CHINESE TECHNOLOGY IN THE SEVENTEENTH CENTURY

T'IEN-KUNG K'AI-WU

SUNG YING-HSING

TRANSLATED FROM THE CHINESE AND ANNOTATED
BY E-TU ZEN SUN AND SHIOU-CHUAN SUN

天工開物

DOVER PUBLICATIONS, INC.
MINEOLA, NEW YORK

Copyright

Published in Canada by General Publishing Company, Ltd., 30 Lesmill Road, Don Mills, Toronto, Ontario.
Published in the United Kingdom by Constable and Company, Ltd., 3 The Lanchesters, 162–164 Fulham Palace Road, London W6 9ER.

Bibliographical Note

This Dover edition, first published in 1997, is an unabridged republication of the work first published by The Pennsylvania State University Press, University Park and London, in 1966.

Library of Congress Cataloging-in-Publication Data

Sung, Ying-hsing, b. 1587.
[T'ien kung k'ai wu. English]
Chinese technology in the seventeenth century ; Chinese technology in the seventeenth century. Sung Ying-hsing ; translated from the Chinese and annotated by E-tu Zen Sun and Shiou-chuan Sun.
p. cm.
Originally published: University Park : Pennsylvania State University, 1966.
Includes bibliographical references and index.
ISBN 0-486-29593-1 (pbk.)
1. Technology—China—History—17th century. 2. Science—China—History—17th century. I. Sun, E-tu Zen, 1921– II. Sun, Shiou-chuan, 1913– III. Title.
T27.C5S9313 1996
600—dc21 96-53237
CIP

Manufactured in the United States of America
Dover Publications, Inc., 31 East 2nd Street, Mineola, N.Y. 11501

CONTENTS

PART III

TRANSLATORS' PREFACE

Nearly ten years ago, in the course of research on a point of Chinese economic history, the translators had occasion to consult *T'ien-kung k'ai-wu*. Its special quality made us realize how limited was our own knowledge of traditional Chinese technology. It occurred to us, as we enthusiastically pored over this work, that others might share our interest.

T'ien-kung k'ai-wu, which may be rendered as "The Creations of Nature and Man," is by no means the only or the oldest Chinese work on technological subjects, but while other works dealt with individual subjects, the present book covers practically all the major industrial techniques of its time, from agriculture, textiles, mining, metallurgy, and chemical engineering, to the building of boats and the manufacture of weapons. Professor Joseph Needham, author of the monumental treatise on *Science and Civilization in China,* calls it "an important book on industrial technology" of the early seventeenth century by "the Diderot of China, Sung Ying-hsing."

The original preface was dated 1637, but only a bare outline of the author's life can be reconstructed from presently available sources.[1] Sung's early years were typical of those of a young man of a scholar-official family. He was born, probably shortly before the end of the century, in Feng-hsin, Kiangsi Province. In his youth he received the usual classical education and passed the public examinations to the level of a provincial graduate in 1615. Thereafter he served in a number of official capacities in several provinces, including his native Kiangsi. It was there, while serving as the Education Officer of Fen-i district, that he wrote the present book. In addition to *T'ien-kung k'ai-wu,* Sung wrote a work on phonology, and a collection of critical essays, but these have been lost. After the fall of the Ming dynasty in 1644 Sung, together with his elder brother Ying-sheng, retired to his native home and led a secluded life. The probable date of his death is placed at around 1660 by V. K. Ting.

In the early years of Manchu rule, as he wrote a short biography of his brother Ying-sheng, who had taken his own life because of acute depression over the fall of Ming, Sung Ying-hsing could hardly have dreamed that it was he himself who was to achieve greater renown three centuries later. Neither could he have foreseen that, decades after the "alien" Manchu rulers had been overthrown, there would be a keen and widespread interest, both in China and abroad, in the technological history of the country, and that *T'ien-kung k'ai-wu* would benefit many scholars other than his book-bound contemporaries whom he so impatiently chided in the Preface.

Although outstanding in scope and factual details, Sung's was not the only book on science and technology produced in his period. At the end of the Ming dynasty (late sixteenth and early seventeenth centuries) many inquiring minds among the literati were rebelling against the futility of the idealist school of thought of Wang Yang-ming, and the result was the upsurge of interest in the materialistic aspects of life and the publication of several works of a factual and technological [2] character that have since become classics in their fields. The seventeenth century as a whole saw the strengthening of this trend, which culminated in the thought and teachings of Wang Fu-chih and Yen Yuan and later in Tai Chen. Yen Yuan, the master of pragmatic philosophy, and Sung Ying-hsing would have found each other congenial; Yen planned for the curriculum of the Chang-nan Academy, of which he was the director for a short period in 1696, courses on military and industrial techniques in addition to those on classical studies.[3] When we compare Yen's plan with the 1637 Preface of *T'ien-kung k'ai-wu,* the similarity of intent is obvious.

Apart from the atmosphere of general reaction against idealist thought, the actual condition of Ming society must have helped inspire Sung's book. Since the latter part of the sixteenth century the decline of Ming power has been accelerated by such familiar manifestations as corruption in government, popular unrest, ineptness in dealing with external foes, and over-all economic maladjustment indicated by measures to change the tax system. In short, it was an era when the more thoughful educated people were casting about for an explanation of the ills of the times and above all for some far-reaching remedy that might help to avert a dimly sensed impending crisis. The first edition of *T'ien-kung k'ai-wu* was published only seven years before the Manchu forces toppled the moribund Ming dynasty, and the disturbed times were probably an added impetus goading Sung to undertake the writing of a factual book on the arts and techniques that went into the making of the necessities of daily life, in an attempt to persuade the vast majority of the scholar-officials that these too were matters that merited attention. The author was apparently a close associate of his elder brother Ying-sheng, though their temperaments were obviously different. While Ying-sheng remained the more conventional Confucian scholar-official, who was later to end his own life as a final gesture of loyalty and patriotism, Ying-hsing expressed his concern over the state of the nation through positive use of the knowledge at his command. The note of conviction rings clear in his preface, in which he makes it plain that he was having *T'ien-kung k'ai-wu* published even though (he adds wryly) this work was not likely to advance any person's official career.

It has often been said that the coming of the Jesuit missionaries to China in the latter part of the Ming dynasty had greatly influenced the intellectual atmosphere of the times, and various works on technology, many of which were translations, are cited as proof of this thesis. While the Jesuits undoubtedly did contribute to the fund of scientific knowledge available in China in the sixteenth and seventeenth centuries, there is little indication in the present work—with the exception of certain items in the chapter on weapons—that they had greatly influenced the writing of Sung Ying-hsing. Seen from the perspective of more than three centuries, it appears that the Jesuits were so readily accepted by the Chinese because the conditions in China were then ripe for the introduction of materialist science, and the late-Ming literati such as Hsü Kuang-ch'i, who collaborated closely with the missionaries in scientific work, reflected the same sort of concern with technological knowledge as did Sung Ying-hsing when he labored over the draft of the present book.

Finally, Sung's home province might have provided much stimulus to his interest in matters industrial and technological. The mid-Yangste province of Kiangsi has been known for centuries for its rich products: rice and other agricultural goods, and above all for its fine porcelain and certain minerals, coal and copper among others. V. K. Ting points out that in Sung's time natives of Kiangsi often worked as miners in neighboring provinces, and recent research as presented in the Yabuuchi volume shows that Sung Ying-hsing did in fact draw extensively upon the data available in his own province in describing many of the techniques of industry and agriculture. Thus, *T'ien-kung k'ai-wu* resulted from both the intellectural climate and an immediate local background in late-Ming China.

Two editions of *T'ien-kung k'ai-wu* appeared before 1644, testimony that the book was well received by Sung's contemporaries. But apparently no new editions were printed during the Ch'ing period, and until recently no copy of the Ming editions had come to light. Various portions of the book, however, were preserved in the eighteenth-century encyclopedia *Ku-ching t'u-shu chi-ch'eng,* in a compendium on agricultural techniques entitled *Shou-shih t'ung-k'ao,* and in an early nineteenth-century edition of the *Gazetteer of Yunnan Province.*[4] During the eighteenth century, one of the Ming copies was also reprinted in Japan. The current "resurrection" of *T'ien-kung k'ai-wu* dates from 1914 when V. K. Ting came upon quotations from it in the *Gazetteer of Yunnan Province* in sections dealing with the metallurgy of copper and silver. The only copy of the book he was able to find, however, was one of the Japanese copies the philologist Lo Chen-yü had brought back to China in the 1880's. In 1927 after collating the Japanese copy with the portions of the book available in the *Ku-chin t'u-shu chi-ch'eng,* Mr. T'ao Hsiang reissued the work in China. A 1929 reprint of the T'ao edition included a Supplement by V. K. Ting. In the early 1930's at least one popular edition appeared, an inexpensive volume published as one of a series of books on Chinese culture designed for use by students.[5] During the 1950's interest in this book was kept alive by a Japanese translation (*Tenkō kaibutsu no kenkyū*—Tokyo, 1953) and critical studies by the seminar of Prof. Yabuuchi and by a reprint in Taiwan, in popular format, of the second T'ao edition. It appears that no one had found any copy of the original book or was aware of the two Ming copies in the Bibliothèque Nationale in Paris—one each of the two Ming editions.[6]

Finally in 1959, the Chung-hua Publishing Company of Shanghai reprinted in three attractive volumes the first Ming edition—322 years after its original publication in 1637. According to the publisher's postface, the original copy had been in the private library of a family named Li in Ningpo, Chekiang, who had recently donated their collection of Chinese books to the National Library of Peking.

The appearance of the 1959 reissue helped to clear up an important question: what exactly did the illustrations look like in the original book? The T'ao edition (as explained by V. K. Ting) incorporated many drawings from the *Ku-chin t'u-shu chi-ch'eng* that were not in the Japanese copy; the added drawings are more ornate and perhaps to some eyes more pleasing, but they add nothing to the technical subjects being illustrated. Moreover, several of the figures in the second chapter (Clothing Materials) are identical with some in another eighteenth century compilation, the imperially commissioned *Keng chih t'u* or *Pictorial Accounts of Agriculture and Sericulture*. This variance led us to the conclusion that the "prettier" drawings in the Ta'o editions are in fact not the authentic Ming illustrations. The figures in the 1959 reproduction of the 1637 version, on the other hand, are striking for their simplicity and clarity, as would have been approved, we believe, by an author whose main concern was didactic rather than esthetic. Although retaining those figures from the Ta'o edition that are not in the 1959 copy, we have indicated the source in the captions.

In our present effort we are deeply indebted to the many scholars who have published works on the very wide spectrum of subjects treated in this book. We wish to express our special appreciation to Professor Lien-sheng Yang of Harvard University, who not only has given us encouragement and helpful suggestions from the beginning, but has also been good enough to read the present work in manuscript form. We are conscious of the fact that, in spite of our application, many imperfections still remain: for these the translators alone are responsible.

E-tu Zen Sun
Shiou-Chuan Sun

October, 1962

The Pennsylvania State University
University Park, Pennsylvania

NOTES

1. The first biographical sketch of Sung Ying-hsing written in modern times was by the geologist V. K. Ting (Ting Wen-chiang), who gathered the data from various historical records and published his sketch, along with a critique of the book, in a supplement to the 1929 reissue: *T'ien-kung k'ai-wu* (in Hsi-yun-hsuan ts'ung-shu, 1929).

This supplement is available also in a new "paperback" version of the 1929 edition that was published by the Chung-hua ts'ung-shu wei-yuan-hui in Taipei, 1955. Sung Ying-hsing's life also appears in Arthur Hummel, ed., *Eminent Chinese of the Ch'ing Period* (Washington, D. C., 1944), pp. 690–691; and in Yabuuchi Kiyoshi, *Tenkō kaibutsu no kenkyū* (Tokyo, 1953), p. 11.

2. Among the works that portray mainly indigenous Chinese technology written by men trained in traditional learning, we may mention the *Pen-t'sao kang-mu* (Materia Medica) by Li Shih-chen, and *Wu-pei chih* (Treatise on Weapons and Military Equipment) by Mao Yuan-i.

3. Joseph Needham, *Science and Civilization in China,* vol. II, p. 915 (Cambridge University Press, 1956), II, 515; Hummel, II, 915.

4. See Ting; Ts'ai K'o-nan, "*T'ien-kung k'ai-wu* tsai Chung-kuo chi-shu-shih shang ti chia-chih," *Cheng-lun chou-k'an* (China Critic Weekly), no. 177 (May 27, 1958), pp. 22–24; and Yuan Yuan, comp., *Yun-nan t'ung-chih* (A Gazetteer of Yunnan Province), 1835 ed., chs. 73, 74.

5. Published by the Commercial Press, Shanghai, in its *Kuo-hsueh chi-pen ts'ung-shu* (Basic Books of Chinese Studies) series.

6. Prof. L. S. Yang called attention to the Bibliothèque Nationale copies in his article reviewing the Yabuuchi volume, in *Harvard Journal of Asiatic Studies,* vol. 17, nos. 1–2 (June, 1954), pp. 310–311.

AUTHOR'S PREFACE TO THE EDITION OF 1637

Under heaven and upon earth, things are to be numbered in the tens of thousands, and there are a like number of phenomena. All details are complete, nothing is overlooked: such cannot have been the fruit of human endeavors alone. Living amidst the myriads of things and phenomena, however, man still knows nothing of them unless one by one each thing is shown before his eyes and he is instructed about it by word of mouth. How much knowledge can one gain in this way?

Of the tens of thousands of things and phenomena in the world, about half are of benefit to mankind and half not. A person who is endowed with intelligence and possesses a knowledge of the natural world is much respected by the multitude of ordinary folk. Yet there are those who, not being able to distinguish between jujube flowers and pear blossoms, would prefer to indulge in speculations about the water-plants of Ch'u; not knowing the measurements and care of cooking pots, would prefer to discourse emptily on ancient sacrificial vessels of Lü; when they paint, they delight in depicting ghosts or monsters but scorn such common subjects as dogs or horses. Even if the talents of these people equal those of Kung-sun Ch'iao of Cheng and Chang Hua of Tsin,* they are still, I say, unworthy of emulation!

We are fortunate to be living in an era of enlightened rule and great prosperity, when the carriages from Yunnan may be seen traversing the plains of Liao-yang [Southern Manchuria], and the officials and merchants from the southern coast travel about freely in Hopei. Within an area of some ten thousand square *li,* is there anything or any phenomenon that may not be seen or heard [by a person with an inquiring mind]? Had one been a scholar of the early Eastern Tsin or late Southern Sung period, then he would have regarded the local products of Yen, Ch'in, Chin, and Yü [the modern Hopei, Shensi, Shansi, and Honan provinces] as "foreign" goods. To obtain a fur hat he would have to resort to the channels of international trade! That indeed would be as if we were, today, trying to obtain the rare arrows made by the Su-shen tribesmen [in modern Kirin]. Furthermore, the scions of noble houses and the imperial princes are born and raised in vast palaces, cut off from the outside world. Yet, while the best rice is cooking in the palace kitchen, perchance one of the princes would wish to know what farming implements look like; or, while the officers of the Imperial Wardrobe are cutting suits of brocade, another of the princes might wonder about the techniques of silk weaving. At such times a volume of drawings that depict these tools and techniques will indeed be treasured as a thing of great value.

During the past few years I have written a book entitled *T'ien-kung k'ai-wu chüan,* or *A Volume on the Creations of Nature and Man,* but it is sadly limited by the author's lack of wealth. It was my wish to purchase some rare artifacts in order that [statements made in the book] might be objectively verified, yet I lacked the funds; and although I wished to gather a group of colleagues to discuss the general subject of the book and to ascertain the truth of its contents, yet there was no meeting place for such conferences to take place. The result, therefore, is that the book has been written entirely out of what I have carried in my own head, based on inadequate observations and scanty knowledge. How worthless such a work must be! However, I have a friend, Mr. T'u Po-chü, who is earnest and sincere in character, and is devoted to the study of natural phenomena. He believes that in any piece of writing, whether by a contemporary or past author, if there be one worthy sentence, or one iota of merit, then this work deserves to be completed and made known, and he himself would work diligently and faithfully to achieve that end. Last year my book *Hua-yin kuei-cheng* was entrusted to Mr. T'u for publication, following which, upon his suggestion, I am having the present work published. Is not this turn of events the handiwork of Fate?

The present book is divided into three parts, the order of their contents arranged in such a way as to indicate my desire to emphasize the importance of the agricultural products and the subordinate roles of metals and gems. In the original draft there were two chapters on astronomy and music, but upon second thought I have decided to omit them before publication, as these subjects require a great deal of special knowledge that does not fall within my own field of interest. An ambitious scholar will undoubtedly toss this book onto his desk and give it no further thought: it is a work that is in no way concerned with the art of advancement in officialdom.

Sung Ying-hsing, written in
the Hall of Inquiry into the Home Arts,
in the fourth month of 1637.

* Kung-sun Ch'iao (Tzu-ch'an) of the Spring and Autumn period (770–403 B.C.), was a contemporary of Confucius. He served as chief minister of the state of Cheng for forty years, during which time Cheng was made strong enough to discourage invasion by its two strong neighbors, the states of Chin and Ch'u. This feat was held as evidence of his great administrative ability. Chang Hua lived in the Tsin dynasty (A.D. 265–420). Learned in a wide range of subjects, he was the author of *Po-wu chih* or *Account of Natural Phenomena and Things.* In later literary allusion he was often paired with Kung-sun Ch'iao.

PART I

Translators' Notes

We have preserved the original order of division of this book into eighteen chapters, which are grouped under three parts. The author explained his reasons for this arrangement in the preface to the 1637 edition.

In the translation, parentheses ( ) denote interlinear notes in the original text, and brackets [ ] indicate words added by the translators to clarify the text.

Reference works consulted by us are listed in the Bibliographies. These works have been useful both as background material for the translation, and as sources for the brief notes at the end of each chapter. For the reader who is interested in pursuing individual topics further, they can serve as the basic bibliography on the historical as well as technical aspects of the subject matter.

1

THE GROWING OF GRAINS

Master Sung observes that, while the existence of the Divine Agriculturist of antiquity is an uncertain matter, the truth denoted by the two words of his name has existed down to the present day. Man cannot live long without the sustenance of the five grains; yet the five grains cannot grow of themselves; they must depend on man to cultivate them. The nature of the soil changes with time, and the species and properties [of the plants] differ according to the geographic environment. But why was it that the classification and explanation of the numerous varieties of grain had to await the coming of Hou-chi, even though a thousand years had elapsed between the time of the Divine Agriculturist and the Emperor T'ao-t'ang,[1] during which interval grain was used as food and the benefits of cultivation had been taught throughout the country? It was because the rich men regarded the [farmer's] straw hat and cape as convict's garb, and in aristocratic households the word "peasant" had come to be used as a curse. Many a man would know the taste of his breakfast and supper, but was ignorant of their sources. That the First Agriculturist should have been called "Divine" is certainly not the mere outcome of human contrivances.

GENERAL TERMS

"Grain" is not a specific term. The "hundred grains" refer to crops in general, while the "five grains" are sesamum, legumes, wheat, panicled millet, and glutinous millet. Rice is not included because the ancient sages who wrote on the subject were natives of northwestern China.[2] Nowadays seventy per cent of the people's staple food is rice, while wheat and various kinds of millet constitute thirty per cent. Sesamum and legumes are used exclusively as vegetables and [for making] oil, although tradition still classifies them among the grains.

RICE

There are numerous kinds of rice. Of the nonglutinous kind, the plant is denoted by one word and the grain by another, both of which are pronounced *keng;* of the glutinous kind, the plant is called *t'u* and the grain *nuo.* (In the south [of the Yangtse River] where no glutinous millet is grown, wines are made from glutinous rice.) A further variety is originally nonglutinous, which yields a late harvest of slightly glutinous grains (commonly known as the "light of Wu-yuan" [3] variety, and is fit only for making gruel). As to shape, there are long-speared and short-speared rice (south of the Yangtse the long variety is called "early Liu-yang," [4] and the short-speared "early Chi-an" [5]), long-grained and pointed-grained, round-top and flat-top grains, and so on. Colors of the grain vary from snowy white, ivory, and red, to semiviolet and dappled black.

Rice should be planted in wet fields not earlier than just before the vernal equinox which is known as *"she* planting" (if the weather is cold the plants are sometimes killed by frost), and not later than the Ch'ing-ming festival.[6] Before planting the seeds are wrapped in straw and soaked for a few days until the shoots appear, then they are planted in the fields. When the shoots are about an inch high they are called "young plants." After thirty days the young plants are thinned and transplanted; this should not be done, however, if the fields are suffering from drought or flood. When past the [transplanting] time the young plants will harden and develop sections [in their stalks], and if left thus in the field they will each produce only a few grains. One *mou* of young plants can provide rice plants for twenty-five *mou* of the transplanted fields. The early variety ripens seventy days after transplantation (of the many local names of this type I shall enumerate only a few: "quell-your-hunger" and "urgent-for-the-throat" for *keng* rice and "silver-wrapped-in-gold" for *nuo* rice), and the latest variety cannot be harvested until winter, since it has a growing period of 200 days. In Kwangtung, where there is neither frost nor snow in the rice region, rice is planted in winter and harvested in midsummer.

Damage from drought is feared if rice plants are short of water for as many as ten days. The variety that is planted in the summer and harvested in the winter needs a field continuously fed by mountain streams; this is a hardy specimen, and this kind of soil is of a cold nature that does not stimulate plant growth. In lake-side fields the transplanting takes place in the sixth month [approximately the month of July in the solar calendar] after the summer innundations are over; meanwhile, the young plants are temporarily kept in high fields to await the proper time. In the southern plains the fields often yield

Figure 1–1. Soil is loosened by ploughing.

two harvests per year. The second crop is commonly called late-glutinous. When the first crop has been reaped in the sixth month, the same field is ploughed over, and the young plants of the second crop are planted. These young plants were sown at Ch'ing-ming at the same time as those of the early crop; yet while the latter cannot survive for one day without water, these plants [of the late variety] could live through the hot sun and dry weather of the fourth and fifth months [approximately May and June] without bad effects. This is a remarkable phenomenon.[7] If there is little rain during the autumn for the second rice crop, then the fields should be constantly irrigated and watered—such are the toils of the farmer for the making of Spring wine! All rice plants die if water is lacking for ten [consecutive] days, yet there has been developed an early nonglutinous strain that can be planted on high hills.[8] This is another remarkable phenomenon. The kind called "fragrant rice" is known for its fragrance and enjoyed by the aristocrats. With a small yield and lacking in nutritious value, this variety deserves no recommendation.

THE CARE OF RICE

The industrious farmer fertilizes his fields and stimulates the [growth of] rice in many ways, for when the soil is lean and depleted the spears of rice become sparse and barren. Commonly used throughout the country as fertilizing agents are: human and animal excretions, dry cakes of pressed seeds ("dry" because the oil has been taken off; of these sesamum and turnip seed cakes are the best, next are those made from rape seeds, followed by the big-eyed *t'ung* seeds [*Aleurites cordata*], and lastly camphor, tallow, and cotton seeds), grass and tree leaves. (In the south where green lentil flour is made, the liquid waste is used in the fields as a very rich fertilizer. When the price of beans is low, soy beans can be cast into the field, each bean enriching an area about three inches square; the cost is later twice repaid by the grain yield.) If the soil contains cold paste,[9] bone ashes should be sprinkled around the roots of young rice plants (the bones of any bird or animal will do). Lime should not be used to cover the plant roots in warm, sunny soil. When the soil is hard, rows should be ploughed and the clods stacked up, then wood is burned under them [to loosen the soil]. This method, however, is not suitable for clayey, carbonaceaous, or sandy soil.

THE CULTIVATION OF RICE: PLOUGHING, HARROWING, ROTARY HARROW, FOOT WEEDING, AND HAND WEEDING

Where no second crop of rice is planted, the land should be turned immediately after harvest, so that the stalks will rot in the ground. The decomposed stalks make a fertilizer twice as effective as manure. If, because of dry weather or laziness, the farmer does not plough the land until the following spring, the benefits will have been greatly reduced. In fertilizing the land,

Figure 1–2. Soil is broken into fine particles by harrowing.

whether using dried cakes or liquid wastes, he must be careful to apply the fertilizer at a time when the application is not likely to be washed away by rain. Therefore, the experienced farmer diligently studies and observes the weather.

The industrious farmer will plough the field two or three times [Figure 1.1] before using the harrow [Figure 1.2], thus breaking the soil evenly and finely and distributing the plant foods therein. Those who do not have work oxen attach a pole to the plough and place it on [the shoulders of] two men who, walking one behind the other, pull it. The result of one day of such ploughing is equal to that done by one ox. If the farmer has no cattle, a rotary harrow that is pulled by two men can be constructed. The result of one day of such harrowing is equal to that done by three oxen. There are only two kinds of cattle in China: the water buffalo and the yellow ox. The buffalo is twice as strong as the ox, but he requires twice as much care as the ox, since he has to be housed in a mud barn during the cold winter and provided with a pond for bathing in summer. Prior to spring the ox must not be exposed to rain while he is perspiring from ploughing, and when it rains he should be driven to the barn immediately. He will be immune to wind and rain after *ku-yü* ["grain rains," a solar period between late April and early May]. In the Yangtse delta area the farmers use hoes instead of ploughs, and do not use working cattle. Considering the cost of the ox and the feed, and the risks of sickness, theft, and death, I think human labor is more suitable [than cattle] for the poorer peasants. Suppose that a farmer with an ox can work ten *mou* of land, and that another industrious farmer, using only a hoe, works half that amount. Since the latter has no ox, he need not concern himself with pasturage problems in his fields after the autumn harvest, when such things as legumes, wheat, sesamum, and vegetables can all be planted, and the additional harvest will compensate the farmer for that part of the land not planted [that is, the five-*mou* difference]. This would seem to be a suitable arrangement.

A few days after transplanting, the old leaves wither on the rice plants and are replaced by new ones; when the new leaves are grown then it is time to weed (commonly known as "beating the plant"). Leaning on a stick [for balance], one pushes the earth around the roots [of the rice plants] with one's feet, at the same time bending the weeds so that they will not grow again [Figure 1.3]. Such weeds as *Backmannia crucaeformis* can be broken by foot, but darnels, tares, and smartweed cannot be killed in this way, and have to be uprooted by hand [Figure 1.4]. Hand weeding is hard on one's back and hands and requires keen eyesight.[10] The plants will flourish after all undesirable weeds are eliminated. Thereafter one needs but to eliminate [excess] water to prevent flooding and irrigate the fields to prevent [damage from] drought, and in due time the harvest will be ready for the scythe.

RICE DISASTERS

The early variety of rice is harvested and stored in early autumn. If it is put away in the barn while the noon sun is shining, so that the heat is carried

Figure 1–3. Foot weeding.

Figure 1–4. Hand weeding.

into the storage bin and closed in immediately, the rice grains will retain the fiery quality of the weather (this misfortune is a common occurrence in the homes of industrious farmers). Then, in the following spring, when the land is fertilized and therefore is hot by nature, with the southeasterly wind to add to the heat, the [soil] will give forth blazing heat waves, which will greatly damage the young shoots [that have germinated from the fiery grains]; this is one kind of disaster. [To prevent this, either] wait until the cool of the evening before putting the grain into the storage bin, or gather a jar of snow and ice water at the time of the winter solstice (water gathered at the beginning of spring is not effective) and sprinkle it on the grain at *Ch'ing-ming* sowing time to the amount of a few bowlfuls per *tan* [of seed grain]. This water will immediately dissolve the hot properties of the grain, so that the young shoots will be unusually handsome even though the warm southeasterly wind prevails (the trouble obviously rests with the seeds, yet people blame it on the spirits and gods!).

A second kind of disaster is that, after the rice seeds are sown into a few inches of water, the grains do not all sink at once, and are blown into one corner [of the field] when a gust of wind comes by. [To avoid this] one should watch until the wind has stopped, then the seeds will sink after sowing and grow evenly into shoots. A third disaster is caused by the birds that eat the shoots. This problem can be solved by erecting a pole and flying an eagle scarecrow from it. A fourth disaster is brought about by prolonged rain before the shoots can fasten their roots in the soil, therefore resulting in loss of more than half of the plants; pray to Heaven for three days' sunshine and all seeds will grow properly. A fifth disaster is plant pest (in the form similar to silkworm cocoons) within the tassles, a result of the combination of rich, wet soil and the steaming effect of warm south wind after the tassels have begun to form. Pray for westerly wind and rain, which will kill the pests and allow the grains to grow.

A sixth disaster is the burning of the rice plants at night by wandering spook fires when the grains are ready for harvest. These fires originate from rotten wood. So long as the wood "originator" containing the fire "offspring" holds together, the fire essence will continue to exist intact for centuries. But in rainy years the neglected graves, having been broken open by foxes and the like, become waterlogged and the wood of the coffins proceeds to rot, and the disintegration of the "originator" releases the fire "offspring" to hover in the air. This fire of the nether world, however, cannot stand sunlight, therefore it will only appear after sundown at twilight, when it rises through clinks [into the air]; and, being unable to rise high, it can only hover aimlessly for a few feet before it has to stop. The grains and leaves of all plants are seared as soon as this fire touches them. Seeing gleaming spots at some tree roots, people who hunt these fires would take them to be ghosts and beat the spots with sticks, and then claim that the ghosts have turned into burnt twigs, not knowing that [the gleaming spots] are spook fires which can be instantly dispelled by lamplight (lights that have not been humanly lighted in lamps are of nether category, therefore disappear upon being confronted with lamplight).

A seventh disaster is the lack of water and drying up of plants, for between the formation of the tassels and the maturing of the grains the water needed [per plant] is three pecks for the early variety and five pecks for the late variety (if the plants are short of one pint of water just before harvest, the grains will still be there, but their size will be small, and are easily broken when husked). Human ingenuity has, however, developed the arts of irrigation and watering to the utmost. The eighth kind of disaster occurs at harvest time, if there should be a strong wind that knocks the grains to the ground or a prolonged rain that dampens the grains, causing them to rot. However, a windstorm does not prevail beyond a distance of thirty *li,* and a rainy spell, 300 *li;* these disasters therefore are of a local nature. As to the grains lost during a windstorm, nothing can be done to retrieve them. [In the case of prolonged rain] the poor families can place the wet grain in a pot, apply heat below, eliminate the chaff and husk by roasting, and eat the roasted kernel to ward off hunger. This would be a way of supplementing [the inadequacies] of Nature.

IRRIGATION: THE CYLINDER WHEEL, OX WHEEL, TREADLE WHEEL, HAND-CRANKED WHEEL, AND COUNTER-WEIGHT LEVER

Water is the most important thing in rice fields. The needs vary according to locality and the nature of the soil: some fields will run dry in three days while others in half a month, and water must be supplied by human effort when rain is lacking.

The cylinder wheel [Figures 1.5 and 1.6] is used along river banks. The river is dammed and channeled [Figures 1.7 and 1.8] so that the water flows onto the lower part of the wheel, resulting in a continuous rotation; the cylinders attached to the wheel are filled with water [as they become submerged], which is subsequently emptied into a receptacle and thence into the fields. This apparatus continues its operations day and night and is sufficient for watering 100 *mou* of land; (when water is not needed, a piece of wood is tied to clog the wheel and stop its turning). [The cylinder wheel can be replaced with a water-turned wheel, as shown in Figure 1.9.]

Near lakes and ponds where there is no flowing water, oxen are employed to turn the water wheel [Figure 1.10], or a few men can work a treadle wheel. Ranging in length between ten to twenty *ch'ih,* the treadle wheel [Figure 1.11], consisting of a wooden paddle chain and a sprocket wheel, [is turned by the weight of one or two persons on the wooden treadle pieces]. Water is carried upward by the paddles [inside a trough, on the upper end of which the water is discharged]. By this method one man can generally water five *mou* of land a day, while the ox wheel doubles the area. Around shallow ponds and small creeks, where lengthy wheels cannot be erected, short [hand cranked] wheels of a few *ch'ih* in length are used [Figure 1.12]. Here a man after cranking all day can irrigate only two *mou.* In Yangchou prefecture [in Kiangsu] the people use another type of wheels that are turned by the wind. They serve to eliminate excess water from the fields before planting rather than to irrigate and are, therefore, not suitable as a remedy against drought. As to other devices, such as the counterweight lever [Figure 1.13] and pulley well [Figure 1.14], they are still less efficient.

WHEAT

There are several kinds of wheat. The ordinary wheat [*Triticum vulgare*] is called *lai* and is the principal species. Barley is called *mou* or *k'uang*. Then there are miscellaneous wheats consisting of oats and buckwheat. All are termed "wheat" because they are sown at the same time as wheat, have similar inflorescense, and are used for their flour in similar fashion. In our empire the staple food of the common people of the Yen, Ch'in, Chin, Yü, Ch'i and Lu provinces [11] consists of fifty per cent wheat, with the remaining fifty per cent composed of millet, rice, and *kao-liang*. [In the southern half of the country] in an area 6,000 *li* square from Yunnan and Szechuan in the west to Fukien, Chekiang, Kiangsu and Hunan in the east, only one-twentieth of the people cultivate wheat. The flour thus obtained is used for pastries and soups, not as a

Figure 1–5. Cylinder wheel.

staple food. Only one-fiftieth of the people plant barley, oats, and buckwheat, which are used only by the poor for meals, but never by the well-to-do.

Shensi is the only place where *k'uang* wheat is produced. Known also as "dark barley," it is barley changed by geographic environment so that the husk is of a blue-black color. The Shensi people use it as horse feed; human beings eat it only during famines. (There is also a glutinous barley, which the people of Honan use for wine-making.)

Oats have very fine panicles, each of which again consists of some dozen or so small seeds. Occasionally they grow wild. Buckwheat actually does not belong to the wheat family, but since its flour is used for food it has become known as a sort of wheat, and so we shall consider it as such.

In the north wheat passes through four seasons [in its growth cycle], being sown in the autumn and reaped in early summer. In the south the period between planting and harvesting is shorter. It is strange that south of the Yangtse the wheat flower begins to bloom at night, while north of this river it starts in daylight. Barley is planted and harvested at the same time as wheat. Buckwheat is planted in mid-autumn and the crop is gathered within two months; frost kills buckwheat shoots, therefore a good harvest is expected if Heaven sends a late frost.

THE PLANTING OF WHEAT: PLOUGHING AND SOWING IN THE NORTH; CULTIVATION

The first stage of ploughing for wheat is the same as that for rice. After planting, however, all the hard labor such as foot and hand weeding is necessary for rice. Wheat needs only hoeing.

Owing to the fact that the soil of the north is relatively sandy, friable, and easy to turn, the method and tools used for planting wheat are different [from those of the south]. In the north the ploughing and sowing are done simultaneously [with a plough-seeder, Figure 1.15]. When an ox is used in turning the soil, the plough's point is removed and in its place are fixed two iron [diggers] on the cross-beam which the local dialect calls *ch'iang*. A small square box containing wheat seeds is placed in the center of the *ch'iang;* five small holes, or "plum blossom holes," are drilled in the bottom of the box through which seeds fall to the ground when the apparatus is shaken as the ox moves forward. When close seeding is desired, the ox is urged to go faster, causing the seeds to fall in greater quantities; when sparse seeding is desired, the ox is made to go slowly, so that fewer seeds will fall. After the seeds are sown a donkey is made to pull two small round stone rollers over the furrows so that the seeds are firmly packed in [Figure 1.16], because wheat will germinate only when firmly pressed into the earth.

In the south the method differs from the north in that, after loosening the soil by much ploughing and harrowing, the [wheat] seeds are mixed with ashes and sown by hand. Next the earth is packed with the heels of the feet [figure 1.17] instead of the donkey-drawn stone rollers used in north China. Planting

Figure 1–6. Elevated cylinder wheel [Ch'ing addition].

Figure 1–7. A water-dam [Ch'ing addition].

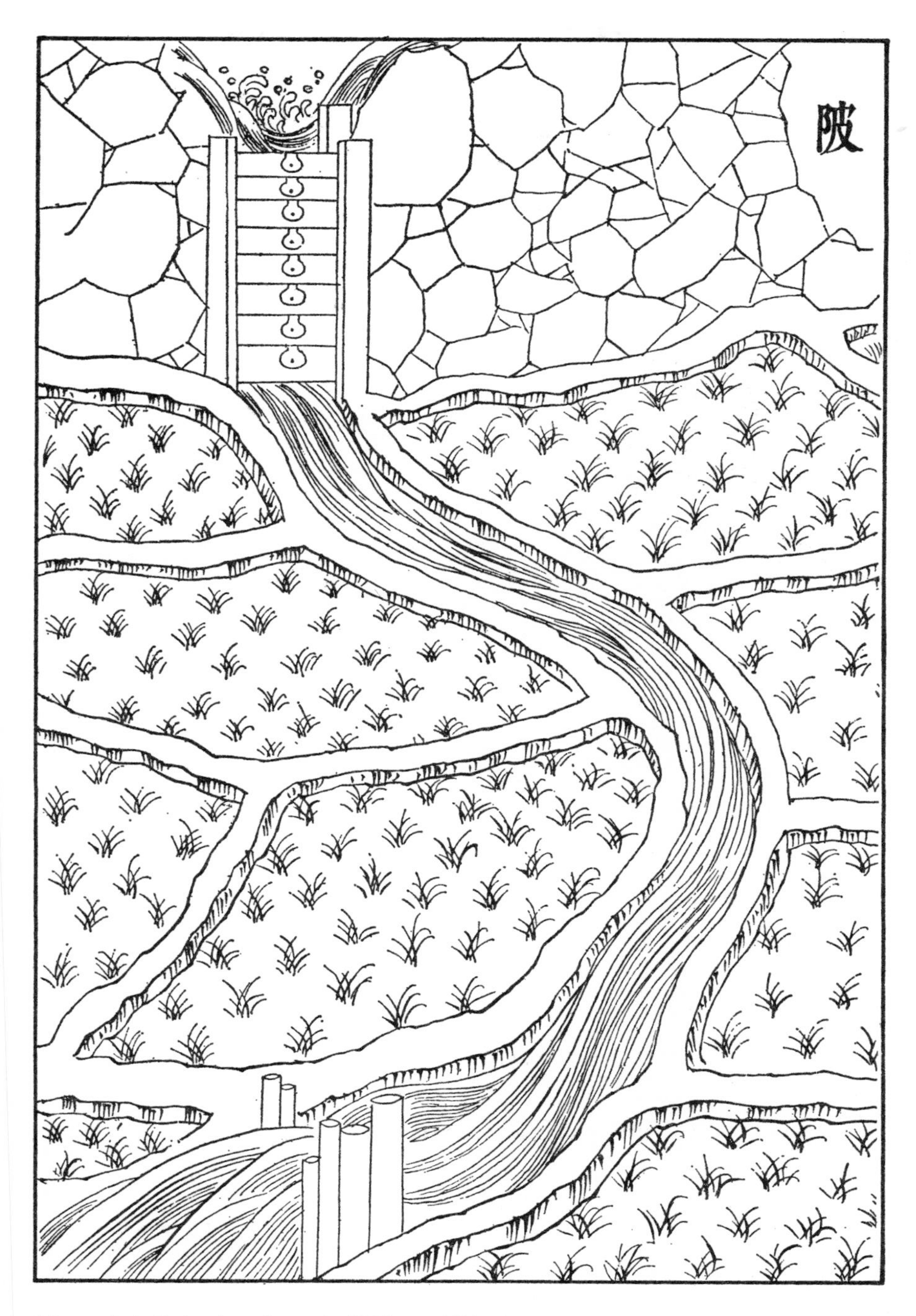

Figure 1–8. Irrigation channels [Ch'ing addition].

Figure 1–9. A water-turned wheel for irrigation [Ch'ing addition].

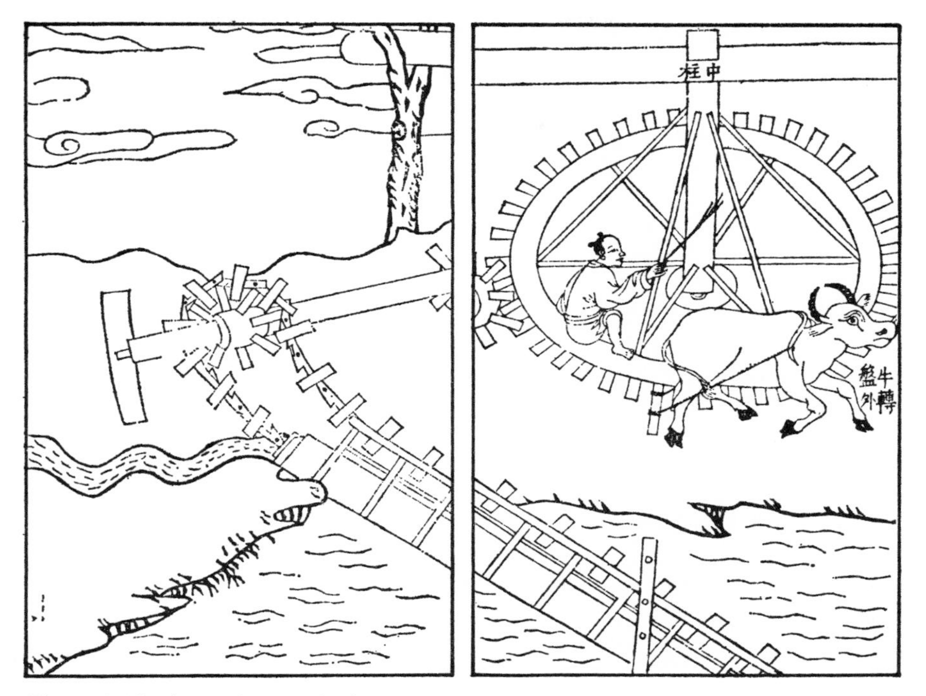

Figure 1–10. An ox-drawn wheel.

should be followed by frequent cultivating, for which purpose the broad-headed large hoe is used [Figure 1.18). There cannot be too much cultivating once the wheat shoots appear (some people do it three or four times); and the entire field can be expected to yield good grain when all the weeds have been killed. "Frequent application [weeding] makes easy weeding" is true of both the north and the south.

Fertilization of the wheat fields must be done before planting, as there is no way to do it afterward. In the Shensi-Honan area some people mix the seeds with arsenic against worms; [for the same purpose] only stove ashes are used in the south (commonly called "ground ash"). There are those who plant "fertilizer wheat" in the southern rice fields; that is, no crop is expected from the wheat, but in the spring, when the wheat and barley plants are good and green, they are ploughed under in the field, and this always doubles the autumn yield of rice.

Other crops can be raised on the land after wheat has been harvested, since there is half a year's time between early harvest and late autumn. The new crops can be anything the farmer chooses to plant that are suitable to the soil. In the south a crop of the late nonglutinous rice is sometimes planted after the barley has been reaped. Heaven will reward in every way the hard efforts of the industrious farmer!

The planting of buckwheat takes place always after the rice harvest in the south, and after the legumes and millet are gathered in the north. Buckwheat absorbs considerable nourishment and thus depletes the soil. Considering yield, however, which would be worth more than half the [regular rice or millet] crop, why should the diligent farmer object to refertilizing the field after it?

WHEAT DISASTERS

Wheat disasters are only one-third as numerous as those pertaining to rice. After the seeds are sown it does not matter whether there be snow, frost, dry weather, or excessive rain. Wheat needs very little water, and in the north one pint of rain in mid-spring will suffice to produce good full kernels of wheat. South of Chin-chou [in mid-Yangtse] and Yang-chou the only fear is prolonged rain, but if at the time of ripening there are some dozen fair days the harvest is more plentiful than the people can use. As a Yang-chou proverb sums it up: "An inch of wheat fears not a foot of water, but a foot of wheat fears only an inch of water"; that is to say, when the wheat plants are young they will survive even though submerged in water, but when the grains begin to ripen even one inch of water will soften the roots, the stalks will bend and fall into the mud,

Figure 1–11. A treadle wheel.

Figure 1–12. Hand cranked wheel.

Figure 1–13. A counterweight lever.

and the grains will rot on the ground. There is a kind of boneless birds in the south which fly in flocks of thousands and live off the wheat fields; they are however not widespread, prevailing only in an area a few scores of *li* in width at a time. But if locusts appear north of the Yangtse, then it will be a catastrophic year in deed.

MILLET AND SORGHUM

There are many kinds of cereal which are used for their grain but not flour. A distance of a few hundred *li* will bring about differences in color, taste, shape and properties of the plants. Though they differ according to the local environment, and are known by an infinite variety of names, yet they are basically the same species of grain. Northerners call the nonglutinous rice, "rice" [*ta-mi,* or "big grain"], while all other cereals are referred to as "millet" [*hsiao-mi,* or "small grain"]. Large-panicled millet belongs to the same group as the small-panicled millet, while sorghum belongs to that of the small-grained millet. The large-panicled millet can be either glutinous or nonglutinous (the former variety being used for wine), but the small-panicled millet is never glutinous. All glutinous millet, whether large-panicled or small-grained, is called *shu* in general; thus, the latter word does not mean an additional kind of millet.

The large-panicled millet is red, white, yellow, or black in color; some people take the black variety to be small-panicled millet, which is not correct. It is close to the truth to say, however, that the small-panicled is an early variety, because it is the first to ripen, therefore more suitable for use as offerings at sacrificial ceremonies. The large-panicled millet appears in the classics under such names as *hsin, ch'i, chü,* and *p'ei;* nowadays it is locally designated as ox hair, swallow's cheek, horsehide, donkey skin, rice tail, and so on. The best planting time is the third month of the year, which gives harvest in the fifth month. The next planting time is the fourth month, harvesting the seventh month. The last planting time is the fifth month, harvesting in the eighth month. In all cases the blossoming and ripening of this millet take place outside the growth period of wheat and barley. The size of the grain depends on the fertility of the soil and the climate. The Sung scholars are wrong in insisting that the musical standards [the twelve semitones] should be determined by the size of millet from a particular area. Sorghum and small-grained millet are called "yellow grain" in general. Wine can be made from a glutinous small grain. Further, there is a "reed millet," known as *kao-liang* because it stands seven *ch'ih* tall and looks like reeds. The sorghum varieties are even more numerous than those of the panicled millet, and they are named either after families, or rivers, or mountains, or shape and size, or the season [of harvest]; they cannot be enumerated here. The natives of Shantung term all of them "grains," making no distinction between sorghum and small-grained millet. All of the four varieties mentioned above are planted in the spring and reaped in the autumn. The methods of ploughing and cultivating are the same as those for wheat; only the planting and harvesting seasons are different.[12]

HEMP

There are only two kinds of hemp [whose seeds] can be used as grain or for oil: hemp and sesame. The latter is generally considered to have been introduced into China from Ta-yuan[13] during the Former Han dynasty [220 B.C.–A.D. 8]. In ancient times hemp was termed one of the five grains, yet this is certainly inappropriate if hemp itself is meant. It is possible that the hemp mentioned in the ancient classics is a species now extinct, or that it was a variety belonging to the legume or millet family, but that in the course of transmission [of the texts] its name was incorrectly rendered.

Sesame is both delicious and nutritious; indeed it would be no exaggeration to say that it is the king of all grains. Hemp, on the other hand, can only produce a little seed oil, and the coarse cloth woven from its bark fibers is of very little worth. As to sesame, a few handfuls [of the seed] are enough to quell one's hunger for a long time; cakes, breads, or sweetmeats, when sprinkled with a few sesame seeds, will have their flavors improved and their values increased. When made into oil sesame can enrich the hair, benefit the intestines, make the strong-smelling [meats] savory, and dissolve poisonous elements. What enormous profits would accrue to the farmers if they devote more of their land to this crop!

Sesame is planted in rows either in the vegetable garden or in the grain fields. The soil must be made extremely fine and all weeds eliminated, then the seeds, having been mixed evenly with slightly damp ground ash, are sown. Planting may take place as early as the third month until just before the Great Heat [in the last part of July] at the latest; those planted early will not mature until the Moon Festival [approximately mid-September]. The hoe is used exclusively for cultivation of the plants. The [seeds] are black, white, or red in color, contained in pods about one inch long. The kind having the four-ridged pod yields less than the eight-ridged, the difference being due to the fertility of the soil rather than to inherent characteristics of the variety. Out of every *tan* of seeds forty catties of oil is made. The dross is used as fertilizer for the fields or, in case of famine, as human food.

LEGUMES

There are as many kinds of legumes as of rice and millet. Their sowing and harvesting times last through the four seasons, and they have been used daily as human food since the beginning of man's need for sustenance was known. One of the legumes is the soy bean,[14] of which there are two varieties: the black and the yellow. These are sown not later than Ch'ing-ming or thereabouts. Of the yellow variety there are three types: "fifth-month yellow,"

Figure 1–14. A pulley wheel [Ch'ing addition].

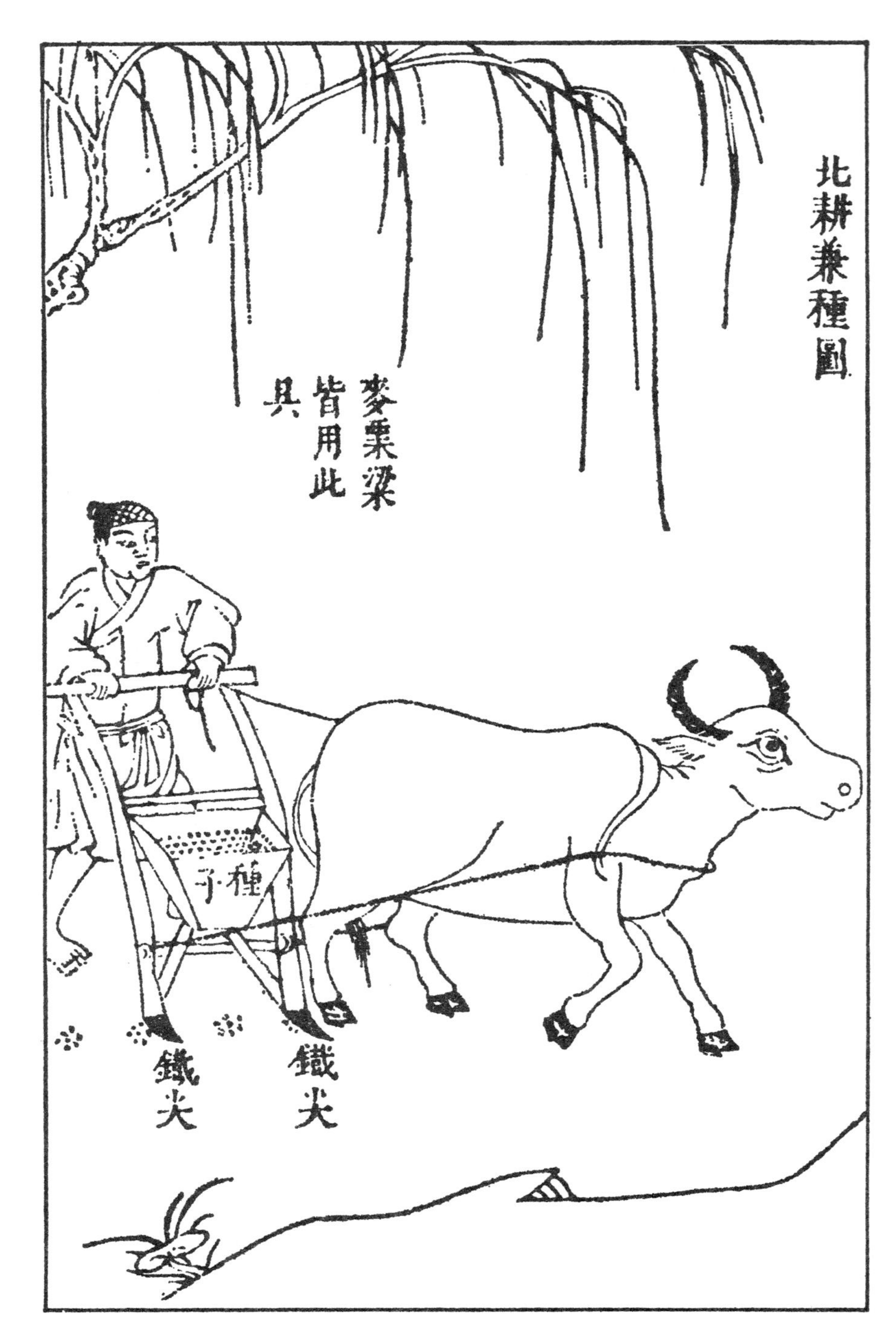

Figure 1–15. A plough-seeder.

Figure 1–16. Stone rollers used in north China for pressing seeds into the earth.

Figure 1–17. Seeds are sown with the fingers, then pressed into the earth with the feet.

"sixth-month popper," and "winter yellow." The yield of the first of these is small, while that of the last is always twice as much. The black variety is harvested invariably in the eighth month. North of the Huai River horses and mules that are used on long journeys must be fed this black soy bean before they can become strong and sturdy. The amount of the yield of the soy bean depends on the quality of the soil, the frequency of cultivation, and the amount of rainfall. All bean jams, sauces, and curds are made from soy.

South of the Yangtse there is another species known as "long-legged yellow," which is planted in the sixth month after early rice has been cut, and is harvested in the ninth or tenth month. The method of planting this bean in Chi-an, Kiangsi, is quite amazing: After the rice stalks are cut the stubble is not ploughed over, but in the open end of each stalk are placed three or four beans, which are pushed down with fingers. The beans are nourished by the dew gathered in the stalk stubs; later when the beans begin to grow the stubs will rot, providing further nourishment for the growth of the new crop. Should the weather be dry after the shoots appear, one pint of water [per plant?] is fed to the plants. In all, one watering and two cultivations are sufficient to bring forth a plentiful harvest. Birds must be kept away after the beans have been planted and before the young shoots appear, and man is the only effective guard against them.

A second kind of legume is the green lentil, shaped small and round like a pearl. It must be planted at the time of the Small Heat [in early July]; if it is planted earlier than that, the stalks will grow up to several feet high and bear few pods; and if it is planted too late the plants, in the Limit of Heat [late in August], will blossom and also bear scattered pods, yielding a small harvest. There are two varieties: "pick green," the pods to be picked day by day as they ripen; and "pull green," the pods ripen at one time and the entire plant is pulled up at harvest. Green lentils are ground into flour, passed through water, collected [dried], and made into chips or noodles, which are delicacies in the diet. The liquid waste from flour-making is a rich fertilizer for the land. For storage the seed beans are mixed with either ground ash, lime, smartweed, or yellow earth, so that no worms will damage them for some four or five months. The industrious farmer frequently puts the seeds in the sun on fair days, which also keeps away the boring worms. If green lentils are planted in the summer or autumn in fields where rice has been reaped, the clods must be finely broken up with a long-handled axe in order to insure a good yield. If it rains heavily within one day of sowing so that the earth is made firm, the green lentils will never germinate. After the shoots have appeared, they must be guarded against flooding by excessive rain, and side ditches can be dug to channel the water out of the fields. In cultivating the fields of soybeans and green lentils, the land should be lightly ploughed, because the roots of the legumes are short and the shoots straight. If the furrows are deep the clods will pile up, preventing half of the seeds from growing. Our ancient agriculturists did not know that deep ploughing was not suitable for legumes.

A third kind of legume is the pea. It is round, like the green lentil, but it has a black spot and is larger in size. The seeds are sown in the tenth month, the

Figure 1–18. Cultivating wheat with broad-headed hoes [Ch'ing addition].

crop reaped in the fifth month of the following year. It can also be planted under trees that are late in putting forth their leaves. A fourth kind is the broad bean with its pod shaped like a silkworm [15] and seeds larger than the soybean. It is planted in the eighth month and harvested in the fourth month of the next year. In western Chekiang the broad bean is frequently planted under mulberry trees. It is a fact that things do not grow under trees where the foliage prevents dew from reaching them, but the pea and the broad bean are already ripe before the foliage becomes shady. Broad beans are produced in large quantities and sold cheaply along the upper reaches of the rivers Hsiang and Han [northern Hupei and southern Shensi] and in usefulness equal millet as staple food. A fifth kind is the small lentil. The red variety is effective when used medicinally, while the white variety (also known as the rice bean) is good as a vegetable. Planted at the time of the summer solstice, this variety is harvested in the ninth month, and is prevalent in the Huai and Yangtse river regions. A sixth kind is the black lentil, which in the old days was a wild plant, but is now a common garden bean in north China. Its flour, made into thin sheets, serves the same purposes as that of the green lentil. In Peking the street peddlers cry their "black lentil sheets" all day long, indicating that the amount produced is considerable. A further kind is the white bean, which grows along trellises and is also known as the "eye-brow bean." In addition, there are long string beans, tiger-spot beans, knife beans [large French beans], as well as the black-skin and brown-skin varieties of soybeans, and so forth, which are too numerous to describe. In all, they can serve as vegetable and take the place of grains in the feeding of mankind. How can students of Nature ignore them?

NOTES

1. *T'ao-t'ang* was another term for the semilegendary Emperor Yao (2356 B.C.–2256 B.C.) during whose reign Hou-chi was said to have served as the agricultural administrator.

2. Generally speaking, the whole of China can be divided into two major parts in terms of principal crops: the wheat region from about 100 miles north of the Yangtse River up through northern Manchuria, and the rice region from the Yangtse River Valley southward, throughout southern and southwestern China.

3. Wu-yuan, a town in Kiangsi province.

4. Liu-yang, a town in Hunan province.

5. Chi-an, a town in Kiangsi.

6. Ch'ing-ming festival occurs in early spring, approximately early April of the solar calendar. On this day family tombs were visited and offerings made; it was also the occasion for a day's outing in the country.

7. Besides the method described here, an "interplanted" method is known to have been practised during the Ming dynasty in Chekiang, Fukien, and Kwangtung: the early variety of rice was transplanted from the seed bed in early July in rows wide apart. Ten days later the late variety was transplanted to the space between the rows. The early rice would be harvested two months after transplanting, and after the period for maturing, the late variety was also harvested. See Yabuuchi Kiyoshi, *Tenkō kaibutsu no kenkyū* (Tokyo, 1953) p. 53.

8. Probably "dry" or upland rice. It was said to have been introduced into China during the second millennium B.C. from northeastern India. In more recent times, a drought-resistant and early ripening rice was imported from Indo-China in the Sung dynasty (960–1279), and proved successful in central and eastern China.

9. The soil containing the so-called "cold paste" is probably a strongly leached soil. Owing to the loss of calcium carbonate, phosphorous, and soluble mineral plant foods, leached soils are usually acid and tend to be sterile.

10. The weeding rake is not mentioned here, but according to Wang Chin, who published his *Nung-shu* (Treatise on Agriculture) in 1313, a new tool for weeding had come into use in Kiangsu and Chekiang, the region where rice cultivation was most advanced in China. This weeding rake consisted of a rectangular piece of wood measuring about one foot long by three inches wide, on one side of which were affixed many short metal nails. A long bamboo pole attached to the piece of wood served as a handle. By using this rake the farmer could weed twice as fast as he could with his hands and feet. Seventeenth-century writings indicate that this tool continued in use during the Ming dynasty.

11. *Yen, Ch'in, Chin, Yü, Ch'i* and *Lu* are the historical designations for the provinces of Hopei, Shensi, Shansi, Honan, eastern and western Shantung, respectively; that is, the area of the north China loess-plain.

12. Maize, the sweet potato, the white potato and peanuts are four newcomers to the sixteenth and seventeenth-century Chinese table, although they are not mentioned in this book. Maize or Indian corn, American in origin, seems to have appeared on the Chinese frontier about 1550. It was brought from the Iberian peninsula to Mecca by way of South Africa, whence it was transported to west China. Some writers believe that maize entered China via northern India, thence to the upper Yangtse valley and then down the river. Other writers are of the opinion that maize was carried by Mohammedan pilgrims from Mecca directly into Sinkiang, and thence into north China, especially Shantung. Sixteenth-century Chinese writings on agriculture referred to maize as "Western barbarian wheat" and "imperial wheat," indicating that it was a novelty to the Chinese in Kiangsu and Chiekiang. In contrast, the mention of cornmeal as an ordinary common foodstuff in the Ming novel *Chin-p'ing Mei* (sections 31, 35, and 74) suggests that maize was planted in Shantung in the later part of the Ming dynasty. By the end of the Ming, however, its cultivation had become quite widespread.

The sweet potato was brought by the Spaniards to the Philippines, whence it was transported to the coastal province of Fukien by Ch'en Chen-lung in 1590. It was first planted on a large scale in that province as a famine relief crop, and within a few decades was grown in many parts of the country. The white potato arrived in Formosa in 1650, and reached the mainland a little later. According to Hommel, several earlier Chinese plants can be loosely classed as potatoes: several species of taro (Colocasia) or yam (Dioscorea), and sweet potato (Ipomoea batatas). The *Materia Medica* (*shih-i* chüan 8 and *kang-mu* chüan 27) indicates that a tuber was probably known and eaten by the Chinese in the southeast coast and islands in the Tsin dynasty (A.D. 265–420). At that time the tuber was called *kan chu,* sweet root, as described in *Nan-fang ts'ao-mu chuang* (written by Chi Han in the third century) and also in *I-wu chih* (written by Ch'en Ch'i-ch'ang). Fan Hsien (fl. 1723–36) states in his *T'ai-wan fu chih* that the original Chinese species is long and white, while the imported one is round and reddish yellow. It can be similarly deduced from *Materia Medica* (Chüan 27) that another plant resembling the potato was probably eaten in the T'ang dynasty (618–906) or at least in the Later Liang dynasty (907–60). At that time the tuber was called *t'u yü,* earth root. The white or Western potato received the name *yang shu,* foreign tuber, because it had been introduced, at least in eastern China, by foreigners. Peanuts were probably brought by the Spaniards to the Philippines, whence they were transported to Fukien by Chinese merchants around 1608. Both potatoes and peanuts can be grown in less fertile soil than other crops.

13. Ta-yuan was a Central Asian state, now within the Uzbek S.S.R.

14. Soybeans first appeared in Chinese written records in the second century B.C., although it is very likely that they had been known and used for a long time before then.

15. The broad bean (also known as horse bean in the United States) is called *ts'an-tou,* silk worm bean, in Chinese. The pods of many kinds of beans are shaped more or less like a full-grown silkworm, and the term had probably come into use because the broad bean ripens at about the time the silkworm season begins.

Figure 2–1. Bathing silk-moth eggs [Ch'ing addition].

2

CLOTHING MATERIALS

Master Sung observes that, Man being the highest of all forms of life on earth, his five senses and the numerous organs are all completely present and should be preserved. Members of the aristocracy are clothed in flowing robes decorated with patterns of magnificent mountain dragons, and they are the rulers of the country. Those of lowly stations would be dressed in hempen jackets and cotton garments to protect themselves from the cold in winter and cover their nakedness in summer, in order to distinguish themselves from the birds and beasts. Therefore Nature has provided the materials for clothing. Of these, the vegetable ones are cotton, hemp, *meng* hemp, and creeper hemp; those derived from birds, animals, and insects are furs, woolens, silk, and spun silk. All the clothing materials [in the world] are about equally divided between vegetable and animal origins.

The ingenious loom was first invented by a divine maiden, who brought the skill to mankind. Weaving progressed from plain to figured patterns, and woven splendor was created out of embroidered stuff. But, although silk looms are to be found in all parts of the country, how many persons have actually seen the remarkable functioning of the draw-loom? Such words as "orderly government" [*chih,* i.e. the word used in silk reeling], "chaos" [*luan,* i.e. when the fibers are badly entangled], "knowledge or good policy" [*ching-lun,* i.e. literally the warp thread and the woven pattern] are known by every schoolboy, but is it not regrettable that he should never see the actual things that gave rise to these words? I shall therefore first describe the raising of silkworms, in order to make known how silk originates. It is to be observed that beauty and plainness, high status and low in the world of man are provided for accordingly by Nature.

[SILK]

Silkworm Eggs

About ten days [after formation of the cocoon], the chrysalis becomes a silkworm moth and emerges by piercing the cocoon. The number of male and female moths are equal. The female stays immobile. The male flutters around, and mates with the first female moth he encounters. The mating continues for a half or an entire day before the two separate, after which the male dies of exhaustion. The female moth immediately begins laying her eggs, which are deposited either on sheets of paper or on cloth, according to local practice (in Chia-hsing and Hu-chou [both in the Lake T'ai region in Chekiang province] thick mulberry bark paper is used; this can be reused in the next year). One moth will produce more than 200 eggs, all adhering naturally to the paper in a single layer across the sheet without the slightest piling up. These eggs are stored by the silkworm raiser for use in the following year.

The Bathing of Silk-moth Eggs

Only the silk-moth [eggs] of Chia-hsing and Hu-chou prefectures are put through the bathing process. In Hu-chou the method consists of using either rain and snow water or lime [water], while in Chia-hsing brine is used. [In the latter instance], for each egg sheet two pints of brine are drawn from a salt storage bin and diluted with water in a bowl. The egg sheet is allowed to float in this solution [Figure 2.1]. (The same procedure is also used for a lime bath.) The bathing begins on the twelfth day of the twelfth lunar month and terminates on the twenty-fourth day, lasting for a total of twelve days. The egg sheet is then lifted out [from the brine or lime water] and dried in the heat provided by a weak fire. It is next carefully stored in a chest or box away from any breeze or moisture, until the time of Ch'ing-ming [i.e. the third lunar month] when the eggs will be hatched.

The time for bathing the silk-moth eggs with rain and snow water is the same as that with brine or lime water. [In the former method, however,] the egg paper is spread out in a woven bamboo tray and held down by small stones placed on the four corners of the paper. The tray [with the egg paper] is placed on a roof top, and left at the mercy of frost, snow, rain, wind, and thunderstorms for twelve full days before the eggs are taken down for storage.

The reason for bathing the silk-moth eggs is that the inferior eggs will die off in the process.[1] Thus, there can be a saving in mulberry leaves, while the yield of silk also will be high. No bathing is necessary for silk-moth eggs of the Late variety of silkworms.

Avoidances for Silk-moth Eggs

[Directly after the silk-moth eggs are laid], the egg papers are held in square frames made of four pieces of wood and hung up in the breeze. The frames should be suspended from rafters and beams inside a house, so as to avoid the sun. Smoke from burning *t'ung* oil and coal fire should also be avoided. In winter the glare of the snow must not reach the egg papers, since the insides of the eggs will shrivel up completely once touched by that light. The egg papers, therefore, should be immediately put away when there is a heavy snowfall. On the next day, when the snow is over, they can be brought out and hung up again. This continues until the twelfth lunar month, when the eggs are put through the bathing process and then stored.

Varieties of Silkworms

There are two kinds of silkworms, the Early and the Late. Each year the [first crop of the] Late variety hatches some five or six days ahead of the Early kind (except for those of Szechuan province, which are different) and also forms the cocoons before the latter does.[2] The cocoons of the Late variety are in weight lighter by one-third. When the Early silkworms begin to form cocoons, the Late worms have already become moths, laid eggs, and so are ready for a new [i.e. late] crop [within the same season]. (The chrysalis of the Late silkworm is absolutely inedible.)

Each of the three ways of bathing the silk-moth eggs should be followed carefully and consistently throughout. In case of error, such as soaking in brine the egg papers that ought to be given the rain and snow water treatment, the result will be that no worms can be hatched!

The cocoons are of two colors only, yellow and white. Only the yellow kind is produced in Szechuan, Shensi, Shansi, and Honan provinces, whereas in Chia-hsing and Hu-chou prefectures only white cocoons are produced. When a white male is crossed with a yellow female, the offspring will make light brown cocoons. Yellow silk will turn white after being washed with hog-fat soap, but still it can not be dyed into light-blue or peachblossom-pink shades.

Cocoons have several forms. The Late-variety cocoon is shaped like a thin-waisted gourd; the cocoons made by rain-and-snow water bathed silkworms are either long and pointed, like a yew nut, or round and slightly flattened like a walnut. Further, there is another kind of silkworm which eats mud-spotted leaves without ill effect. This is known as "lowly silkworm," and its yield of silk is particularly abundant.

In accordance with their appearance, silkworms have been designated as pure white, tiger stripes, pure black, spotted, and so on. But they all produce silk in the same way. Recently some small silkworm raisers have crossed an Early male with a Late female, thus hoping for the production of an excellent breed. This is an unusual occurrence worth noting.

In places like Ch'ing-chou and Yin-shui [both in Shantung province] there is a wild silkworm [3] that makes the cocoon [without human care], its natural habitat being old trees. Garments made of this silk are rainproof and dustproof. The moth flies away immediately upon emerging from the cocoon, and its eggs are not preserved on paper sheets. Wild silkworms are found but rarely in other localities.

Hatching and Raising the Silkworms

Three days after Ch'ing-ming the silkworms will hatch out of the eggs without being huddled in cloth wrappings. The silkworm room should face southeast, all cracks [in the walls] should be papered to prevent draft, and the room ought to be furnished with a ceiling. If the weather is cold, the room is kept warm by means of a charcoal fire. Newly hatched silkworms are fed mulberry leaves that have been cut into shreds. No damage will be done to the knife blade if the leaves are not wrapped in straw while being cut. Freshly picked leaves are put in earthen jars so as to prevent wilting.

Before the silkworms have had their second moulting, they should be picked up with small, round-pointed chopsticks when they are being changed from one basket to another, but after the second moulting they may be moved with the fingers. The frequency of basket changes depends on the diligence of the silkworm raiser. The result of infrequent changes is that the silkworms are often pressed to death by the weight of both the old thick leaves and the droppings.

Silkworms [of the same crop] will moult at about the same time, going to sleep [4] after emitting some threads of silk. If the baskets are changed at this time, care must be taken that all traces of the old leaves are cleaned out. Should any leaves, entangled with bits of silk and left in the basket, be eaten by the silkworms after coming out of the moulting, the worms will die of the "swelling" sickness.

In case of very hot weather after the third moulting, the silkworms should be moved to an airy and cool place that is free from wind and draft. After the final moulting the silkworms are to be given twelve feedings before changing baskets, more frequent changings will later result in producing coarse silk.

Avoidances for Silkworms

Silkworms are afraid of both fragrance and bad odors. They die if exposed to such smells as from the burning of bones, from the cleaning of latrines, and often also from the frying of fish or other oderiferous meats next door. They likewise will die if exposed to such smoke as that from coal stoves, incense, or sandalwood burners. They are also injured by the smell of night pots that some indolent woman may carelessly shake up. The southwest wind is dangerous: when it is too strong an entire trayful of silkworms are known to have stiffened and died. When bad odors approach, leftover mulberry leaves should be burned, and the smoke will ward off the odors.

Mulberry Leaves

Mulberry trees are grown everywhere. In Chia-hsing and Hu-chou [new trees are planted] by pressing the branches [of old trees] into the ground. The fresh side-branches put forth by a mulberry tree this year are gradually pulled downward to the ground with bamboo hooks, then in the winter they are covered with earth. In the following spring each of these branches will have taken root at every joint, whereupon they are cut apart and planted separately. Such mulberry trees do not flower or bear fruit, but all the essence of the tree is concentrated in the growing of leaves. These leaves can be cut with scissors whenever they are needed. When a tree reaches seven or eight *ch'ih* in height the top is cut off, so that [branches will] luxuriantly spread sideways, and the leaves can be easily reached for picking without ladders or climbing.

Another method consists of planting with seeds. At the beginning of summer [5] the ripe purple mulberries are crushed and rubbed in yellow mud water, after which the entire pulp is poured to the ground. By autumn of the same year the seedlings will be over one *ch'ih* high, and in the following spring they are transplanted. These can also grow into luxuriant trees if they are manured frequently and diligently. Some of them, however, will bloom and bear fruit, in which case the leaves will turn out to be sparse and thin.

Still another type of mulberry tree is called the "flowering mulberry," of which the leaves are so thin as to be useless. However, when [twigs of the leafy type are] grafted onto it, this tree is also capable of growing thick leaves.

In addition, there are three kinds of *che* trees, of which the leaves are used to supplement the mulberry leaves. *Che* trees are seldom seen in Chekiang,[6] but are most abundant in Szechuan. There the poor families cultivate the Chekiang variety [of silkworms] using *che* leaves as feed when mulberry leaves are insufficient. The physical properties of *che* leaves are the same as those of mulberry leaves. The cocoons of *che*-fed silkworms are called "tough cocoons," their silk is considered to be the most durable, and is used in the manufacture of lute strings, bow strings, and the like.[7]

Leaves are picked with shears. The sharpest iron shears are the products of T'ung-hsiang district in Chia-hsing prefecture [in Checkiang province], unequalled elsewhere. Cutting off the leaves together with the twigs attached to them will enable a second full growth of leaves and branches to take place in the following months; thus a plentiful supply of leaves is insured at a minimum cost of labor. When the leaves of this second growth are gathered in midsummer to feed the [second crop of the] Late variety of silkworms, only the leaves are picked, but the twigs are not cut. After this second growth has been used, still a third growth of leaves will mature in the autumn. The Chekiang people allow them to fall naturally with the coming of frost, then sweep them up as sheep feed, gaining therefrom good profits from the making of woolen materials.

Care in Feeding

Silkworms are allowed to eat wet leaves after their final moulting. The leaves picked on a rainy day can be fed directly to the worms; if picked on a fair day, however, the leaves should be sprinkled with water before feeding. This will make the silk glossy and lustrous. Before the final moulting takes place, however, the leaves gathered on rainy days should be hung by strings under the eaves and occasionally shaken, so as to enable them to dry in the

Figure 2–2. Transferring mature silkworms [Ch'ing addition].

breeze. If the leaves are dried by patting them with the palms of hands, they will lose moisture, so that later the silk will also have a dried-up appearance.

Just before each moulting period the silkworms should be well fed, then no harm is done if fresh leaves are given to them a few hours late after they emerge from moulting. The most unhealthful leaves are the damp ones gathered while there is fog. Leaf-picking should never be done on foggy mornings, but one should wait until the fog has lifted and, whether rain or shine, the leaves can then be gathered. Likewise, dew-covered leaves should not be picked until they have dried.

Silkworm Diseases

A previous section [i.e. avoidances for silk-moth eggs] has already described the sickness and destruction of silk-moth eggs. After hatching, the young silkworm is allergic to dampness, heat, and pressure, and these should be the care of the silkworm raiser. When the box of silkworms is changed during the first moulting period, the box should not be closed if it is coated with Chinese lacquer, otherwise the worms will emit foam and sicken.

These are the symptoms of sickness: if the head of a silkworm becomes shiny, its entire body yellow, and its head increases in size while the tail part contracts; or if when a moulting period is due a worm remains wakeful, wandering about but eating little; such are the signs of sickness, and the affected silkworms must immediately be eliminated so as not to infect others.[8] The healthy and strong silkworms always stay on top of the leaves while moulting. Those that stay covered under are either weak or lazy, and their cocoons are also thin. There are other silkworms which do not know how to spin a closed cocoon but, emitting their silk in every which way, produce wide open nests [instead of the regular cocoons]. These worms are not lazy but merely stupid.

Mature Silkworms

When silkworms have completed their feeding and reached maturity, they will only await the proper hour [in order to commence cocoon-making]. As the young worms generally hatch from the eggs between eight o'clock in the morning and twelve noon, so the mature worms also [begin to] make their cocoons during those hours.

The maturity of a silkworm is indicated by its throat, which turns transparent when it is fully mature [and thereupon it is caught and transferred to the bamboo screen and rice straw where it will spin the cocoon]. If the silkworm is caught too soon the yield of silk will be small, but if too late some silk will have been already spun out, so that the cocoon will be thin. The best result would be that every single worm is caught at exactly the correct time, and this will require good eyesight as well as deft fingers [Figure 2.2]. Black silkworms are the most difficult to catch at the proper time, as the transparent parts in their bodies are not easily seen.

The Spinning of Cocoons

The method used in Chia-hsing and Hu-chou for the spinning of cocoons is absolutely the best. People of other provinces, not knowing how to use [the controlled heat from] fires, allow the silkworms to take a free hand in spinning cocoons, some of which are even formed deep inside stacks of straw, or within crates and boxes, where the silk is not tempered by either wind or heat. That is why the pongee of north China and Chang-chou [Fukien] and the silks of Honan and Szechuan deteriorate so easily, whereas a garment made of Chia-hsing or Hu-chou silk, on the other hand, will remain intact even after numerous launderings.

The [Chekiang] method is as follows: split-bamboo screens equipped with straw cocks are placed on wooden supports about six *ch'ih* from the ground, under which braziers of burning charcoal are placed (the charcoal should be explosion-free) some four or five *ch'ih* apart from one another. [Figure 2.3]. The fire is kept low when the silkworms are first put on the straw cocks. As the worm starts to spin out the silk, he is induced by the warmth of the fire to stay on at the same spot instead of wandering around. Then when the outlines of the cocoons are defined, the fires are increased by the addition of half a catty [of charcoal] into each brazier. Thus, the silk can be dried as soon as the worm spins it out; that is why such silk is long-lasting.

The cocoon room should not have a low ceiling, because while the lower part of the room must be warm, the upper part should be kept cool. Cocoons that are formed directly above the fires are not to be used for producing silk-moth eggs; the breeding moths should come from cocoons further away from the heat.

The straw cocks on top of the bamboo screens are made by twisting bundles of rice or wheat straws of equal lengths, and then securing them onto the screens. Persons who make the straw cocks best are those with a strong and sure hand. Some straw should be spread over the surfaces of the screens, so as to prevent the silkworms from tumbling through the cracks between the bamboo slots, to the ground or into the fire.

Gathering the Cocoons

The cocoons are gathered from the bamboo screens three days after they are formed. The loose silk or floss, also known as stripping, on the surfaces of the cocoons is usually bought at a low price (100 cash per catty) by old women in Hu-chou, who spin it into yarn with the use of copper coins [as spindles], and weave it into the "spun-silk" of Hu-chou. After the floss has been taken off, the cocoons are spread out in large trays and placed on shelves to await reeling [Figure 2.4]. The silk fibers will decay and break if the cocoons are put away in covered containers.

Pests

Birds, rats, and mosquitoes are the three enemies of silkworms. The worms are safe from birds, however, in the cocoon stage, and the Early variety

Figure 2–3. Mature silkworms spinning cocoons on split-bamboo screens.

Figure 2–4. Gathering of cocoons into large trays [Ch'ing addition].

Figure 2–5. Separating single cocoons from the double cocoons and the multiple-worm cocoons [Ch'ing addition].

Figure 2–6. Reeling silk fibers.

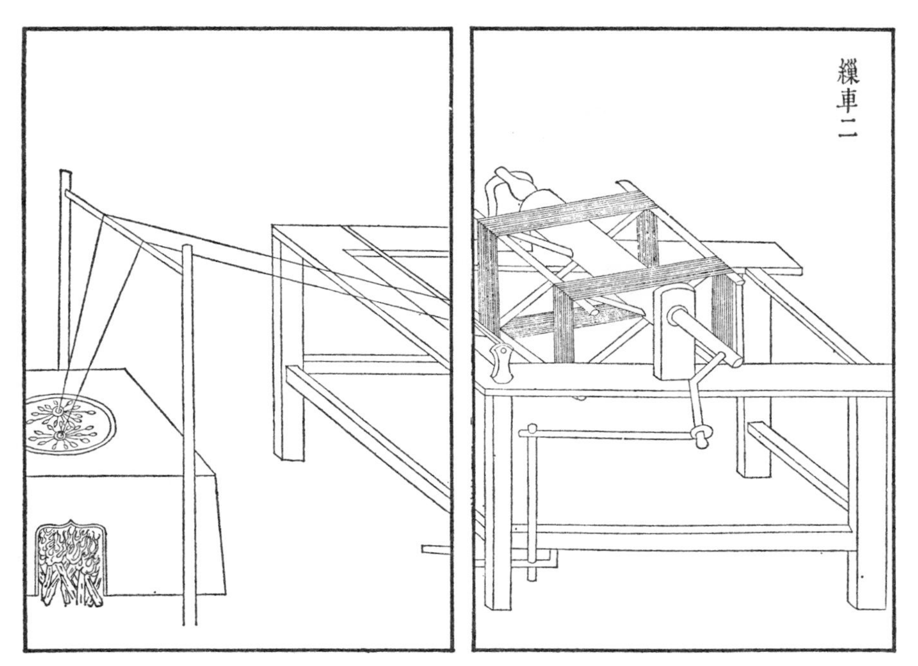

Figure 2–7. A silk reel used in south China [Ch'ing addition].

Figure 2–8. Another silk reel used in north China [Ch'ing addition].

silkworms are not troubled by mosquitoes, but the menace of rats is with the worms from beginning to end. There are many different methods of preventing or getting rid of these three pests, depending on the ways of particular individuals. (A silkworm will immediately die and decay, if it eats leaves soiled with bird droppings.)

Sorting the Cocoons

For reeling silk, only the properly oval-shaped single cocoons should be chosen, because then the fibers will not become tangled. The double cocoons and the four- or five-worm cocoons are sorted out and put aside as a separate lot [Figure 2.5] for making spun or florette silk; if these defective cocoons are reeled, the resultant silk will be extremely coarse.

Making Silk Wadding

The fibers of such materials as double cocoons, remnants in the pot after reelings, and pierced cocoons from which the moths have emerged, are all either entangled or broken. They cannot be reeled into ordinary silk, but are made into silk wadding [in accordance with the following procedure]. The [cocoons are] first boiled in an aqueous solution of rice stalk ash (lime solution is not suitable), and then transferred to a basin of clean water. After clipping his thumb-nail, the worker picks up a group of four cocoons at a time, and bores a hole at one side of each cocoon with his thumb. This operation is continued until four groups of fours or a total of sixteen cocoons are treated. The thumb-pierced opening of each cocoon is further enlarged by pressing through it with the worker's fist. These sixteen cocoons are then put under small bamboo bows [for a still further stretching]. This is what [the ancient philosopher] Chuang-tzu called "washed spun silk." The secret of the whiteness and purity of Hu-chou products lies in nothing else but superior skill. When a cocoon is put under the bow, it ought to be done very quickly and deftly, so that it can be easily stretched while still wet. If the handling is slow and the bowing is done when the water has already drained off, then the silk fibers will become matted, and the product will not be pure white.

The silk wadding made from the remnants of silk reeling is known as "pot-bottom" silk. Sewn inside the linings of garments or bedclothes, silk wadding will ward off the intensest cold and is called quilting.

Making silk wadding requires eight times as much labor as silk reeling, the result of [one man's] work for an entire day being only four ounces of wadding. The Hu-chou silk woven from yarn spun from this silk floss fetches a rather high price. A further variety is the stuff woven on draw-looms and called "figured spun silk," which is even more costly.

Reeling the Silk Fibers

For reeling silk, a reeling machine must be made ready first, of which the design and dimension are given in the accompanying illustrations [Figures 2.6,

2.7 and 2.8]. A vat of boiling water [is used to kill the chrysalis and to soften and dissolve the sericin]. The coarseness or fineness of the reeled thread depends on the number of cocoons thrown into the water at one time. One reeler is able to reel thirty ounces of silk per day, but if the silk is intended for weaving kerchiefs, then the amount is reduced to twenty ounces, because the fibers have to be considerably longer. For the reeling of damask for clothing and gauze silk, twenty cocoons are thrown into the water at the same time, but for that of kerchief silk only about a dozen are put in.

As the cocoons boil, the surface of the water is stirred with bamboo sticks, by this means the ends of the silk fibers are made visible. The ends are first taken by hand, then are passed through the bamboo eyelets [of the reeling machine]; they are next placed over the guide rolls (made of bamboo cylinders, resembling incense-stick containers) and guide rings, then fixed to the thread-passing rod and thence to be wound by the winch. In case one fiber breaks off during the mechanical process of reeling, the loose end is found and put in place; it is not necessary to tie it. The even distribution of the silk fibers on the winch is attributed to the [skillful application of the] guide rolls and the thread-passing rod.

The method of silk reeling in Szechuan is somewhat different. Here the reeling machine is placed directly over the vat, and is fed with four or five silk fibers gathered from the vat at a time. Two persons are employed to look for the fiber ends in the vat [Figure 2.8]. This is not as good a method, however, as that practiced in Hu-chou [as described above].

The fuel wood used in silk reeling should be extremely dry and smokeless, so as not to damage the lustre of the silk. There are [two phrases consisting of] six Chinese words that constitute the secret of preparing high-quality silk. The first is "dry out of mouth"; that is, drying the silk with a charcoal fire as the cocoons are being spun by the silkworms. The second is "dry out of water": as the silk is being wound onto the winch of the reeling machine, a small charcoal fire is made in a brazier and placed about five *ts'un* away from the winch. As the latter rotates and stirs the air nearby, [the silk] is dried in the heat as it passes through. This is called "dry out of water" (the fire is not needed if the day is sunny and there is a good breeze).

Spooling the Silk Fibers

The first step toward the preparation of silk fibers for weaving is spooling [Figure 2.9]. At a well-lighted spot under the eaves, a skein frame is set up by afixing four bamboo sticks to a wooden board, which is set on the ground. The [reeled] silk is stretched on the frame. At the point eight *ch'ih* high on a pillar nearby a device is set up which consists of a small, semicircular bamboo hook hanging at an angle, through which the silk fibers are passed. The ends of the fibers are attached to a hand spool which is held and rotated by the [operator's] hand, so that [the silk is wound and] ready to be converted into weft or tram yarn. In case some fiber breaks off during the spooling process, [the hook] can be lowered by means of a lever that consists of a small bamboo rod with a suspended stone at its end [to serve as a weight].

[*Spinning Silk Fibers into*] *Weft Yarns*

After spooling, the silk threads are made into warp and weft yarns [by means of spinning]. The warp yarn consumes less silk than the weft yarn, and the general proportions being four-tenths for the warp and six-tenths for the weft. The spools of silk fibers that are intended for making weft yarn are first sprinkled with water, then twisted with a spinning wheel [Figure 2.10] carrying a rotary ring, and finally spun onto the bamboo rod or bobbin (made of slender "arrow bamboo").

Warp Frame

After spooling, the silk threads are drawn into warp strands for weaving. [In the process of drawing,] a straight bamboo rod is pierced with thirty holes, each of which is connected with a split bamboo ring, known as "slippery eye." The rod is held in a horizontal position by fixing its two ends separately on two supporting pillars. The silk thread of each spool is first drawn through a separate bamboo ring, then passed through one of the holes of a palm-shaped "warp guiding-rake" or *Chang-shan,* and finally wound on a warp rack [Figure 2.11].

When a sufficient amount of the silk strands or *roverings* has been wound up, they are [taken down from the rack], and wrapped on [the roll of] a warp frame [Figure 2.12]. These threads are separated alternately one up and one down into two groups, by means of two pieces of bamboo. After that, the threads are first drawn through a combing harness (this is not the one used in weaving), [and then fastened to the warp beam]. When this is done, the beam is placed at a distance of fifty to seventy *ch'ih* away from [the roll of] the warp frame. If the threads are to be sized, the sizing is carried out at this point; if this is not required, then the threads are rolled onto the warp beam and thus made ready for weaving.

Sizing the Silk

The material commonly used for sizing is the starch obtained from the gluten of wheat. Sizing is necessary for gauze [*lo*] and thin gauze [*sha*], but for damask and plain silk fabrics sizing can either be used or not. If the thin gauze is [later] to be dyed in such colors as not to show any of its natural hues, a solution of ox glue is used in sizing, and the product is called "thin-glue gauze." The sizing liquid [feeder] is held on top of the combing harness, which is moved back and forth through the silk threads in order to size and subsequently dry them. On a clear and sunny day the [sized threads] can be dried in a few moments, but on cloudy days a breeze would be needed for drying.

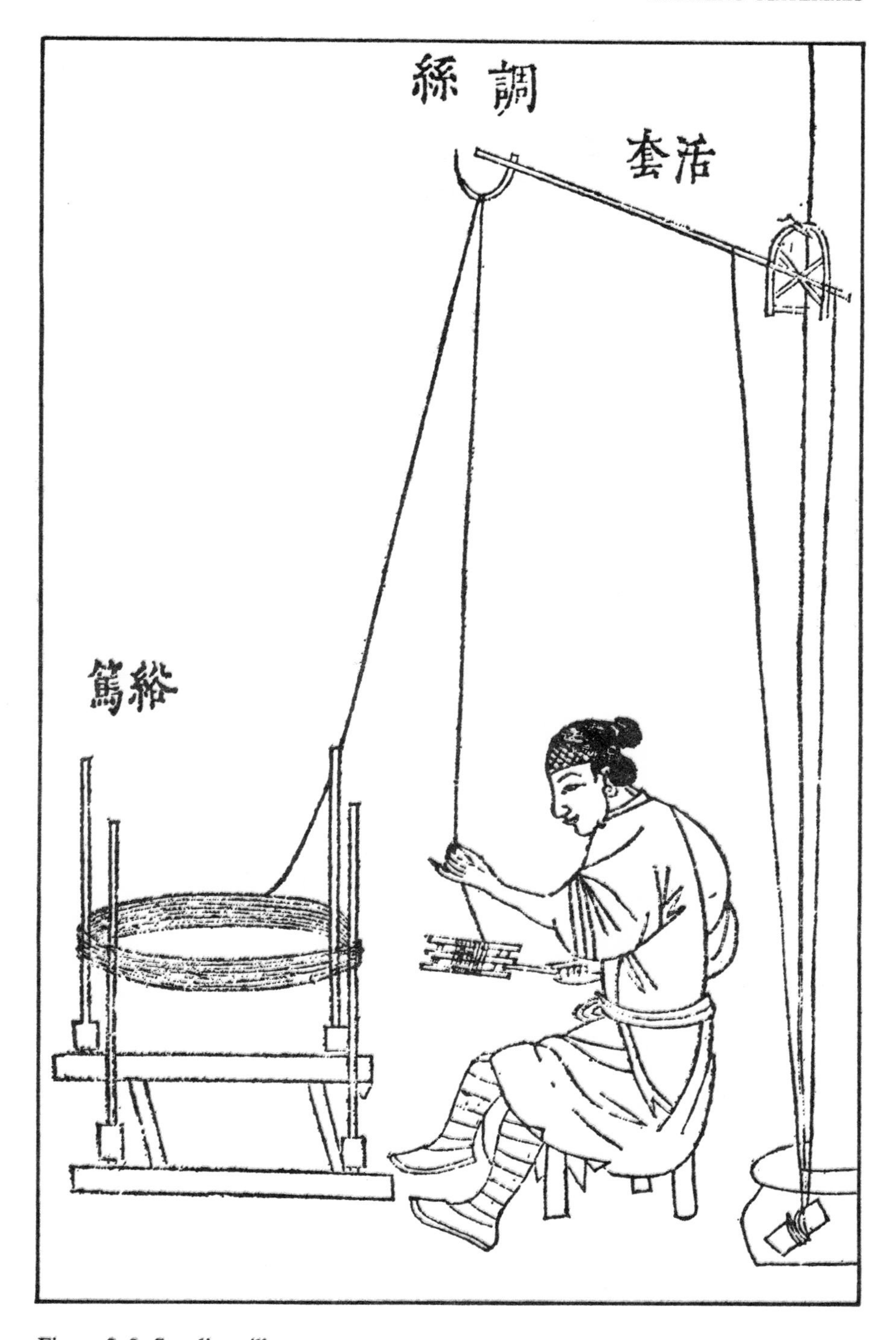

Figure 2–9. Spooling silk.

Figure 2–10. A spinning wheel for making yarns.

Figure 2–11. Drawing silk threads into warp strands.

The Border of Silk Fabrics

All silk fabrics, whether damask or gauze [or any other kind], must have specially woven borders. These borders measure twenty-odd [warp] threads on each side. The threads used for weaving borders must be first sized and then dried by [the above-described method] of using a combing harness. In order to reduce the number of warpings, which is a laborious process, the warp threads for the weaving of damask and gauze are usually three hundred to five or six hundred *ch'ih* in length. In order to show that the fabric is up to its proper length in each bolt, ink marks should be stamped on the border [at regular intervals].

The warp threads for weaving the borders of the silk fabric are not rolled on the warp beam but are wound on a different cylinder of the loom.

Warp Count

The harness used for weaving gauze and thin gauze usually consist of 800 "teeth" or heddles, whereas that for damask and pongee contains 1,200 "teeth." Through the "hole" or eyelet at the central part of each "tooth" four [un-sized] warp threads pass at the same time. In the case of sized threads, four of them are combined into two warp yarns and subsequently drawn through an eyelet. It follows, therefore, that the total number of warp threads

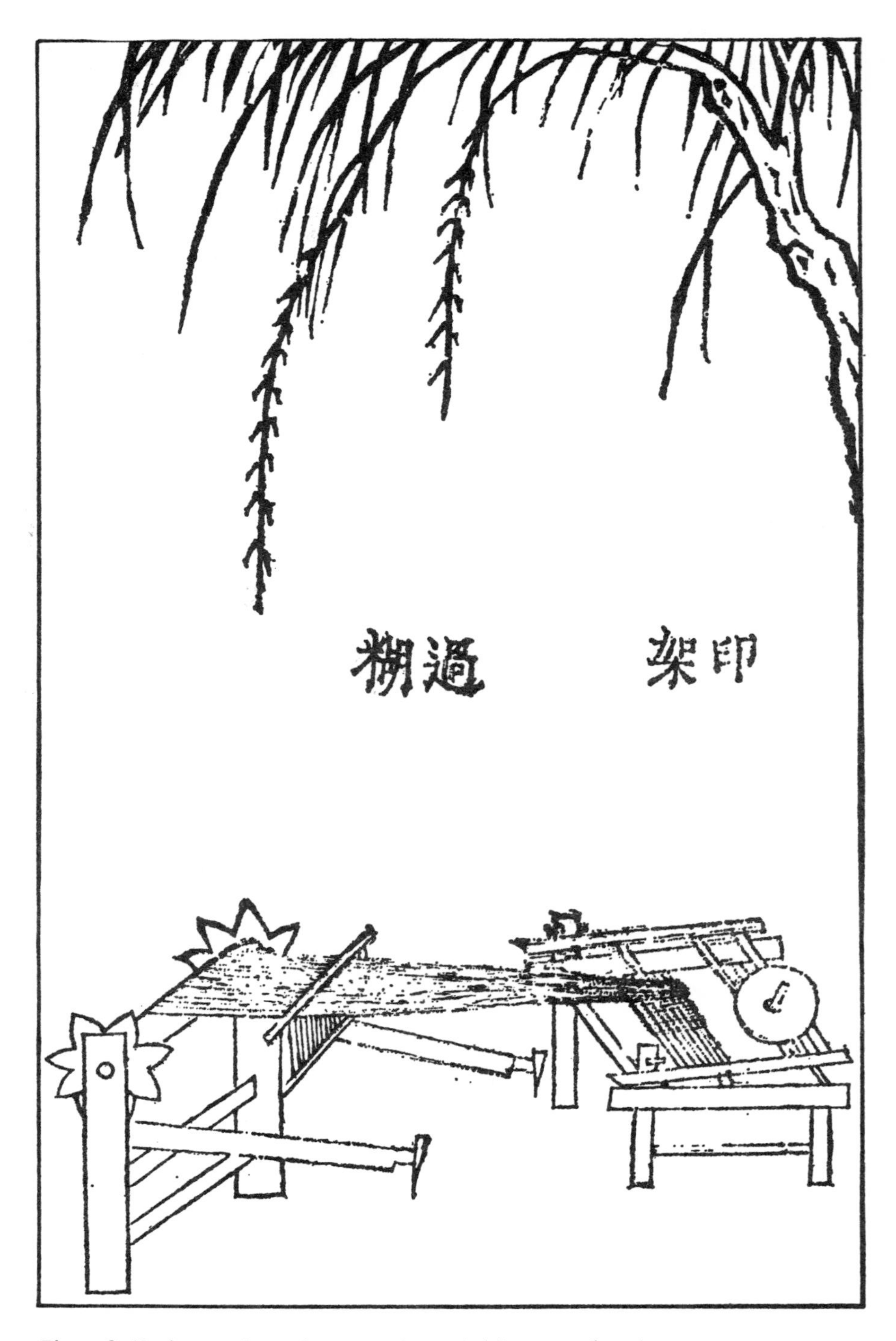

Figure 2–12. A warp frame for separating and sizing warp threads.

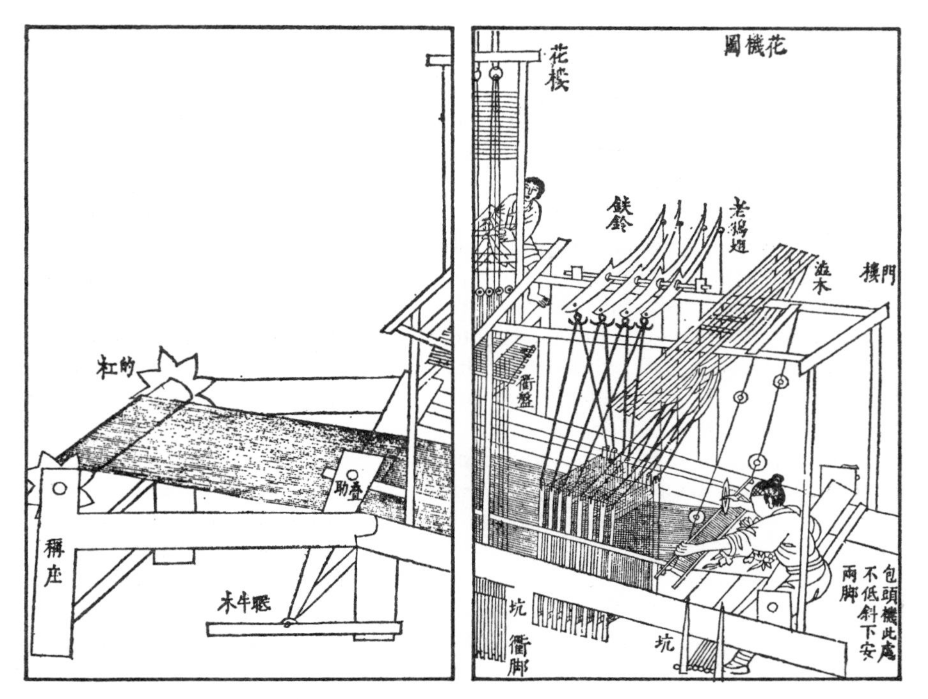

Figure 2–13. A drawloom for figure-weaving.

for weaving gauze and thin gauze is 3,200, and that for damask and plain silk fabric is 5,000 to 6,000. In ancient [Chinese] writings eighty warp threads were termed as one "strip"; the thick damasks and pongees of today, therefore, are the equivalent of the "sixty-strip fabric" of olden times.[9]

For weaving figured fabrics, the "dry out of mouth" and "dry out of water" silk fibers from Hu-chou and Chia-hsing must be used as the foundation or warp, because they will not break off in the repeated processes of lifting and shedding. Silk fibers produced from other provinces can barely stand the strain of figure-weaving, and their use will result in an inferior product.

Drawloom

The drawloom frame [Figure 2.13] has a total length of sixteen *ch'ih*. At the upper part of the frame a *hua-lou or* "figure tower" is located. Below it a *ch'ü-p'an* or "drawer board" is placed [for keeping the many heddles of the weaving harnesses in their proper positions]. [On the lower end of each heddle] hangs a separate *ch'ü-chiao* or "rigid rod" or weight (the latter is made of water-polished bamboo rod, numbering 1,800 to a loom). A pit of about two *ch'ih* deep is dug in the ground at a spot directly beneath the "figure tower" to make room for the "rigid rods" (where the ground is damp the pit can be replaced by the construction of a two-*ch'ih* high frame) [on which the loom rests]. A cross-plank is provided in the figure tower for the drawboy to sit or stand on.

The warp threads are wrapped [in parallel order] on a warp beam, placed in the front of the loom. [From here the threads pass to the back of the loom where they are attached to the cloth roll]. Toward the middle part of the loom, two *tieh-chu mu* or "driving shafts" are connected separately with two wooden poles four *ch'ih* long, which in turn are attached separately to the two ends of a reed. For weaving gauzes, the shafts should be lighter by about ten catties in weight than those used for damask and pongee. For weaving plain gauze or soft thin gauze, or damask and pongee showing small, scattered designs, only two cross beams need be added to the [lighter loom] used for weaving plain gauze, and one weaver operating the loom with treadles is enough to finish the task; in this case no drawboy is needed to man the "figure tower," nor need the "drawer board" and "rigid rods" be set up.

The frame of the [draw] loom consists of two sections. [See Figure 2.13 again.] The front section is level. The [back] section declines one *ch'ih* from the "figure tower" to the weaver's body [or cloth roll], thus increasing the force of the driving shafts. For weaving kerchiefs and other small, delicate items, however, a [small] loom of level surface is used; and the loom treadles are operated by the two feet of a seated weaver. This is because the very fine silk fibers used for making such fabrics may not be able to withstand the force of the driving shafts.

The Waist [*or Small*] *Loom*

The drawloom need not be used in weaving such silk fabrics as "Hang-chou" pongee, "gauze-weave" pongee, light silk, plain silk, "silver stripe" thin gauze, and gauze for making caps, hats, etc.; a small loom is used instead [Figure 2.14]. The artisan works sitting down, with a square piece of cured leather placed under the seat.[10] Because the motions of weaving depend entirely on [the strength of] the weaver's waist and lower spine, the apparatus is therefore called a waist loom. When used in weaving hemp or cotton cloths, it produces better looking and more durable fabrics [than does the draw loom]. Unfortunately, however, the practice has not yet become widespread.

Figure Designing

The artisan who makes the figure design for weaving is the most ingenious person. The pattern and color of a fabric design are first painted by an artist on a piece of paper. The artisan takes the painted design and translates it, precisely, in terms of the silk threads used, down to the last thousandth of an inch, and makes a pattern for weaving. This pattern is hung up in the "figure tower" of the drawloom; [it guides the drawboy] to lift the correct "rigid rods" of the heddles. Even if the weaver does not know what the figure and color on the fabric will turn out to be, he has only to follow the figure design and interlace the weft and warp threads according to specifications. Then, lo and behold, when the shuttle passes the desired figure appears on the fabric.

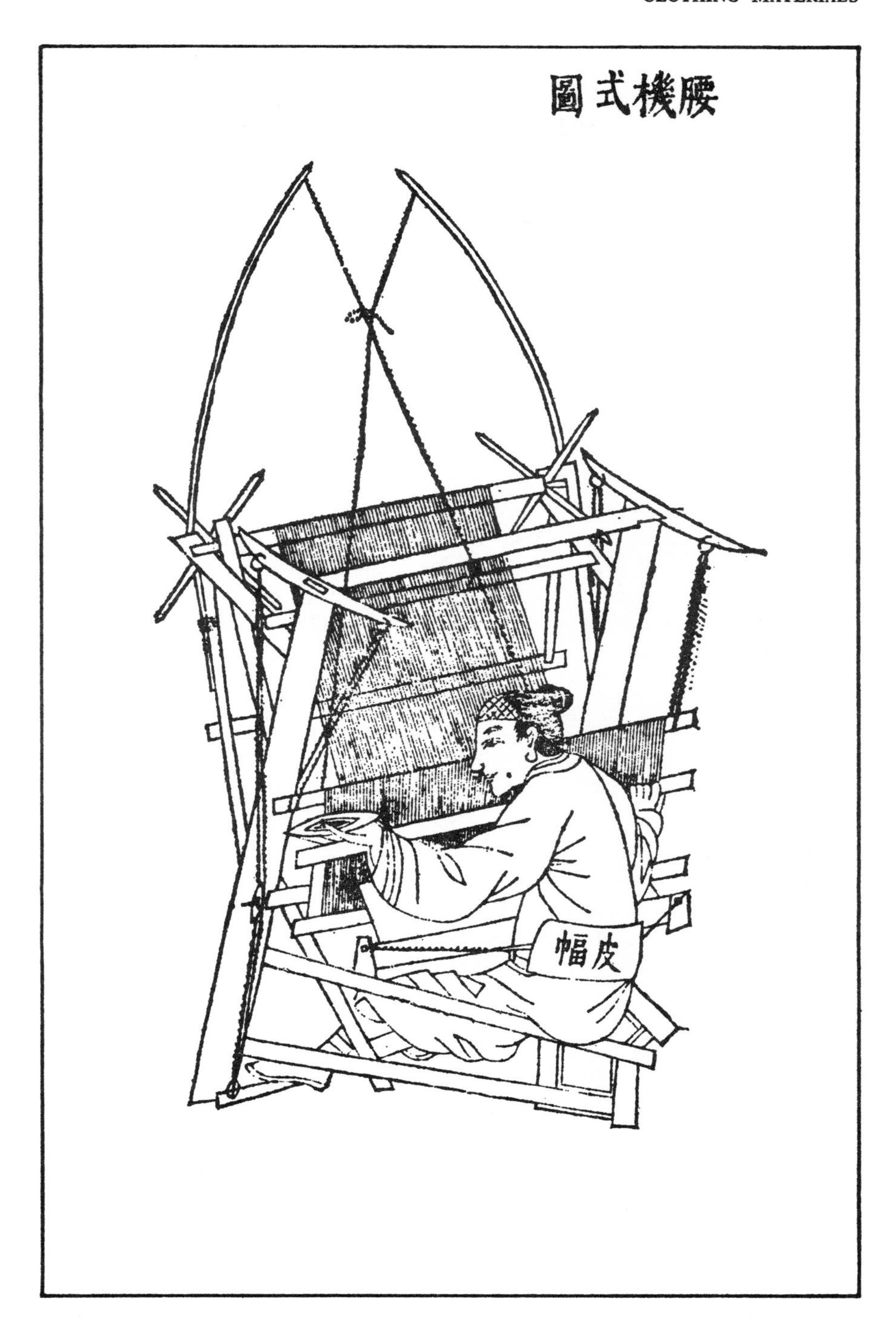

Figure 2–14. A waist [or small] loom.

The figure-patterns of damasks and pongees are formed by floating the warp threads over the filling or weft threads, while those of gauzes and thin gauzes are produced by combining [two or more] single threads or yarns in each weft strand. In the weaving of damasks and pongees the [warp] is lifted at every pick of the shuttle, but in the case of gauzes it is lifted only at every other pick.

The looms first devised by divine maidens have indeed been brought to perfection through human skill.

Warping

The [laborious process of] passing each warp thread or yarn through [a separate heddle-eye of] the weaving harness is performed by four persons, sitting alongside each other. The man in charge of passing the warp threads through the heddle-eyes holds the harness frame ready and waits for the threads to arrive. When the threads have been passed through [the proper eyelets of] the harness, they are held together with two fingers until some fifty or seventy threads are gathered; then these are tied together [to prevent slipping or tangling]. The key to keeping the threads in good order without tangling lies in the use of a separation bamboo rod. In case a thread breaks, it needs only to be pulled out a few inches; when it is released after the ends have been tied, it will snap back into position. Such is the natural good property of silk.

Classification of Silk Fabrics

In a *lo* [11] or net fabric many tiny open ribs are woven for ventilation and coolness. The key to this weaving lies entirely in the function of the "soft" harness. The loom has two harnesses, one being "soft" and the other "hard," and each harness is manipulated by a cam-type lever, [as exemplified by the *lao-ya-ch'ih* and *se-mu* of Figure 2.13]. After the shuttle is passed for five or three times, (a seven-time passage being used for weaving the thickest *lo* fabric), the two sets of warp yarns are individually twisted with each other through the raising of the "soft" harness. The warp yarns [in the close vicinity of the twisting] are not interlaced with filling, thus creating many open roads. If an open-mesh effect is obtained by interlacing the entire length of warp yarns at a considerable reed spacing, then the resultant fabric is called *sha* or gauze. This weaving depends also on the two cams [for manipulating the harnesses]. The replacement of the two cams by eight cross beams [probably a dobby arrangement] takes place only in the weaving of figured damask and pongee. [When the warp yarns are interlaced with two oppositely twisted weft yarns], which are carried separately by a left-hand shuttle and a right-hand shuttle, the resultant fabric is called crinkled gauze or crepe.

The single-strand yarns are used for the ground warp of *lo* or net, the two-strand yarns for *chüan* or pongee, and the five-strand yarns for *ling* or damask. The structural figure of fabrics is classified as plain-weave and damask-weave. The former gives a dull appearance, while the latter exhibits a smooth surface with a characteristic luster. When silk filaments are [first twisted into yarns], then dyed, and finally woven, the resultant *tuan* fabrics are known as [warp-faced] satin and [filling-faced] sateen. (Silk yarns are also dyed prior to the weaving of plain pongee in north China.) In the weaving of *ch'iu-lo* or "autumn net" on a silk-pongee loom, the warp yarns are interlaced twice with a light shuttle carrying the thin filling yarn and once with a heavy shuttle carrying the thick filling yarn. The same interlacing pattern [is repeated throughout the entire length of the fabric] at a considerable reed spacing, thus creating many open roads. This weaving technique had its origin in recent times. The *ch'iu-lo* fabric manufactured in Kiangsu and Chekiang as well as a similar fabric made in Fukien and Kwangtung, which is known as *huai-su,* are all used for making the summer garments of high officials. The plain pongee, being inferior to the brocaded and/or embroidered silk fabrics, is used as the clothing material for the provincial and minor officials.

Boiling Off [*the Gum from Raw Silk*]

The silk in the woven cloth is still in its raw state, which will become "souple" after boiling. It is first boiled with an aqueous solution of rice-stalk ashes and then steeped overnight in a solution of lard soap. Next the silk fabric is rinsed in hot water, and its dazzling luster will appear. Some people use [the aqueous extract of] smoked Chinese plums [as a degumming agent], and the result is a slightly duller luster. When the warp yarns of the fabric are made of Early silk, and the weft yarns Late silk, there is a loss in weight of three-tenths after boiling. If both the warp and the weft yarns are made of fine quality Early silk, then the loss in weight amounts to only two-tenths.

After being boiled, [the fabric] is quickly spread out to dry in the sun. Next, in order to bring out all the luster in the silk, it is thoroughly polished with a large, round, smoothly ground piece of clamshell.

Dragon Robes

In our Dynasty the dragon robes [12] for Imperial use are woven in factories in Soochow and Hang-chou. [The drawloom used for such figure weaving is equipped with] a "figure tower" of fifteen *ch'ih* in height. Two highly skilled artisans [stay in the "figure tower" to] manipulate the "drawer board" in accordance with a prepared figure design. The shape of the woven dragon changes after every few inches of weave—[the design] being the result of the collective effort of all the [government] weaving establishments and not the work of any individual designer. The silk is dyed yellow before weaving. The weaving equipments are actually not different [from those used in ordinary figure weaving], but the special care of the artisans as well as the sums expended [on the dragon robes] are scores of times more than the ordinary. Thus do subjects show their loyalty and respect [for the Emperor]. It is not possible to know the minute details involved in the weaving of these robes.

Japanese Satin

Originating in the country of the Eastern barbarians, the manufacture of Japanese satin has spread to the [Chinese] coastal areas of Chang-chou and Ch'üan-chou. The silk used in this fabric is produced in Szechuan, whence merchants bring it to sell in exchange for pepper, which they take back on the return journey. The method of weaving [this satin] also came from the Japanese.[13] The silk is dyed before weaving, and silk floss is concealed between the interlaced filling yarns. After every few inches of weaving, the warp yarns [showing on the surface of the cloth] are napped to produce a [soft hairy] fabric of dark luster. This [flannel-type] satin was very popular with the northern barbarians at the market places, but it is extremely nondurable and easily soiled. If it is made into hats or caps, they will accumulate dust in no time; and when it is made into garments their collars will wear out in a matter of days. Nowadays it is reported as of little value by both Chinese and foreigners. Since it will become an unwanted commodity in the future, there is no need to note down the method by which this satin is woven.

COTTON TEXTILES

Rich and poor alike use cotton clothing in cold weather. In ancient writings cotton [14] was called *hsi-ma* or "nettle-hemp," and was widely grown in the country.

There are two species, the tree cotton [*Ceiba pentandra*] and the cotton plant [*Gossypium indicum*]. [Of the latter the] flowers are of two colors, white and purple, the white being nine-tenths and the purple one-tenth of all the cotton planted.

Cotton is planted in the spring, and the bolls form in the autumn. The bolls are picked day by day as they mature and split open; therefore, not all cotton can be picked at the same time. The cotton seeds, which are tightly fastened to the fibers in the bolls, are eliminated by means of a cotton gin [Figure 2.15].[15] Then the clean fibers are bowed [Figure 2.16] (and subsequently used as cotton padding for garments and bed clothes. This is the last step in the process.)

After bowing, [the cotton fibers] are straightened and made into long slivers by rolling them with wooden boards [Figure 2.17], so as to be rendered ready for the spinning wheel [Figure 2.18], where the slivers are drawn out to the desired size and twisted into [single-strand] yarns. These yarns are spooled [and used directly as weft in weaving]. [Figure 2.19 seems to suggest that four spools of the single yarns are combined into two multiple-strand warp yarns through the use of a foot-operated spinning wheel.] The resultant warp yarns are wrapped in parallel order on the loom beam, and then interlaced with the weft yarns to form fabrics. A skillful spinner can hold a device consisting of three "tubes" in one hand [for guiding] the spinning of yarns onto the wheel. (The resultant yarns will be weak if the spinning is done too quickly.)

Figure 2–15. A cotton gin for separating fibers from seeds.

Figure 2–16. Bowing of cotton fibers.

Cotton fabrics are produced in all parts of the country. The best weaving is done, however, in Sung-chiang [Prefecture], and the best sizing and dyeing in Wu-hu. The tightly woven fabric is strong and long-lasting, while a loosely woven fabric is easily breakable. The best stone rollers [used for calendering the fabrics to give a smooth finish] are made of the cold-natured and fine-textured stones produced north of the Yangtse River (the cost being more than ten taels of silver for a high-grade piece). [In the process of calendering], the stone rollers will not become heated and the tightly woven fabrics will not become loose. The large firms in Wu-hu are best known for the good stone [rollers] they use. A great deal of cotton fabric is produced in Kwangtung in the south; but here the people also import the stones from afar [i.e. from north of the Yangtse], indicating that the good qualities of the latter have been truly proved. Cotton garments are often washed, and the same attention is paid to the pounding stones on which the laundering is done. The above method [for making cotton fabrics] is also used in foreign countries and Korea. However, the methods employed in the West are not known, since no information on the workings of the western looms is available.

The patterns used for the figure-weaving of cotton fabrics are "cloud," "twill," "elephant eye," and so on. These patterns are modeled after those used in [the weaving of silk fabrics by the] drawloom. But, since the fabric is only made of cotton, a plain unfigured weave would actually be good enough. It is not necessary to give illustration of the loom for weaving cotton fabrics, as it is to be found in every tenth household.

Cotton Padding

Out of a hundred persons who wear quilted garments in winter, only one [can afford to] use spun silk padding; the rest all use vegetable fibers. In ancient times hemp-quilted clothing was known, and nowadays the common practice is to use cotton-padded garments. After the cotton has been bowed, it is shaped to fit the garments and filled into the latter. The newly made clothes are light in weight and warm to the body, but the warmth is gradually lost with the passage of time and when [the padding] becomes hard and thin. To restore it to its previous state, the cotton should be taken out and bowed again. It will be as good as new when put [back] inside the garment.

SUMMER CLOTHING

Ramie, known in China as *Chu-ma,* is grown in all parts of the country.[16] It can be planted by either sowing the seeds or burying root cuttings. (In the Ch'ih prefecture [Kiangsi] the heads of roots are covered each year with a mixture of grass and manure, under which the root stumps will grow to the height of the manure covering. In contrast, the green ramie in Kwangtung province is planted by casting the seeds in fields.) There are two varieties of ramie: the Green and the Yellow. Each year [the stems of the plants] can be cut two or three times; from these, fibers are obtained for the making of summer garments, curtains, and [mosquito] nets.

After being peeled from the [woody portion] of the stem, the tissues of ramie fibers are stacked to dry in the sun, as they will rot if exposed to moisture. In order to separate the individual fibers [through the removal of gum], the dried fibers are steeped in water, but the length of steeping is not to exceed twenty quarters in time,[17] as prolonged soaking without separation will also cause rot. The natural color of ramie fibers is light yellow, but it is bleached into pure white. (The fibers are first boiled in an aqueous solution containing the ash of rice stalks and lime, then repeatedly rinsed in flowing water, and finally dried in the sun until the pure white color is achieved.)

Those who are skillful workers use treadle-powered wheels for spinning ramie yarns, and with this machine one female worker can produce [in one day] as much yarn as that produced by three men [using ordinary hand spindles]. However, the separation of the fibers [is a time-consuming process]. The entire day's effort [of one person] will yield only some three or five ounces [of separated fibers]. The loom used for weaving ramie fabrics is the same as that for cotton cloth. Ramie fibers are always used in the threads for sewing cotton garments and leather shoes.

[There is a similar species called] *ke* plant [*Pueraria thunbergiana*], which is a creeper whose fibers are several *ch'ih* longer than that of ramie. The fabric woven from the fine fibers of the *ke* plant is a costly commodity. [Another similar species] is *meng* hemp [*Abutilon avicennae*], which can be woven into a very coarse fabric. The extremely coarse grade is used in making mourning clothes. It might be mentioned that the crudest fabric made from [the regular] ramie is used by the lacquer-ware manufacturers as "ash cloths," while in the Imperial Palace it goes into the construction of torches. In addition, there is a kind of palm-fiber gauze, which is made in Fukien Province from the fibers of plantain bark. This cheap fabric is thin and light, and deteriorates rapidly. It is not suitable as a clothing material.

FURS

"Furs" is a general term for garments made from the skin of fur-bearing animals. There are hundreds of kinds of furs classified according to their values, ranging from the costly sable and fox to the inexpensive sheep and deerskin.

The sable is a native of Manchuria and is found in the territory of Chien-chou as well as in Korea. This animal is fond of pine nuts. Therefore the native hunters, waiting in absolute silence, shoot it at night under the [pine] trees. The skin of one sable is less than a square *ch'ih* in size, so that over sixty sable skins will go into the making of just one fur garment. Dressed in a sable gown, a man standing in a snow storm will feel warmer than if he were indoors; and any sable hair that gets into one's eye is easily removed. For these reasons it is most expensive. Sable furs are of three colors: one is the white kind, called

Figure 2–17. Straightening and rolling bowed cotton fibers into long slivers with the aid of wooden boards.

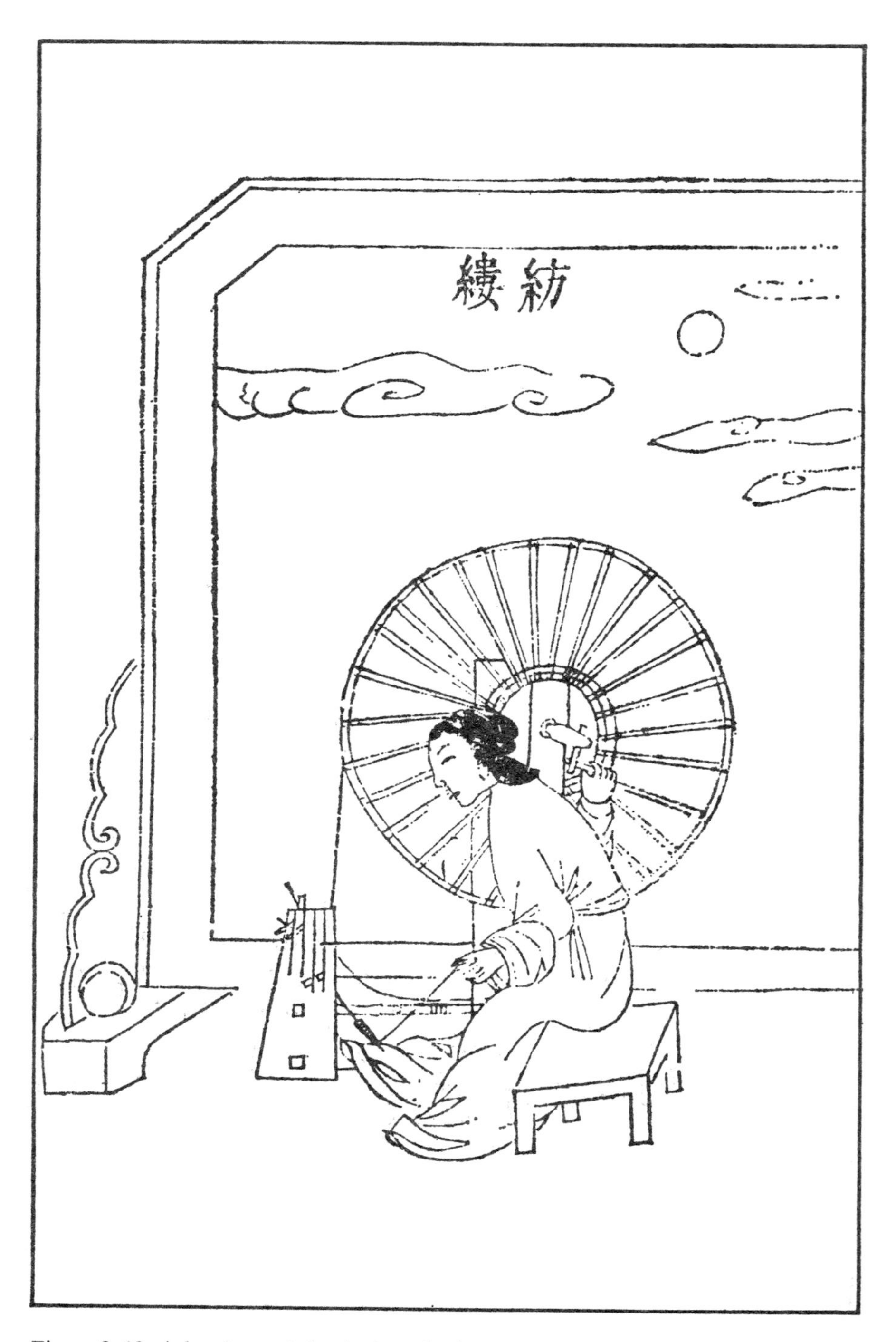

Figure 2–18. A hand-operated spinning wheel.

silver sable; another is jet black; and the third is dark brown. (The long-haired variety of black sable is so valuable that merely an over-cap made of it is now worth some fifty taels of silver.)

Fox furs are also produced in [north China such as in] Hopei, Shantung, Southern Manchuria, and Honan. A gown lined with pure white fox fetches almost as high a price as sable, but yellow and brown fox furs are equal to one-fifth the price of sable. Fox furs are second to sables as protection against the cold. The underfur of Manchurian foxes is dark bluish in color, while that of foxes in China Proper is white. By blowing on the pelt and separating the overhair, the buyer [ascertains the origin of the fur and] grades it on the basis of the color of the underfur.

As to sheep furs, that from the mother sheep is cheaper than that from the lamb. [The fur of] the lamb taken before it was born is called "unborn lamb" (with slight beginnings of curls in the hair); that of newly born lambs is called "suckling lamb" (with curled hairs like the ends of earrings); that of three-month olds is "running lamb"; and seven-month olds, "walking lamb" (with gradually straightened hairs). Fur clothes made of "suckling" and "running" lamb are free from sheep odor. [Such] lamb garments were worn exclusively by government ministers in ancient times, and even nowadays the high gentry of the northwestern provinces still greatly value them. The skin of adult sheep, however, are cured and made into heavy garments that are worn by the lower classes.

All the furs named above are prepared from the [long-haired] sheep skins. The short-haired [goat] skins produced in southern China are only valuable for their hides. These hides are cured and made into parchment, which is as thin as paper and used in the manufacture of painted lanterns. Those who are accustomed to wearing sheepskin will cease to notice its odor as time goes on. The southern Chinese, who are not used to it, however, can not bear the smell; but since the cold weather is not severe in the south, there is actually little need there to wear sheepskin.

After the removal of hair, the deerskin is cured and turned into a kind of soft leather for making jackets and trousers. This leather affords good protection against wind and is comfortable to wear; particularly good are the socks and boots made from it. In addition to being produced in large quantities in Kwangtung, deerskin is also prepared in Hunan and Hupei, where an important distribution center for it is the Wang-hua Mountain. Another property of deerskin is that it wards off scorpions. In north China, in addition to being clothing material, it is also cut into strips and used as a border for bed coverings, which naturally keep scorpions away.

The strikingly marked leopard and tiger furs are used in making the military outfits for generals; the cheap dog and pig skins are made into shoes for the laborers. The otter fur is highly valued by the western tribes in making clothes, and is also used for decorating collars of clothing worn by officials and gentry. The hunters in Hsiang-yang and Huang-chou [in Hupei] travel over mountains and across great distance [to the places] where they can shoot otters whose furs they sell in a distant [market] at enormous profits. Further, there

Figure 2–19. A foot-operated spinning wheel [Ch'ing addition].

are various kinds of rare furs from exotic places, such as "golden-haired ape" fur for making the Emperor's over-caps and *"ch'e-li* monkey" fur for making the Emperor's gowns. These furs are not Chinese products.

The above is a general sketch of the animal furs that have been used for clothing, not counting the numerous varieties of local products. Birds also furnish material for clothes; there are people who make use of the small, inner feathers from the eagle's breast and those under the wild goose's wings. This stuff is called "swan's-down." Tens of thousands of birds have to be killed in order to obtain enough feathers for a single cloak. It isn't worth the trouble.

WOOLENS AND FELT

There are two kinds of sheep. One is known as woolly sheep. Its wool fibers are made into felt or downlike fabric that is subsequently fashioned into hats, socks and the like, which are worn all over the country. In ancient times, before the "Western-Region" [18] sheep was introduced into China, the hair of this [woolly] sheep was also woven into coarse wool fabrics for the poor people's clothing. These fabrics lacked any fine quality. Today some of them are still made from the wool of this sheep. Woolly sheep flourish in all areas north of Hsü-chou and the Huai River. South [of the Yangtse] sheep are raised only in Hu-chou, where they are sheared three times each year (in summer the hair is sparse). Each sheep yields enough wool every year for making three pairs of socks. Furthermore, each pair of sheep will produce two lambs in each lambing season. Therefore if a north China household keeps 100 sheep, the income it derives from them will be some 100 taels [of silver] per year.

The other kind of sheep is *yü-t'iao* sheep (this is a foreign term), which was introduced into China from the Western Regions only toward the end of the T'ang dynasty. The outer hair of this sheep is not very long, but the inner hair is fine and soft and is separated and woven into pubescence fabrics. The Shensi people call this animal [cashmere] goat in order to distinguish it from the [woolly] sheep. The former species was first brought from the Western Regions to Lin-t'ao [in Kansu, an old name for Lanchou]. At present Lanchou is best known for the raising of [cashmere] goat. The high-grade wool fabrics are all manufactured in Lanchou and are called "Lanchou woolen" or *"Ku-ku* woolen," according to its original foreign name.

[This cashmere] goat hair is classified in two kinds. One, termed "combed wool," is obtained by combing [the coat of the goat]. The resultant hair fibers are first spun into yarns and then woven into such fabrics as serge, *pa-tzu* [possibly twill], and so on. The other, termed "picked wool," is obtained by hand-picking the finest inner hairs, one by one. This is also made into yarn and then woven. Rubbed against the face, this fabric feels so smooth and silky as silk itself. A spinner working for a whole day can produce only about 0.1 ounce of yarn; and one bolt of cloth [made of this "picked wool"] will require half a year to complete. The amount of "combed wool" that can be spun in a day, however, is several times that of "picked wool."

To make [a fine and even] wool yarn, a lead weight is suspended from the lower end of the strand of wool fibers, which are rolled and twisted by the spinner's two hands. The loom used for weaving wool fabrics is larger than the cotton looms. Eight cross beams are employed on the wool loom; the wool yarns are placed on the warp beam and passed through their respective harness slots. [The harness is lifted by] stamping the four treadles under the frame. This is to separate the warp yarns [and to form the shed] so that the filling goes over two warp yarns, [then under one, over two, under one, and so on, with a progression of one at the point of interlacings]. The result is that [a twill weave or] a diagonal pattern is created across the face of the finished fabric. The shuttle is 1.2 *ch'ih* in length. Both the loom and the goat were brought to China in past ages by barbarians who came [to China] (but whose names are not known); to this day, therefore, the weavers are all members of this race. No Chinese takes part in this work.

Of the wool sheared from the [woolly] sheep, the coarse fibers are used in making felt,[19] and the finer ones, in weaving woolens. For making felt, boiling water is first prepared. Into it the wool is thrown and subsequently washed and agitated until the fibres have matted together. The resultant crude felt is spread over a wooden board [or mat] that has been fashioned in the shape of the thing to be made. Then the felt is finished by pressing a roller back and forth on it. The natural colors of felt are black and white; all the other colors are created by dyeing. As for such terms as *ch'ü-yü* carpets, *p'ang-lu* rugs, and the like, they are simply the names derived from many dialects both Chinese and foreign. The coarsest kind of blankets or rugs are made with horsehair and shoddy mixed with sheep's wool.

NOTES

1. The bathing of eggs in rain and snow also prevents them from hatching until the next spring when young and tender mulberry leaves sprout.

2. The Early and Late varieties of silkworms refer to the *annual* (one crop) and polyvoltine (two or more crops per year) classes respectively. Although the first generation of the latter is hatched somewhat earlier than the annual variety, the last bit of work

involved with it would be finished later than that of the one-crop worm, therefore it is here called the late variety.

3. Wild silk is produced by caterpillars other than the mulberry silkworm, or *Bombyx mori*. The term "wild" implies that these silkworms are not capable of being domesticated and artificially cultivated like the mulberry worms.

4. Moulting of the silkworm is expressed in Chinese by the word *mien,* meaning "sleep," because when the worm is about to shed its skin, it stops eating for a time and becomes motionless. Under normal conditions the worm sheds its skin four times prior to spinning the cocoon.

5. "Beginning of summer" or *li-hsia* refers to a solar term in the traditional calendar year. It occurs approximately in the first half of May.

6. In Chekiang, that is, around the silk-producing region of Chia-hsing and Hu-chou.

7. In his *Up the Yangtse in 1891* E. H. Parker recounted that in Chiating silkworms are frequently fed at first with the leaves of *che* or "silkworm oak"; however, when the worms grow older, the diet is changed to mulberry leaves. The silk is known as "hard silk."

8. *Grasserie* seems to be the silkworm disease described in this book. It is present when the worms become restless, bloated, and yellow in color; if punctured they exude a fetid matter filled with minute granular crystals. The disease is neither contagious nor hereditary. Its chief cause is mismanagement of the worms at moulting periods and uneven feeding.

9. The warp count of Chinese fabrics underwent a series of changes with the progress of time. For example, Fan Wen-lan states that in the Chou dynasty kerchiefs were woven with 170-warp yarns in every inch of width. Lao Kan relates that in the Han dynasty an eight-thread "strip" of warp yarns referred only to hemp and tree-cotton fabrics, while the silk fabrics were measured by weight, usually in terms of so many "ounces" of silk. In the Ming dynasty, however, the term eighty-thread "strip" was broadened to express the warp count of silk damasks and pongees, as indicated in the present chapter.

10. The piece of cured leather seems to be a sort of bolster for supporting the waist and back of the weaver. It is not clear why the text states that it is placed under the weaver's seat.

11. Gauze fabrics have long been known in China as *lo* and *sha*. The complex *lo* fabrics were much more popular than the simple *sha* fabrics in the Han and T'ang dynasties (206 B.C.–906 A.D.). In the Sung dynasty (A.D. 960–1279), however, the popularity of *lo* declined, while the manufacture of *sha* was considerably increased. From that time onwards *lo* was gradually replaced by *sha* and fell into disuse after the overthrow of the Ch'ing dynasty in 1912, chiefly because *lo* was more expensive than *sha* to manufacture.

12. Dragon robes had a long tradition in China dating back to the Sung dynasty. During the Ming and Ch'ing periods they generally served as an insignia of rank through the numerous variations of design and color. The imperial dragon robes, described briefly in the present section of the book, were worn by the emperor and those of highest ranks in his immediate family, while the lesser dragon robes or *mang p'ao* were used as court dress by officials down to the seventh rank. For an informative study on the subject see Schuyler Cammann, *China's Dragon Robes* (New York, 1952).

13. The so-called "Japanese satin" or *wo-tuan* is a Chinese misnomer. It was first brought into Japan by the early Spanish merchants and was subsequently introduced into China. The Chinese manufacture of the fabric, begun in the Ming dynasty, has centered at Chang-chou in Fukien province, thus giving the satin the other name *Chang tuan*. This material was widely used by the noblewomen of the Ch'ing dynasty (1644–1911) for making their richly ornamented dresses.

14. *Ceiba pentandra* was known in China as *mu-mien* or "tree cotton" in very ancient times. This has led to a confusion of terms. The real cotton plant, *Gossypium indicum,* is believed to have been introduced into China from India via Persia in the early Han period,

ca. 138–126 B.C. For centuries, however, it seems to have been grown only as an ornamental shrub. Not until the Southern Sung (1127–1279) and the Yuan (1279–1368) dynasties was cotton widely cultivated for the fiber of its fruit, and the two main areas of production were the lower Yangtse valley centering in Sung-chiang and Shensi in North China.

15. The cotton gin used by the Chinese was very similar to the one invented by Eli Whitney in the United States in 1793.

16. Ramie or "China grass" is an Asian perennial plant, *Boehmeria nivea.* It is known in China as *chu-ma,* in Cochin China as *cay-gai,* and in Bengal as *kankura.* The shrub reaches four to six feet in height, and is very hardy. It is cultivated largely in China and to a limited extent in Japan and India. The prehistory of ramie in China is still obscure, since a large number of bast fibers, including hemp, flax, and ramie, are all referred to as *ma* or "hemp" in ancient Chinese writings. One thing is certain, however, that *ma* was the first fiber plant of the Chinese and was used for clothing almost exclusively before the arrival of cotton. The use of *ma* as a textile was gradually replaced by cotton during the Ming and Ch'ing dynasties.

17. Twenty quarters in the old Chinese way of calculating time are equal to ten hours.

18. The "Western Region" sheep, as described in the present chapter, was probably introduced into China during the Han dynasty. It is related in Chinese history that, on a campaign against the Hsiung-nu (121–119 B.C.), the army of the Martial Emperor of Han defeated or killed 19,000 of the enemy and captured a million sheep. In addition to wool fibers, the hairs of the camel and the rabbit were also used in the T'ang dynasty (A.D. 618–906) for making soft and fine wool cloth. Another important development at that time was the introduction of the cashmere goat into China proper from Tibet and northern India. The soft, silky inner hair of the animal was used by the nomadic Chinese in the Western Region for making wool cloth. However, for the next seven centuries most Chinese continued the ancient tradition of using long-fiber silk and hemp as textile materials and made little use of the short-fiber wool. Matthew Ricci writes of the Chinese wool industry at the end of the sixteenth and the beginning of the seventeenth centuries as follows:

> They shear sheep, but in the use of sheep's wool, they are not nearly as adept as the people in Europe, and though they place a high value on imported woolen cloth they do not know how to weave wool into cloth for clothing. True, they do weave a woolen cloth of light weight for summer use which is much in demand by the poorer classes for hats and for the carpets which they use as sleeping mats. These carpets are also used in the performance of their social rites. Woolen cloth is more in demand in the northern parts of China where the cold is almost as biting as it is in the northern parts of Europe. The natives gather the skins of foxes and of Scythian weasels, which they make into garments to ward off the rigorous frosts.

19. Felt was widely used by nomads, particularly Tibetans and the Mongols. Their tents had to be impervious to wind and water and their saddlebags and other containers strong, light, and dustproof. Felt met these qualifications, and consequently became an essential item among the wandering people. In contrast, the use of felt by the ordinary Chinese has been limited to the making of pallets, runners, hats, and shoes.

3

DYES

Master Sung observes that in the sky the clouds are of different hues, and on earth the flowers and leaves are dissimilar in appearance. What Heaven has established as model, the sages then followed. Dyes of various colors therefore were invented, which are derived from the five [primary] colors. Who can say that the Emperor Shun has not deliberated carefully in this matter?

Among the many birds of the air the phoenix [i.e. king of birds] sports a red color, and among the multitudes of animals the unicorn [i.e. king of beasts] is blue-green. The same principle underlies the decision that vermillion and yellow [i.e. colors of the imperial palaces] should be the objects of obeisance by the multitudes of dark-robed literati. Lao Tzu once said, "The bland will absorb a mixture of tastes, and the white will absorb the rainbow colors." All the silk, hemp, fur, and woolen stuffs are naturally plain, yet when dyed in various colors their value can be much enhanced. Is there one who contends that the Power of Creation had not exercised great care in its undertakings? I certainly would not believe him.

DYES OF VARIOUS COLORS

Crimson. The ingredient for this color is safflower cakes, boiled in the juice of smoked Chinese plums, and then levigated in an aqueous solution of caustic soda. The same result can be obtained by substituting for caustic soda the ash of rice stalks. The more times the substance is decanted, the more brilliant will be the color. Some dye works, however, dye [the materials] first with Venetian sumach as a base in order to save expenses.

Safflower red is most allergic to garu-wood and musk [i.e. perfumes]. If [safflower-dyed] garments are stored together with these scents, the coloring will be ruined within a few months. If, after a piece of cloth has been safflower-dyed, a removal in color is desired, the cloth needs only be moistened with a small amount [lit. "a few dozen drops"] of the aqueous solution of caustic soda or rice stalk ash, and the red color will be completely withdrawn from the cloth and returned to its original form. This liquid is stored in green lentil flour and can be released again for dyeing with no loss. Information about this process is not obtainable from dyers, who regard it as a trade secret.

Lotus pink, peach-blossom pink, silver pink, and clear pale pink. For dyeing these colors the sole ingredient is also safflower cakes, the different shades resulting from variations in the amount of dye used. These colors will not show on yellow silk; therefore, white silk must be used.

Wood red. Dye for this color is made by boiling sappanwood [1] in water, with gallnuts and alum added.

Purple. This color is achieved by using sappanwood as a base, then dyeing again with green vitriol [ferrous sulfate, $FeSO_4$].

Earth brown. The method for dyeing this color is not clear.

Canary yellow. This color is dyed first with the aqueous solution of boiled yellow berberine wood and then soaked in the [seeth] water of indigo.[2]

Golden yellow. This color is achieved by first dyeing with the aqueous solution of boiled Venetian sumach [3] wood, followed by shampooing with the alkaline solution of water-leached hemp ash.

Tea brown. Dyed with the aqueous solution of boiled lotus-seed shells and then rinsed with the aqueous solution of green vitriol.

Dark green. Dyed first with the juice of boiled "huai" flower and next soaked in indigo solution. Alum is used [as a mordant] for both the light and darker shades of this color.

Bright green. Dyed first with the liquid of [boiled] yellow berberine and then soaked in indigo solution. Nowadays there is a shade called bright grass-green, which is obtained by using the liquid of boiled small-leafed *Polygonum tinctorium* plant [after dyeing with yellow berberine]; this results in a very brilliant hue.

Light green. This is slightly dyed with "huai" flower [liquid] and soaked in green vitriol [solution].

Deep sky-blue. This color results from first dyeing the material lightly in a vat of indigo and then washing with sappanwood solution.

Grape blue. First the material is deeply dyed in a vat of indigo and then washed with concentrated sappanwood solution.

Egg-shell blue. This color is obtained by first dyeing [the cloth] with yellow berberine solution then soaking it in indigo vats.

Peacock blue and sky blue. Both are dyed with indigo, the difference being only one of shade.

Black. It is first dyed a deep blue with liquid indigo then soaked in the liquids of boiled Venetian sumach wood and afterwards [in those of boiled] myricaeceae bark. According to another method, the tender indigo leaves are first soaked in water; next, green vitriol and gallnuts are added, and [with the cloth] are soaked together in this liquid. Cloth dyed in this fashion, however, will deterioriate rapidly.

Pale blue and light blue. These two colors are both achieved by dyeing slightly in an aqueous solution of indigo. A new method nowadays consists of dyeing in the slightly boiled liquid of *Polygonum tinctorium.*

Ivory color. This is obtained by dyeing slightly in the liquid of Venetian sumach, or in [an aquaeous solution of] yellow earth.

Mauve. Dyed first in a dilute sappanwood solution, followed by soaking in the dilute liquids of lotus-seed shells and of blue vitriol, respectively.

Supplement [1]:The dyeing of black kerchiefs. This black color is not obtained from indigo. [The material to be dyed] is first boiled for a day in a mixture of water and acorn shells or lotus-seed shells, and strained off; and is again boiled for one night in a pot containing the aqueous solution of iron ore and green vitriol. This results in a deep black color.

Supplement [2]: Dyeing navy-blue cloth. For many centuries Wu-hu [on the Yangtse in Anhui] was famous for its dark blue cloth. Its blue sheen, a result of [this locality's] starching and smoothing process, was highly valued by peoples of countries far and near. Yet human nature eventually tired of this material, therefore a method for dyeing the navy-blue color has been developed in recent times. This new method consists of the following: the high-quality [cotton] cloth of Sung-chiang [in Kiangsu] is first dyed a dark blue and is dried in the air without the starching and smoothing treatments. The cloth is next rinsed in an aqueous solution of glue and bean milk. This is followed by dyeing the cloth slightly in a vat of the highest grade of indigo, called "standard vat." Thus the finished product will show a red iridescence, and it has become much prized.

INDIGO

There are five kinds of indigo plants, and they all yield indigo. "Tea indigo" is another name for the variety *Isatis tinctoria,* a plant that is propagated by sprig-planting. Others, the *Polygonum tinctorium,* "horse indigo" [*strobilanthes flaccidifolius*], and "Kiangsu indigo" [*Indigofera kiangsu*] are all seed-grown. There has recently been developed a variety of *Polygunum tinctorium* having small leaves, commonly called "Pigweed indigo" [*Amarantaceae tinctorium*], which is a still better species.[4] The "tea indigo" plant is harvested in the eleventh month of the year. All the leaves are cut off and put into a pit for the manufacture of indigo. The tops and bottoms of the stems are trimmed off so that only a few inches near the roots are left, which are dried in hot air and then covered with earth for storage. In the spring the soil of the hilly country is made extremely fertile and friable by burning [the winter's grass and leaves?]. In this prepared soil holes are then made at a slanting angle with an awl-hoe (this hoe, about eight inches in length, has an end that curves toward the holder), and the indigo stems are inserted into them. They will live and grow into new plants without trouble.

As to the other varieties of indigo, their seeds are gathered and sown in prepared seed beds. Young shoots of the plants will appear in late spring, in the sixth month the fruits are formed, and in the seventh month [approximately early August to early September, solar calendar] the plants are cut and harvested for making indigo.

In making indigo, large amounts of [the harvested] leaves and stems are placed in pits; for smaller amounts barrels or vats are used. The plants are soaked for seven days in water to extract [and hydrolyze] the [indigo] juice [i.e., indican].[5] Five pints of lime are added to each *tan* of this solution, which is then stirred and agitated several dozen times [to secure air for oxidation]. The indigo will precipitate and settle in the bottom [of the vat] when the agitation of the water ceases. In recent times the people of Fukien province have cultivated mostly "tea indigo" on their hills, the proportion being several times the amount of all the other kinds of indigo planted [there]. [The harvested plants] are packed in wicker baskets in the mountains and brought [to markets] by boats. [In the process of soaking the plant in water], the foamy substance on top of the liquid surface, that has been skimmed off and dried, is called "indigo florets." Before being put into the vat [for oxidation] the indigo [plant extract] must first be mixed with the aqueous solution of rice-stalk ash, and [after it is in the vat] it must be stirred innumerable times every day with a bamboo stick [to secure air for oxidation]. The finest grade [of indigo] is called "standard vat."

SAFFLOWER

Safflower is planted by sowing the seeds in fields at the beginning of the second month [early March, in the solar calendar]. If it is planted too early, a kind of black antlike insect will appear when the shoots are about one foot high and attack the plant roots, causing the plants to die immediately. In fertile soil the shoots will grow to be two or three feet tall. Stakes should be fixed along each row and strings tied across them, in order to prevent the plants from being broken by strong winds. But this procedure need not be followed if the safflowers are planted in poor soil and the plants are under one and one-half feet high.[6]

The safflower begins to bloom in early summer. [The calyx] under the flower is ball-shaped and covered with thorns, and the flower rests on top of the ball. Safflower must be picked early in the morning while it is still moistened with dew; as the sun rises high and the dew dries, the flower will close up into a solid ball, and it is no longer fit for picking. But on a cloudy or rainy morning, when fewer flowers will bloom, it is safe to pick them even though the dew drops have evaporated, since there is no sunshine. Safflower will continue to bloom every day for an entire month. Those flowers destined for use in medicine need not be made into cakes. Those that are to become dyes, however, must be put into cakes before they can be used: in this way the yellow juice [in the flower] is eliminated and a true red color obtained. The seeds of safflower can be pressed for oil. A silver-leafed fan will turn golden after having been brushed with this oil and dried over a fire.

The Manufacture of Safflower Cakes

The safflower blossoms are picked while they are still moistened with dew, and are thoroughly pounded. After the pounded mass is put into a cloth sack and soaked in water [for a certain time,] it is squeezed to remove the yellow juice. The solid residue is again pounded and further purified [by decomposing its yellow coloring matter] with soured millet or rice juice. [The acid-treated mass] is first put in a sack, then soaked in water, and finally squeezed to remove its decomposed yellow matter. Then for one night this solid residue is covered with [branches of] *Artimisia apiacea,* after which it is shaped into cakes, dried in the shade, and stored. When dyers know the correct method [of preparing safflower cakes], "brilliant will be our red color." [7] This is the color known as scarlet (safflower cakes are also necessary in dyeing red paper for ceremonial uses; otherwise the color would be quite pale).

Supplement: Rouge

In ancient days the best rouge was made by tinting cotton-wool with litmus [*tzu kuang* in the text],[8] while that tinted with the juice of safflower or mountain pomegranate flowers ranked second. Nowadays in the Chi-ning area [in Shantung], however, it is made only from the dregs of safflower left over from dye works and is worth very little. The dried [safflower] dregs are called "purple powder"; it is sometimes used as a color by artists, but dyers discard it as waste matter.

"Huai" Flowers

A huai tree [9] will not bloom and bear fruit until it is ten-odd years old. The immature, unopened flowers are called huai flower buds, which are a necessary ingredient in dyeing cloth green, just as safflower is in dyeing it red. To gather the buds, bamboo mats are spread closely under [the tree] to receive them [as they are picked]. The buds are then boiled once in water, strained, and fashioned into cakes; they are now ready for use by dyers. The mature flowers will gradually turn yellow. After being gathered they are mixed with a little lime, [dried] in the sun, and stored for use.

NOTES

1. Sappanwood is the hard wood of the *Caesalpinia sappan*, a tree common to the warmer regions of Asia. According to Li Shih-chen, sappanwood was used by the Chinese for dyeing fabrics in the Western Tsin dynasty (A.D. 265–316) and possibly earlier. It is described in *Nan-fang t'sao-mu chuang* (A.D. 264–367) and also in *T'ang Pen-ts'ao* (A.D. 600). It constituted an important item of foreign commerce in the Ming dynasty, being imported chiefly from Malaya and the East Indies.

2. Yellow berberine wood is from the *Berberis thunbergii*, a tree which the Chinese have used for its yellow dyestuff ever since the Han dynasty (206 B.C.–219 A.D.), if not earlier. "Seeth" water is the liquid from which the indigo precipitate settles out at the end of the oxidation process. When seeth water is evaporated, yellow-brown to deep brown residues are obtained.

3. Venetian sumach or Turkish sumach is a small tree whose wood and leaves yield the yellow dyestuff known as "Young Fustic."

4. The mention of indigo plants in *Erh-ya* and *Kuang-chih* suggests that they might have been used as dyestuff or medicine in the Chou dynasty (1122–256 B.C.) or at least in the Ch'in dynasty (221–206 B.C.). The cultivation, preparation, and use of indigo as a dye are fully described in *Ch'i-min yao-shu* written by Chia Ssu-hsieh in the fifth century. The commonly used indigo plant in China has been and is the *Polygonum tinctorium*, chiefly found in the central and northern parts of China as well as in Manchuria. The *Isatis tinctoria* or wood plant, at one time widely grown in south China, is now used in limited quantity to blend with indigo in the dyeing of fabrics. The so-called "tea indigo" plant of the Fukien province may belong to one of the many species of *Indigofera*, such as *tinctoria, sumatrana* (the Indian plant), *arrecta* (the Natal plant), *Paucifolia* (Madagascar plant), *secundiflora* (Guatemala plant), *argentea, disperma*, and others.

5. The extraction of indican by hot water is mentioned in the older *Ch'i-min yao-shu* but not in the present book. It is possible to recover 80–85 per cent of the indican by hot water extraction, as against 40–50 per cent by ordinary steeping.

6. Safflower or bastard saffron is an annual thistle-like plant belonging to the *Cynarocephalae*. A native of Southern Asia, it has been cultivated and used as both a red dye and a drug in China since ancient times. According to Li Shih-chen, *Chung-hua ku-chin chu* states that safflower juice was first used for making face-paint rouge in the time of King Chou (1154–1122 B.C.) in the Shang dynasty. *Po-wu chih* relates that the seeds of a "barbarian safflower" were brought to China from the Western Region by Chang Ch'ien in 126 B.C. The cultivation, preparation and utilization of safflower were fully described in *Ch'i-min yao-shu*, written long before the present book.

Modern research discloses that safflower is a very weak dyestuff, four ounces are necessary to dye one pound of cotton light pink, eight ounces for a rose-pink, and about one pound for producing crimson. It is now rarely used as a dyestuff, though it is still employed in China and other countries in the making of cosmetics.

7. A quotation from the *Book of Poetry*.

8. Litmus, assuming that this is the substance meant, is well known to the modern chemists and is derived from various species of lichens. It is one of the tints called "lichen purple dye" and is sometimes used in dyeing crimson.

9. Sophora japonica or "huai," a large and beautiful tree belonging to the *Leguminosae* grows abundantly throughout China. It can be seen from *Erh-ya* and the *Rites of Chou* that in the Chou dynasty (1122–256 B.C.) huai was considered the fittest tree for the

imperial palace as well as for the meeting places of highest officials. The cultivation of huai trees and the use of their flower-buds as a yellow dyestuff are respectively mentioned in *Ch'i-min yao-shu* (fifth century) and *Pen-ts'ao shih-i* (eighth century). The method of dyeing, not described in the present book, consists in simply boiling mordant-treated silk for one to one and one-half hours in a decoction of the flower buds. When applied to wool, the huai buds give a dull orange color with chromium as the mordant, a yellow of moderate brilliancy with aluminum, a bright yellow with tin, and a dark olive with iron.

Figure 4–1. Beating rice grains into a wooden barrel in the wet field.

4

THE PREPARATION OF GRAINS

Master Sung observes that, as Nature creates the five grains to nourish mankind, it places the essential part of the substance inside covers as though wrapping these cereals in yellow robes. Rice is enclosed in chaff, wheat is covered by bran, and the millet and sorghum grains are hidden behind featherlike spears. It would seem as though the method of obtaining the fine and polished parts is to be a veiled mystery. For those who are discriminating in food nothing can be too refined, and the work of polishing and grinding provides a livelihood for thousands of people. This [desire for refinement] results in excessive use of small and humble tools such as the mortar and pestle. Is it not true that those who invented these implements were really divine forces in human disguise?

POLISHING RICE

For the separation of rice grains from the stalks after reaping, half of it is done by beating sheaves of the plants by hand [against a receptacle], while the other half is accomplished by spreading the rice plants on the ground and passing over them a stone roller drawn by an ox. If the former method is used, the sheaves can be beaten against either a wooden barrel or a slab of stone. If during harvesting time the weather is often rainy, and the fields and rice plants are both wet so that nothing can be spread on the ground, the grains are beaten into a wooden barrel right in the fields [Figure 4.1]. But if the weather is fair and the rice plants are dry, a slab of stone will serve the purpose very nicely [Figure 4.2]. Rolling off the grain with an ox-drawn stone roller requires one-third the human labor required by hand-threshing [Figure 4.3], but there is the danger that the tips of seed grains may be damaged so greatly that they will not germinate. In the south, therefore, in households where large amounts of rice are produced, the ox [drawn roller] is used to separate rice grains [from the stalks] for ordinary use, but seed rice is obtained by hand-beating against stones.

In the best rice crop, nine-tenths of the heads are full of kernels and one-tenth, empty. Owing to unfavorable weather or improper cultivation, however, the proportion may be six-tenths to four-tenths. In the south the winnowing machine [1] is used throughout for the elimination of husks [Figures 4.4 and 4.5]. In north China, on the other hand, where less rice is produced, it is generally winnowed by tossing [the unseparated grains and husks in the wind]. The same method is used for wheat and millet, but it is not so efficient as the machine.

The husks of rice grains are removed with a hulling mill, while the bran is eliminated with either a pounding mill or a rolling mill. A water-powered pounding mill can serve the dual purposes of both pounding and hulling. The hulling process can be eliminated also by feeding dry grains into a rolling mill. There are two kinds of hulling mills. The first is made of wood. The wood (mostly pine) is sawed into pieces one foot long and built into the shape of large millstones. The [inner] faces of the two circular blocks are both marked with diagonal grooves, and the protruding [center] of the lower block is inserted into the upper; in the center of the upper block there is a large hole into which the grain is fed [Figure 4.6]. A wooden mill is good for hulling over 2,000 *tan* of grain before it is worn out. It will not pulverize the grain even though the latter is not entirely dry, therefore it is used for preparing the myriads [of *tan*] of rice set aside for government taxes, army provisions, or tribute. The second kind of hulling mill is made of earth. A round frame is first constructed of bamboo, into which clean yellow earth is packed solidly; to both the upper and lower piece is then affixed a number of bamboo teeth.[2] The rice grains are fed through the upper piece [Figure 4.7]. The capacity of the earthen mill is twice that of the wooden one. The slightest moisture in the grain will result in broken kernels if it is put through this earthern mill, which is worn out after hulling 200 *tan* of rice. Strong men are required to operate the wooden mill, but even women and children can work an earthen one. Rice hulled by the latter provides the daily food of the common people.

After hulling, the husk and the bran are eliminated by means of a winnowing machine. The grain is then poured onto a sieve and agitated with a rotating motion [Figure 4.8]. Kernels whose husks are not yet broken will rise to the top, and they will be returned to the hulling mill [Figure 4.9]. A large sieve has a circumference of five feet, a small one is half as big. The center of the large kind rises in a hump, and it is operated by able-bodied men. The small kind, with a level center surrounded by sides two Chinese inches high, is worked by women and children.

Figure 4–2. Beating rice grains against a slab of stone on the dry ground.

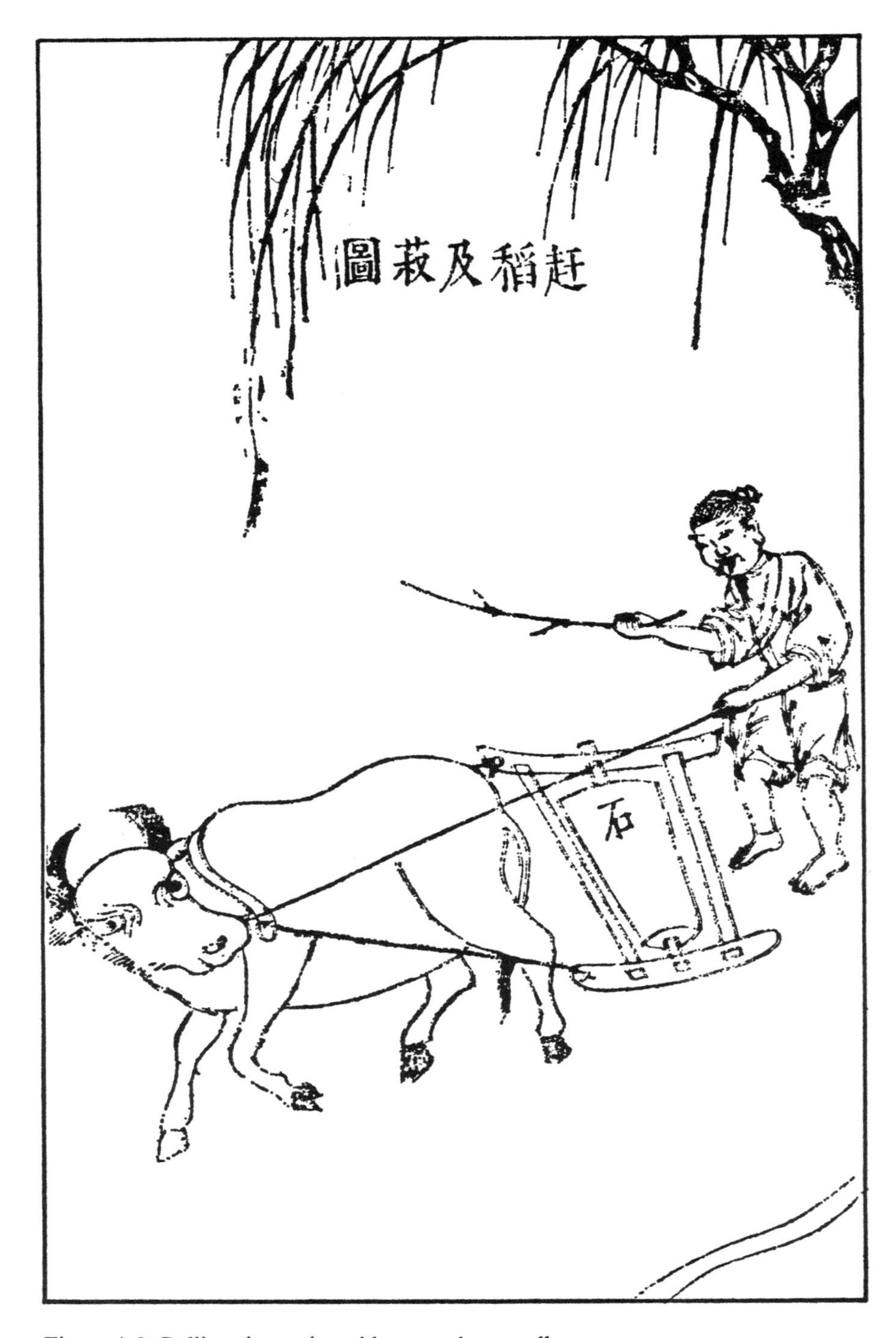

Figure 4–3. Rolling rice grains with an ox-drawn roller.

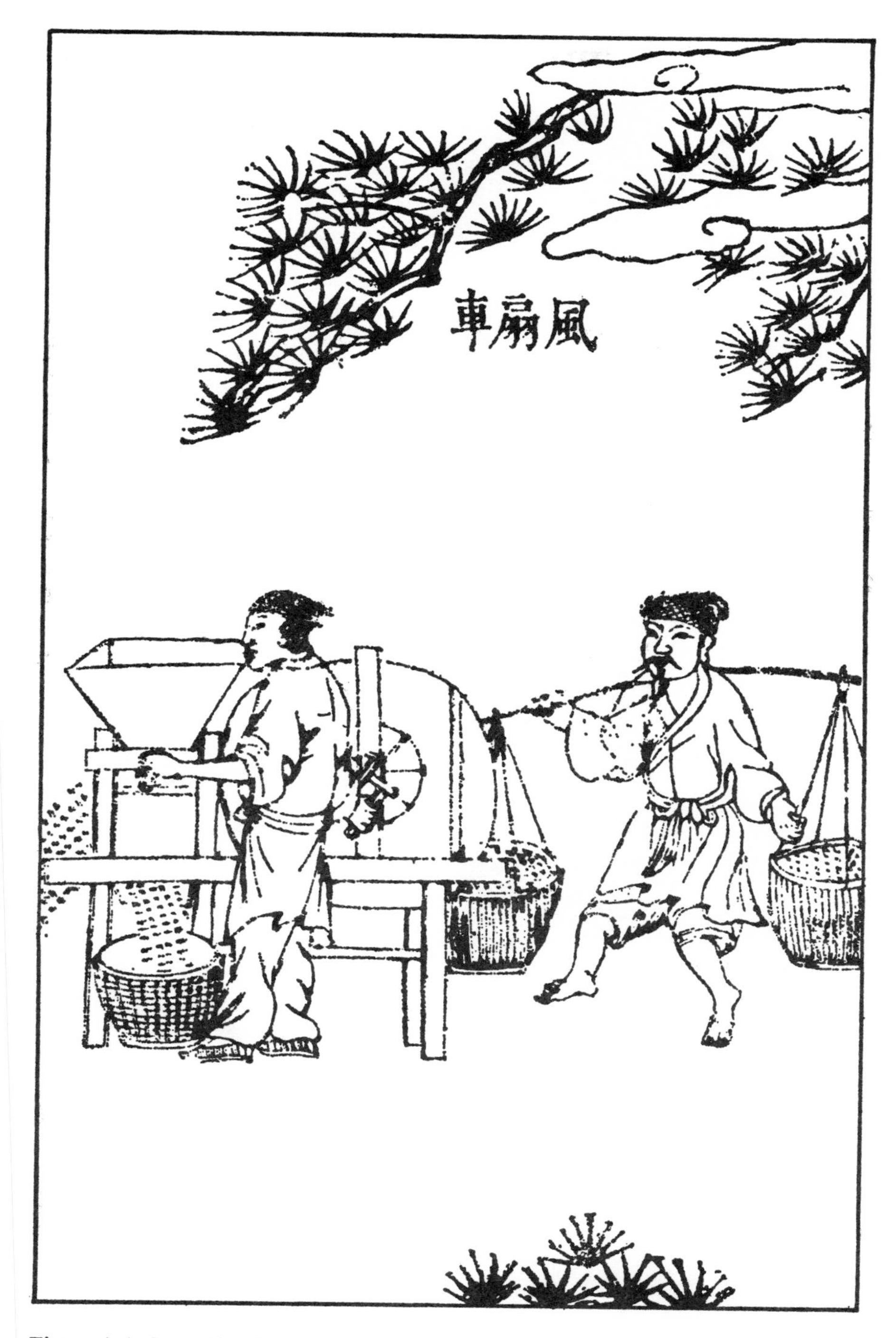

Figure 4–4. Separating husks with a winnowing machine.

Figure 4–5. The rotating fan of a winnowing machine [Ch'ing addition].

Figure 4–6. A hand-operated wooden hulling mill.

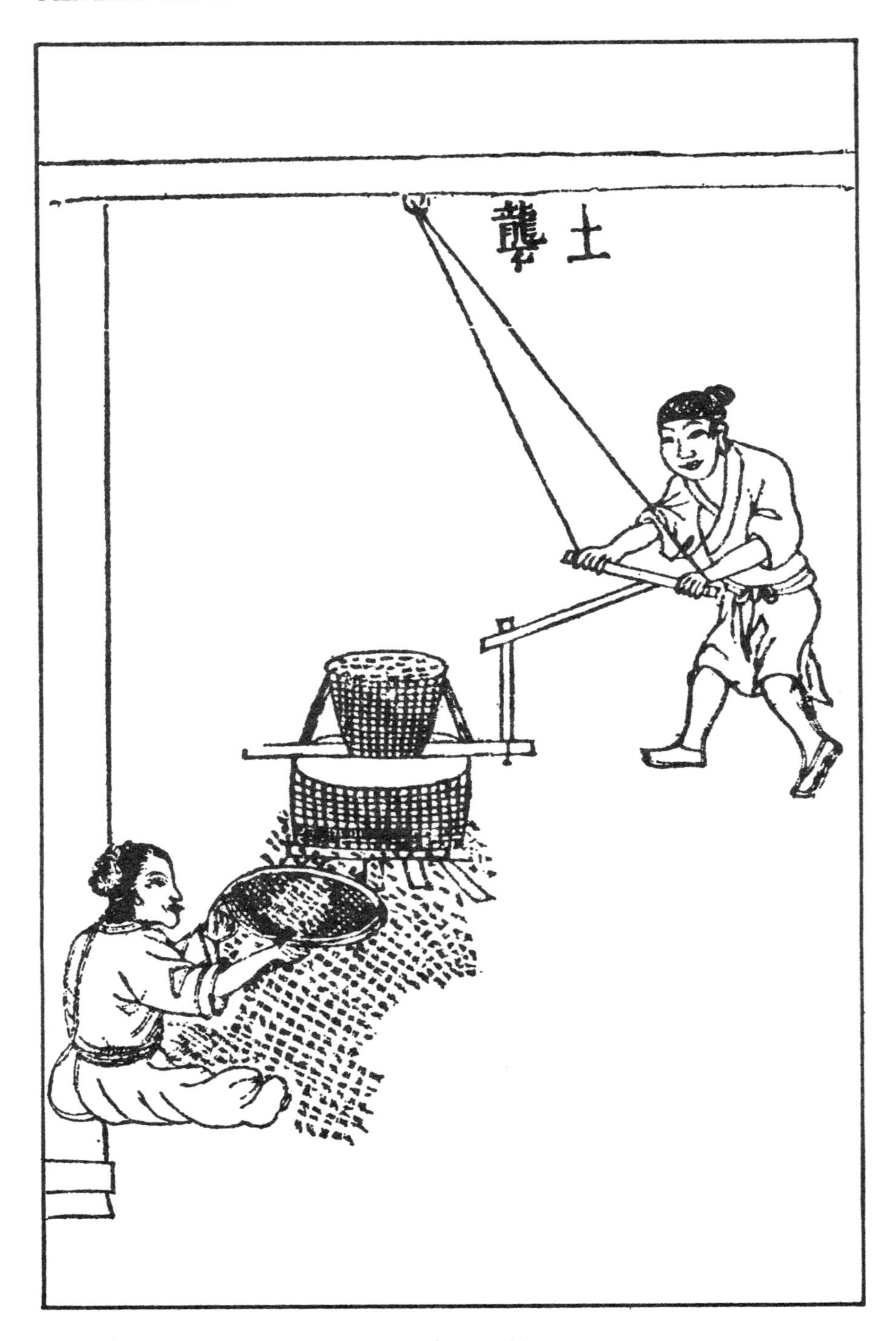

Figure 4–7. A hand-operated earthen hulling machine.

Figure 4–8. Separation of the husk-free grains by sieving [Ch'ing addition].

Figure 4–9. Unhusked grains being sieved out and returned to the hulling mill [Ch'ing addition].

Figure 4–10. A foot-operated pounding mill [Ch'ing addition].

Figure 4–11. Hand- and foot-operated pounding mills.

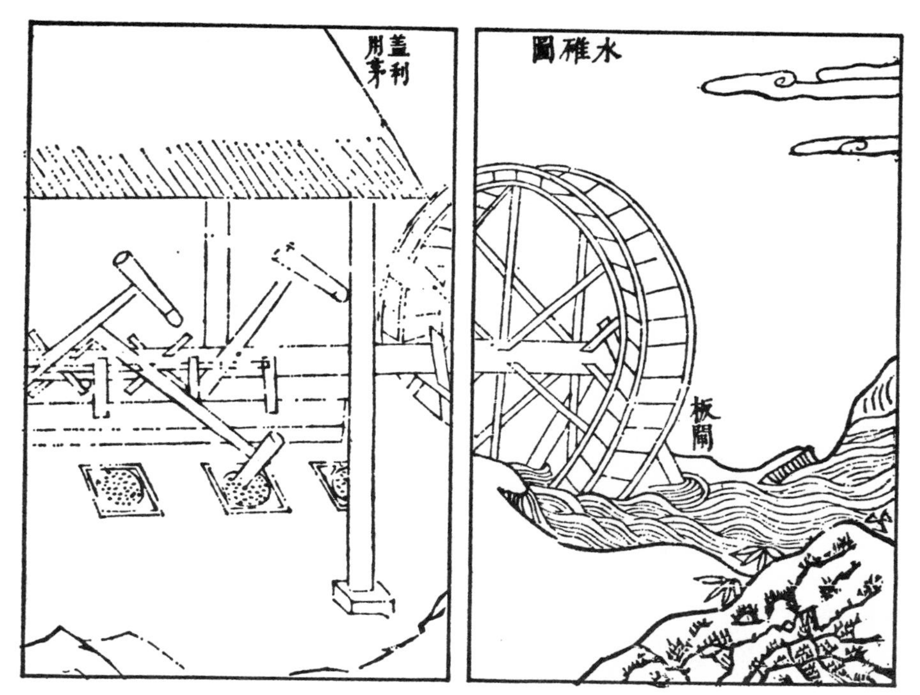

Figure 4–12. A water-powered pounding mill.

After being [hulled] and sieved, the rice grains are next subject to pounding in a mortar, of which there are also two varieties. For a large family of more than eight members, the stone mortar is settled in a depression dug in the ground. A large mortar will contain five pecks of grains, a small one half as much. [A few feet away] the pestle is affixed to a cross beam (the tip of the pestle is made of iron and glued [to the wooden part] with vinegar dregs), and is operated by foot [Figure 4.10]. Too little pounding will result in unpolished grains, and too much, in fragmentary ones. The most refined rice is produced in this fashion. For [smaller] households that need not cook large meals, a wooden hand-pestle is used in pounding against a mortar of either wood or stone [Figure 4.11]. After pounding, the chaff becomes pulverized, and is known as fine bran. It is [commonly] used as dog and pig feed, but it also serves as human food in times of famine. When this fine bran is winnowed and blown away by the wind, all chaff and dirt are eliminated, leaving behind the polished rice.

Water-powered mills [3] are used by people who live beside rivers in mountainous country. These mills are very popular as they save ninety per cent of human labor when used in pounding rice. They are constructed on the same principle as that of cylinder wheels for irrigation, [the main element being] the proper channeling of water [Figure 4.12]. The number of mortars [to each water mill] varies, ranging from two or three mortars in places where the water supply and land area are limited, to as many as ten, the latter set up where the volume of water is large and land plentiful. An extremely ingenious way of building water-powered pounding mills prevails in Kuang-hsin prefecture [in Kiangsi province] of southern China. The chief difficulty in the construction of such a mill [is the elevation of the land]: if the land is low where the mortars are set in the ground, they may be damaged by floods; if, however, the land is high, it would be difficult to make use of the water currents. In Kuang-hsin the people use a boat as the base ground. It is secured by wooden pilings that surround it, and is filled with earth in which the mortars are set. Across the current of the stream a low stone dam is erected [to provide water power]. Thus is a water-powered pounding mill completed without having to build embankments of wood, earth, and so forth.

There is still another type of water mill that is made to do threefold work simultaneously: When the swift current sets the water wheel in motion, it turns a flour mill, a pounding mill for polishing rice, and a device for drawing water to irrigate the rice fields. Such a machine can only be invented by unusually clever minds. People who live along rivers and use water mills sometimes never cast their eyes on a [wooden or earthen] hulling mill all their lives. For the elimination of husk and bran they only see the [water-powered] pounding mill. However, they winnow grain as is done in other parts of the country.

[One kind of] hulling mill is built of stone; stone is also used for making pivot poles and driving wheels. The stone hulling mill can be powered by either ox or horse [Figure 4.13], the work done by an ox in one day being equal to that of five men. Only thoroughly dry grains can be hulled and/or polished by this mill, for moistened grains will be crushed into fine fragments.

GRINDING WHEAT FLOUR

The essential substance of wheat is flour. As the utmost in refined grains is the twice-polished rice, so the utmost in pure substance of grains is the double-bolted wheat flour. When wheat is harvested it is put in bundles and the grains are separated from the stalks by beating, as is done with rice. In North China, where the winnowing machine is not widely known, the elimination of husks is accomplished by hand-winnowing. This must be done away from the house when there is wind. Winnowing cannot be undertaken on windless or rainy days [Figure 4.14]. After winnowing the wheat is washed thoroughly in water, dried in the sun, and then run through the mill. There are two species of wheat: the purple and the yellow, the former being the superior variety. The yield from good wheat is 120 catties of flour per *tan* of wheat, while that from the inferior kind is less by one-third.

There is no fixed size for grinding mills.[4] The large ones are turned by strong oxen, over whose eyes the shells of wood-oil seeds [*Aleurites cordata*] are placed as blinders, so as to prevent dizziness. A wooden pot is suspended under the belly of the ox to receive its excretion, otherwise [the wheat and the flour] will become soiled.[5] Smaller mills, powered by donkeys, [Figure 4.15] can take fewer catties of wheat. The smallest kind of mill is operated by men. An ox-powered mill can grind two *tan* of wheat in a day; a donkey-driven mill, half of that amount; while a strong man is able to grind three pecks [i.e. three-tenths the amount done by an ox], a weaker one can do [only] half as much.

Water mills have been described in detail in the section on polishing rice. The same system is employed [for grinding flour, Figure 4.16], and is three times more efficient than an ox-powered mill. Above an ox-, donkey-, or water-powered mill a cloth sack, wide at the top and narrow at the bottom end, is suspended; [a small hole is made in the bottom end and] in it are put several pecks of wheat grains, so that the receiver in the millstones is constantly supplied. This is not necessary when the mill is hand-operated.

There are two kinds of millstones, and the nature of the stone determines the quality of the flour. The lack of pure white, top-grade flour south of the Yangtse River is due to the fact that [mills there are provided with] sandy stones. In the process of grinding, the rough surfaces of the millstones are heated by friction and for this reason they pulverize [instead of simply pressing] the bran of the ground wheat. This results in the mixing of the dark-colored bran particles with the flour, from which they cannot be separated by screening. North of the Yangtse, however, the stones are cold and smooth; those produced on Chiu-hua Mountain in Ch'ih-chou [in Anhui] are especially fine. Buhrs made of this stone will not become heated, and the bran, though pressed into the thinnest slivers, will not break into fragments. Therefore, no black particles are present in the flour, which will be pure white. The teeth in southern millstones are worn out after twenty days of use, but those in northern mills will last for six months. Where 100 catties of flour, including the ground-up bran, are obtained in the south, only eighty can be gotten from northern mills. The price of top-grade flour therefore is two-tenths higher [than that of the inferior grade]. However, as this [high quality] flour is also the material out of which gluten of wheat and starch are manufactured, it would seem that in the final balance the profit is even greater.

After grinding, the flour is passed through the flour-bolter several times; for the industrious worker there can not be too many repetitions of the process [Figure 4.17]. The screen of the bolter is made of silken gauze woven for bolters. That which is woven of Hu-chou silk will remain in good condition after bolting some 1,000 *tan* of flour, but those [screens] made of yellow silk from other places cannot outlast 100 *tan*. When the final product is obtained, it will keep well in storage for three months in the cold season, but during spring or summer flour will spoil within twenty days. For best eating, therefore, flour has to be used while it is fresh.[6]

Barley is usually polished to rid it of chaff, boiled, and eaten in grain form; less than one-tenth is made into flour. As for buckwheat, it is first slightly polished, and then pounded or ground into flour before it is used as food. Buckwheat ranks far below wheat, both in quality and in price.

Figure 4–13. An animal-driven hulling mill [Ch'ing addition].

Figure 4–14. Winnowing by tossing [Ch'ing addition].

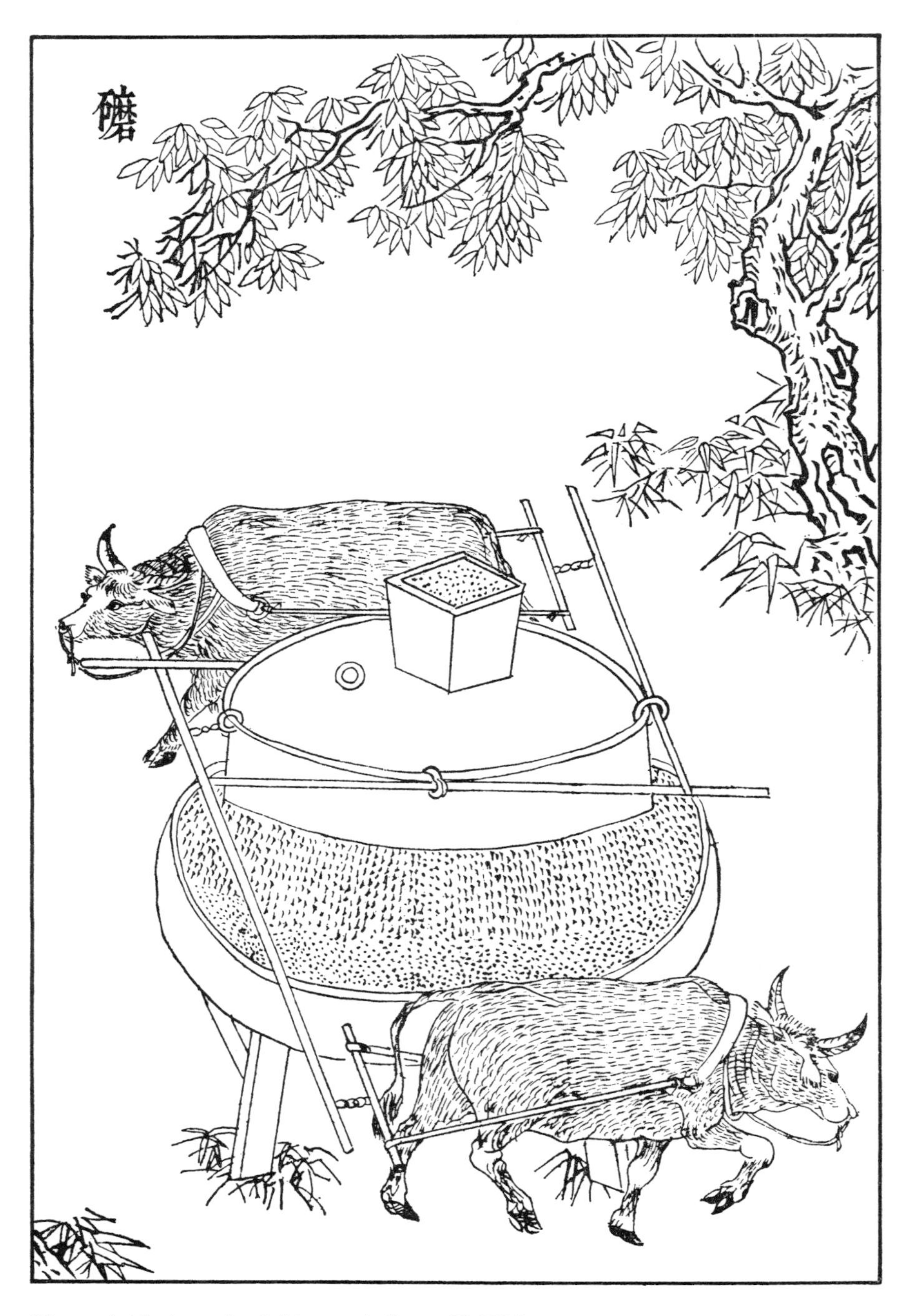

Figure 4–15. An animal-driven grinding mill [Ch'ing addition].

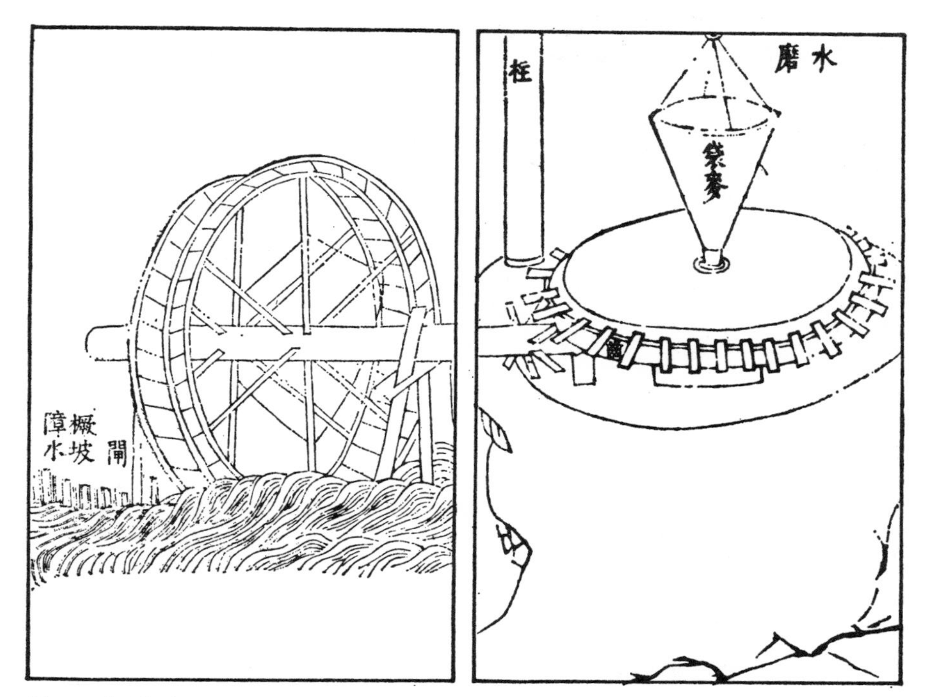

Figure 4–16. A water-powered grinding mill.

PREPARING MILLET, SORGHUM, SESAMUM, AND BEANS

[The threshed] millet is winnowed to obtain the grain, pounded to eliminate the husks and chaff, and ground into flour. Besides using the winnowing machine and the tossing method there is still another way [of winnowing millet], that is by shaking it with winnowing riddles. For this, round riddles are woven of split bamboo, on which the millet grains are spread and shaken into the air. The lighter particles [i.e. the empty husks etc.] will fly forward and fall to the ground, while the heavier particles that remain behind, are the good, full kernels.[7]

All the methods and implements for pounding, grinding, and winnowing millet [are the same as those for the other grains and] have been described in the sections on rice and wheat. In addition, however, there is an implement called the small rolling mill [Figure 4.18]. In every North China household that has millet to prepare there is a stone block, the top of which has a center rising higher than the edges. No opening is made along the edges. Millet is put on the block, and is rolled by two women who stand facing each other across the block and pass the rolling stone back and forth between them. The rolling stone is cylindrical in shape and is similar to that of the ox-pulled rolling mill [Figure 4.19]; but a wooden handle is attached to each end [of the small

Figure 4–17. A flour-bolter.

Figure 4–18. A hand-operated small rolling mill.

Figure 4–19. An ox-driven rolling mill.

Figure 4–20. A water-powered rolling mill [Ch'ing addition].

Figure 4–21. Separating sesamum grains from their stalks by striking them against a stone [Ch'ing addition].

Figure 4–22. Separating beans from the pods by beating with a flail.

roller]. Any grain that falls toward the edge of the block is promptly swept back with a little brush. In many a household this small roller is used to the exclusion even of pestle and mortar. [A water-powered rolling mill is illustrated in Figure 4.20].

After sesamum is harvested, [the cut plants] are dried in very hot sun and gathered into small bundles. Then, holding one bundle at a time with both hands, [the farmer] hits it [against a stone], causing the seeds to fall on a woven bamboo mat that has been prepared as a receptacle [Figure 4.21]. Sieves for screening sesamum resemble small rice sieves, except that the former is five times finer than the latter. The seeds will pass through the holes, while bits of leaves and other waste matter will remain on top and can be discarded.

[To separate the beans from their pods], a small crop may be simply beaten with a flail after harvesting [Figure 4.22]. For a large crop, however, the labor-saving method is to spread it on the ground, let it dry under bright sunlight, and then roll off the beans by means of a stone roller pulled by an ox [Figure 4.3]. The flail for beating the beans consists of a handle made of a bamboo or wooden stick, through one end of which a round hole is bored. Another piece of wood about three feet long is inserted in this and tied. The bean pods are spread on the hard ground, and the flail is applied by swinging the handle. After the beans have been flailed, the pods and leaves are blown off with the winnowing machine. This is followed by sieving. Thus, the good beans are made ready for storage in the barn. It is therefore said that pounding and grinding do not apply to sesamum, and that hulling and rolling are not to be used on beans.

NOTES

1. The winnowing machine with attached rotary fan was invented by the ancient Chinese, according to Needham. For a description of this machine, see R. P. Hommel, *China at Work* (Boylestown, Pa., 1937), pp. 74–77.

2. The hand rice-hulling mill with clay disks and bamboo teeth was used in the Chou dynasty (1122–256 B.C.) and is still used in China today. The principle is the same as the wooden mill: the dried clay serving merely to hold the rubbing surfaces, the bamboo strips, in place.

3. Water-powered mills for preparing cereal grain were probably used in China as early as the first or second century A.D., and became common in the period of the Three Kingdoms (A.D. 220–64).

4. It is difficult to ascertain exactly when wheat was first ground into flour in China. The *Tributes of Yü* in the *Book of History* lists "grinding stones" (*li-ti*) as one of the native products to be sent annually to the government from Chin-chou, in the modern Hupei-Hunan-Kiangsi area. But these were probably whetstones for sharpening blades.

5. According to one account, Ch'ü, a clever miller of the Yuan dynasty, solved the sanitation problem with a two-story millhouse. On the ground floor animals drove shafts which powered millstones upstairs.

6. Flour made in this traditional way spoiled quickly, chiefly because of the high moisture content.

7. According to Hommel, the ordinary grain-riddle in Chekiang province is made of split bamboo strips. For the binding around the edge, strips of cane are used. The square meshes measure 0.1875 or 0.125 inch, and the diameter of the whole riddle is 21.5 inches. On a windy day one man holds up with outstretched arms a riddle filled with grain and shakes it in such a way as to throw the contents up in the air. The wind carries off the chaff and the grain drops straight down into a shallow basket about four feet in diameter and with edges about nine inches high. Another man scoops up some more grain from a large basket and fills the riddle whenever it becomes empty.

For winnowing threshed rice which has not yet been hulled, the meshes of the riddle are equilateral triangles, each about 0.75 of an inch. The riddle is made entirely of split bamboo strips, and its diameter is 22 inches. One man holds it up and another, stepping upon a little bench about a foot and a half high, pours into the riddle a scoopful of grain which runs through with little shaking, the bits of straw staying behind. The first man must keep the riddle at a certain height in order that the swiftly running grain may have the full benefit of the wind.

Figure 5–1. Spreading ashes for "planting" sea salt (lower) and collecting sun-dried raw salt (upper).

5

SALT

Master Sung observes that, as there are five phenomena in weather, so are there in the world five tastes; [1] this being the truth first imparted to King Wu of Chou [eleventh century B.C.] by the saga Chi-tzu. A man would not be unwell if he abstained for an entire year from either the sweet or sour or bitter or hot; but deprive him of salt for a fortnight, and he will be too weak to tie up a chicken and feel utterly enervated. Is it not because this taste, deriving from [salt] water, which is the primary Creation of Heaven, is the main life-giving source for human beings? In all parts of China there are always some places where certain grains or vegetables can not grow, yet east and west, across the land, salt is everywhere obtainable in various forms. Who can tell the whys and wherefores of this?

THE SOURCES OF SALT

There is no uniformity in salt sources. They are generally classified as six different kinds: sea, lake, well, earth, rock, and gravel salt. This does not include the "tree-leaf" salt of the Eastern barbarians and the "bright" salt of the Western barbarians.[2] Of all the salt produced in China, sea salt constitutes eighty per cent, while well, lake, and earth salt make up the remaining twenty per cent. Some kinds of salt are produced by human labor, others are obtained in their natural state. In brief, as soon as it appears that there might be difficulties in transporting salt [from one place to another], Nature then will devise some other way to cope with the situation [so that no region will lack salt].

SEA SALT

Sea water contains salt. High ground on the sea shore is called a tidal mound and low ground, marshes. They both produce salt. Methods of extracting salt differ widely, though sea water is the same everywhere. One method, employed on high ground which is not submerged by tides, is to "plant" salt. Each "planter" has his own area staked out, on which no encroachment is allowed. When the weather is predicted to be fair, ashes of rice and wheat stalks, as well as those of reeds, are widely scattered on the first day on the ground until [the layer is] about one inch thick; then the ashes are rolled smooth. When the dew begins to evaporate on the following morning, the salt will rise rapidly [through the ashes]; after it has been exposed to the noonday sun the salt is swept up together with the ashes to be boiled and crystallized [Figure 5.1]. A second method, employed on lower land that is slightly submerged by tides, does not [require the] use of ashes. After the tide recedes, the [brine-wetted] ground is exposed to the sun for half a day, when salt frost will appear on the surface. It is immediately swept up and refined. Yet a third method, employed in low ground where the tides are deep, makes use of deep pits dug in the ground. Some wooden sticks are laid across the opening of the pit; it is then covered with a reed mat, and over that sand is spread. When the tide comes in and covers the ground, the brine will drip through the sand into the pit. The mat and sand are then removed and a lamp is held over the pit. The presence of briny vapor is indicated when the lamp goes out. The brine is then taken out and refined.

In all these methods the important factor is fair weather, and prolonged spells of rain will create a "salt famine." Also, at the Huai salt factories [Northern Kiangsu] there is a kind of salt that crystallizes on the ground and is shaped like horse's teeth. It is called "sun-dried salt," and is usable as soon as it is swept up. Still another variety is "straw salt" which is obtained from boiling straws and grass drifted ashore from the sea.

For draining and crystallizing salt two pits are dug, one shallow and the other deep. The shallow one is about one foot in depth, with the opening covered by reed mats resting on a wooden frame. The raw salt swept [from the ground] is then spread on the mat (the method of draining is the same whether or not ashes are mixed with the raw salt), with the edges made higher than the centre like a dyke; next, sea brine is poured over the top [of the spread salt] and allowed to drain down into the shallow pit. The deep pit is some seven or eight feet in depth; it receives the drained brine from the shallow pit, which is then evaporated in a pan for crystallization [Figure 5.2].

In ancient times, the pans used for crystallizing salt were termed "strong pans," of which there were two types. They have a circumference of several dozen feet, and a diameter of about ten feet. The type made of iron is constructed of iron plates held together by iron rivets; their bottoms are level, and the sides are 1.2 feet high. When the seams of a pan are filled with the solidified brine, they will become watertight for all time. The furnace under one single pan is equipped with a number of doors [at which fires are lighted],

Figure 5–2. Sea brine passed through raw salt in a reed mat and allowed to drain into a shallow pit.

Figure 5–3. Boiling salt brine.

Figure 5–4. Weighing and storing salt [Ch'ing addition].

Figure 5–5. Introducing lake brine into plots (lower); collecting sun-dried salt (upper).

ranging from seven or eight to about a dozen [Figure 5.3]. Another type [of pan], employed in Kwangtung, is made of bamboo. The latter is woven into pans ten feet across and one foot deep. The bamboo pan is plastered with clamshell lime and is attached to the inside of a large [iron] pot. When fire is lighted under the pot, heat is transmitted [to the bamboo pan, and the brine] boils, subsequently crystallizing into salt. These salt pans are not as suitable for the purpose as the iron ones.

If the boiled brine is slow to crystallize, a mixture of the ground pods of *Tsao-chiao* (*Gleditschia chinensis*), millet grains, and chaff should be added to the boiling liquid, and salt can be obtained shortly afterward. The role of the *gleditschia* pods in the crystallization of salt is similar to that of plaster-of-Paris in the solidification of bean curds. The salt produced at the Huai River and Yang-chou factories [3] is heavy and dark, while that produced in other places is light and white. Compared in terms of density, if one peck of Huai salt weights ten ounces, [an equal volume of] the product of Kwangtung, Chekiang and Ch'ang-lu will weigh only six or seven. The production of straw salt is irregular, since [the drift-straws] may come at intervals of either once in a few years or several times within a single month.

Salt dissolves immediately in water and is quickly converted into brine by the wind, but it can be made more solid with fire. No closed barns are needed for the storage of salt. As the thing to avoid is wind but not moisture, a three-inch layer of straw on the ground will suffice to protect the salt from dampness. If around this [stored salt] a wall of sun-baked brick and mud plaster is erected and covered with thatch about one *ch'ih* thick, the salt will remain in good condition indefinitely [Figure 5.4].

LAKE SALT

Lake salt is produced in two places in the Empire: one is Ningsia, which provides for the needs of the frontiers; the other is Chieh Lake [4] in Shansi, which supplies the various districts in Shansi and Honan provinces. Situated amid the districts of An-i, I-shih, and Lin-chin, Chieh Lake is surrounded by a brick wall which serves as a protective boundary. The lake brine is dark green in color where it is deepest. The people who "plant" salt there plough the land near the lake into plots separated by dykes, and clear brine from the lake is introduced into the plots. It is important that no turbid brine be allowed to contaminate the clear, lest the salt brine vein will be covered by the settled impurities. This step of the operation should be done in the spring. After a period the brine will turn red,[5] until late summer or early autumn when the south wind blows hard across this area, so that the [brine evaporates entirely and] salt crystallizes overnight [Figure 5.5]. This salt is called "grain salt," also termed "large salt" in ancient records, because while sea salt is finely granulated, this lake salt is formed in big crystals, hence the term "large." It can be used as table salt as soon as it is swept up. The salt producers, however, have to deliver it to the authorities, and receive only a few dozens of copper cash for each *tan*.

At Hai-feng and Shen-chou [in Kwangtung and Hopei, respectively] where sea water is introduced into ponds [on the shore] and crystallized by sun, the salt can also be used directly as table salt without refining, which is the same as Chieh [lake] salt. It differs vastly from the latter, however, in the season of production and in the fact that [the crystallization of salt in these localities] is not dependent on the south wind.

WELL SALT

The provinces of Yunnan and Szechuan, situated far from the seacoast and on high elevations, are not easily reached by boat or cart. Here salt is buried underground. In the rocky mountains of Szechuan, salt wells can usually be drilled not far from rivers. The diameter of the well shaft is only a few inches [Figures 5.6 and 5.7], and the opening on the ground can be covered amply with a small bowl. The well must be more than 100 feet in depth, however, before the salt water vein can be reached, hence the drilling of a salt well is a difficult and costly undertaking. The tool used is an iron drill with a very hard and sharp tip shaped like the blade of a chisel, which bores holes in the mountain rocks. The drill is suspended and held in place by a bundle of split bamboo strips, fastened together with ropes [Figure 5.8 and 5.9]. Each time the rock has been penetrated a few feet, the bamboo-suspender is lengthened by attaching to it another section of bamboo. Up to the first ten feet or so of the well, the drilling equipment can be operated by stepping on it with one's feet, like the motion of pounding rice in a mortar. When the well becomes deeper, the drilling equipment is operated by hand. The rock fragments that result from the drilling are scooped up with an iron vessel attached to a long bamboo pole.[6] The time required for completing a salt well ranges from over a month for a shallow one to some six months for a deep one. [The reason for the small diameter of the shaft is that] a wide shaft allows the brine to degrade, through dissipation of its vapor, and become unsuitable for making salt.

When the brine level in the well is reached, good bamboo stalks about ten feet in length are brought in, and from them all the inner section partitions, except the one in the bottom end, are removed [Figure 5.9], and a valve which permits the brine to enter is set in that partition. This bamboo is lowered into the well with a long rope [Figure 5.10]. After it fills with brine, it is raised to the surface by means of a pulley fixed over the well and [a windlass] turned by an ox-powered wheel [Figure 5.11]. The brine so obtained is then poured into a pan for evaporation and crystallization (only a medium-sized pan is used, not the "strong" pan) [Figure 5.12]. Salt of a very white color will quickly crystallize from it. [The transportation of salt in Szechuan and Yunnan provinces is illustrated in Figure 5.13.]

A highly amazing phenomenon is the fire wells[7] of western Szechuan. These wells actually contain cold water, without the least appearance of fire.

Figure 5–6. Opening ground for drilling a salt well [Ch'ing addition].

Figure 5–7. Lining a drilled shaft with stones [Ch'ing addition].

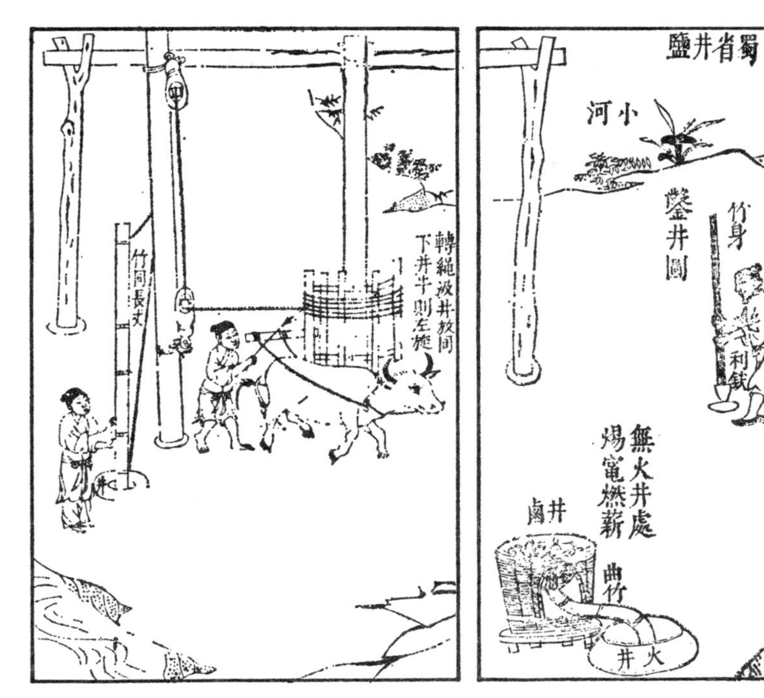

Figure 5–8. Drilling a salt well in Szechuan.

Figure 5–9. Preparing bamboo stalks [Ch'ing addition].

Figure 5–10. Lowering a bamboo stalk into the bottom of a salt well [Ch'ing addition].

Figure 5–11. Raising brine from the bottom of a salt well [Ch'ing addition].

Figure 5–12. Boiling and crystallizing salt brine [Ch'ing addition].

Figure 5–13. Transporting salt in Szechuan and Yunnan [Ch'ing addition].

But, split open a long bamboo stalk and take out its inner section partitions, then put the stalk together again and securely wrap it with varnished cloth. Next, place one end of this bamboo into the well, while the exposed end is connected with a curved section [of bamboo] to reach the bottom of a [salt] pan filled with brine. It will be seen that heat radiates [from the mouth of the bamboo] to boil the brine in the pan [lower right of Figure 5.8]. Yet if the bamboo stalk is opened and examined, not a bit of charring or burning evidence can be seen. To use the essence of fire yet seeing not the fire—this is indeed one of the strangest things in the world!

It is extremely easy for [the operators of] the salt wells in Szechuan and Yunnan to avoid paying the salt tax by covering up the wells. It is impossible to track down all these people.[8]

POWDERED SALT

Salt can also be obtained from alkaline soils. Aside from the powdered salt [so produced] in Ping-chou [northern Shansi], the natives of Ch'ang-lu [eastern Hopei] also scrape up the earth and derive salt from it. This salt is of a mixed dark color and is rather unpalatable.

ROCK SALT

In such places as Chieh [in Kansu] and Feng [in Shensi] in the western provinces, where neither sea salt nor well salt is available, a kind of rock salt is produced from the cliffs and caves. This natural salt has the color of red earth, is freely obtainable by scraping, and is [sold] without refining.

NOTES

1. The five phenomena of weather are fair, rainy, hot, cold and wind. The five tastes are sweet, sour, bitter, hot and salty.

2. "Tree-leaf" salt, called "wood salt" in Pei shih (*History of the Northern Dynasties* (A.D. 286–581), and also in *T'ang shu,* is presumably formed by evaporating ground

brine on trees and is still used by the aborigines of Formosa. "Bright salt" is simply a salt of large crystals.

3. The Huai River region and Yang-chou were the largest sea-salt production areas and distribution points of eastern China. The salt industry of this territory was started by Liu P'i, otherwise known as the prince of Wu, of the Former Han dynasty (ca. 206–193 B.C.), and was fully developed by a salt-and-iron minister named Liu Yen (ca. A.D. 756–66) of the T'ang dynasty. In comparison, the salt industry of Kwangtung reached its maturity in the Han dynasty, and the salt yards of Ch'ang-lu in Hopei province, according to one modern author, started in the Chou dynasty (1122–256 B.C.). Other sea-salt producing areas, not mentioned in the present book, are the seacoasts of Shantung, Manchuria, and Fukien. The salt industry of these three regions dates from the Hsia, Chou, and T'ang dynasties, respectively.

4. The Chieh Lake in Shansi, approximately 50 *li* long and 15 *li* wide, was the most famous salt-producing lake in China. References to this salt can be found in many ancient Chinese records, including the *Tso Chronicles*. It should be noted that most of China's salt lakes are in the western and northwestern provinces, such as Shansi, Shensi, Ningshia, Kansu, Sinkiang, Chinghai, Tibet, and Mongolia.

5. The red color is believed to be caused by numerous small insects in the lake brine.

6. The technique of drilling deep wells dates from the Warring States period (403–221 B.C.) or at least the Former Han dynasty (206 B.C.–25 A.D.).

7. Natural gas from the "fire wells" or *huo-ching* of Szechuan was known in the Han dynasty (206 B.C.–220 A.D.). *Hua-yang-kuo chih* says that the Han Chinese used this natural gas for various purposes, including boiling salt brine.

8. From a historian's point of view, the real interest attached to Chinese salt lies in its taxation and administration, as evidenced by the numerous Chinese records. The system of collecting tribute salt from the salt-rich districts was probably started in the Hsia dynasty and continued in the Shang (1766–1122 B.C.) and Chou (1122–256 B.C.) dynasties. Although tribute salt can be considered as a sort of salt tax, the taxation of salt was not proclaimed until the Spring and Autumn period. Chinese records show the Kuan Chung (696–681 B.C.) advised Duke Huan of Ch'i to levy taxes on salt and iron. The Ch'in dynasty (221–207 B.C.) collected twenty times more salt tax than the previous dynasties. Salt was continually taxed in the early part (206–127 B.C.) of the Former Han dynasty and became a government monopoly in the reign of the Martial Emperor (126 B.C.). At the beginning of the Later Han dynasty (A.D. 25), the government monopoly of salt was replaced by taxation, which was generally continued in the Southern and Northern dynasties and in the early part of the Sui dynasty. The salt tax was eliminated by emperor Wen-ti (A.D. 584) of the Sui dynasty but was restored by emperor Hsüan-tsung (A.D. 723) of T'ang. About thirty years later (A.D. 756) another emperor of the same dynasty again instituted a government monopoly of salt, which continued for more than seven hundred years. In 1618, the government monopoly was modified into a monopoly by government-licensed merchants, a system which lasted until the end of the Ch'ing dynasty.

6

SUGARS

Master Sung observes that pleasant scents, fresh colors, and mild, sweet taste are the general desires of man. Yet Nature is able to enhance these qualities to their extreme states, so that the pleasant become fragrant, the fresh become brilliant, and the mildly sweet become very sweet. Of the material from which sweet things are made, eight-tenths come from vegetable and wood sources, while winged insects also exert their efforts to the utmost, gathering [nectar from] many flowers and producing a delicious concoction. Who was it that [first] discovered the use [of honey], so that people in all parts of the country might benefit from it?

PLANTING OF SUGAR CANE

There are two kinds of sugar cane, most of which comes from Fukien and Kwangtung. The other provinces altogether produce but one-tenth the amount produced in these two. The kind that is large and resembles bamboos is fruit cane: it is cut into sections and chewed fresh for its sweet juice, but is not used for making sugar. The other kind, smaller and resembling reeds, is sugar cane [*Miscanthus sacchariflorus*]. People do not dare eat it, because its sharp fibres will cut the mouth.[1] This is the material from which both white and brown sugar are made.

The manufacture of sugar from cane juice was not known to the ancient Chinese.[2] It was during the Ta-li reign [A.D. 766–88] of the T'ang dynasty that the art of sugar-making was first introduced at Sui-ning in Szechuan by the monk Tsou, who had come from the west. The large quantities of cane produced today in Szechuan owe their origin also to the Western Regions.

In early winter just before the coming of frost the sugar canes are cut down, stripped of tops and roots, and buried under earth [3] (avoid low ground or places where water gathers). These are dug out on sunny days the following spring, five or six days before the Rain Water period.[4] The bark is peeled off, and the canes are chopped into sections about half a foot long, usually leaving two joints to a section. These are laid on the ground close to each other and lightly covered with earth, the ends of the sections overlapping like fish scales. [Each section must be so placed that] both buds on it are laid in a level position. The buds must not be placed so that one points up while another points down, resulting in difficulties in germination. When the shoots are about one or two (Chinese) inches high, they are repeatedly watered with the supernatant liquid of manure. [The young plants] are transplanted when they reach six or seven inches in height.

Sugar cane must be planted in sandy soil, the best choice being the alluvial soil along rivers. To test the soil, the earth is dug to a depth of about one and one-half feet, and a sample of this sandy soil is given a taste test. Cane should not be planted where the soil tastes bitter in the mouth. Deep in the mountains and along the upper reaches of rivers, however, no cane should be planted even if the soil tastes sweet, because the cold-retaining climate of the mountain will later give a bitter taste to the sugar. [The location of cane planting therefore] should be at a distance of some forty or fifty *li* from the mountains, on a plot of good earth on level alluvial soil (but yellow-mud bottom land should never be used).

Furrows for sugar cane are ploughed four feet apart and four inches deep. The cane shoots are placed in the furrows, spaced at about three clusters for every seven feet, and covered with about one inch of earth. Fewer shoots will grow if the covering earth is too thick. When three, four, or half a dozen or so shoots have emerged, more earth is gradually added while the rows are cultivated; this thicker earth at the base will insure the growth of tall plants and deep roots, without the danger of toppling. The more frequently the cane is cultivated, the better, while the amount of manure to be added is determined by the original richness of the soil. After the plants are about one or two feet high, the residue of sesamum or rape seeds, from which oil had been extracted, is soaked in water and poured between the cane rows. Later when the cane has reached [a height of] two or three feet the earth between the rows is turned with an ox-drawn plough twice a month. First the earth is turned and the side roots are cut by the plough, then the earth is heaped [around the base of the plants] for the protection of the plant roots. At the beginning of the ninth month the roots should be protected by [a further] covering of earth, so as to prevent harm from frost or snow after the harvest.

VARIETIES OF CANE SUGAR

Three varieties of sugar are made from the sugar cane: rock sugar, white sugar, and brown sugar, the differences being determined by the age of the cane juice. The latter's nature is such that in autumn it turns a dark red color; after the winter solstice, it changes from red to brown, and finally to white.[5] In regions south of the Five Ranges [6] where there is no frost, the cane is left standing [in the fields until winter] in order to obtain white sugar. North of Shao-chou and Nan-Hsiung [in Northern Kwangtung], however, frost appears in the tenth month [i.e. approximately through November], and will immediately destroy the sugar element in the cane. The plants therefore cannot wait to produce a white color [cane juice] in winter, and should be cut down earlier for making brown sugar. To do this, the entire harvesting process must be accomplished ten days [before the first frost]. Before the ten-day period, the cane will not have matured sufficiently, while after frost the crop may be worthless. For this reason every household that plants ten *mou* [i.e. 1.7 acres or more] of sugar cane is equipped with its own cane crusher and juice boiler in order to carry out the work with dispatch. In the frost-free southern Kwangtung, man can set the time [of harvesting the cane] early or late, as he prefers.

THE MANUFACTURE OF SUGAR

The [vertical] roll crusher used for crushing cane in sugar manufacturing [Figure 6.1] is erected as follows: Take two wooden boards measuring five feet long, two feet wide, and one-half foot thick. At the ends of each board holes are bored to accommodate the supporting posts [for the upper board, which is placed horizontally several feet above the lower]. The upper ends of the posts protrude a little bit above the upper board, while the lower ends extend two or three feet below the lower board and are buried in the ground, so that [the whole apparatus] will be stable. At the center of the upper board two openings are made, through which two large rollers (made of very hard wood) are placed one next to the other. It would be best to have the rollers measuring seven feet in circumference. One of the rollers is three feet long, the other four and a half feet. Affixed to the protruding end of the latter one is a curved pole made of bent wood, measuring fifteen feet in length; this is attached to an ox that pulls it in a circle [and thus turns the rolls]. The surfaces of both rollers are deeply corrugated or cogged; the cogs of one roll fit into the grooves between cogs of the other. The sugar cane which passes between the two toothed rollers is pressed and crushed. This is on the same principle as the cotton gin. As the cane is pressed the juice runs off. The residue is fed into the crusher again through the receiving hole on the rollers, and after having been crushed three times all the juice in the cane will be pressed out. The bagassee is used as fuel.

On the lower board two depressions one and one-half inches deep are hollowed out to receive the bottom ends of the rollers. The board is not perforated, and therefore it catches the juice. An iron pellet [7] is affixed to the center of the bottom end of each roller to facilitate rotation. The cane juice flows into an earthenware jar through a trough that has been cut in the board.

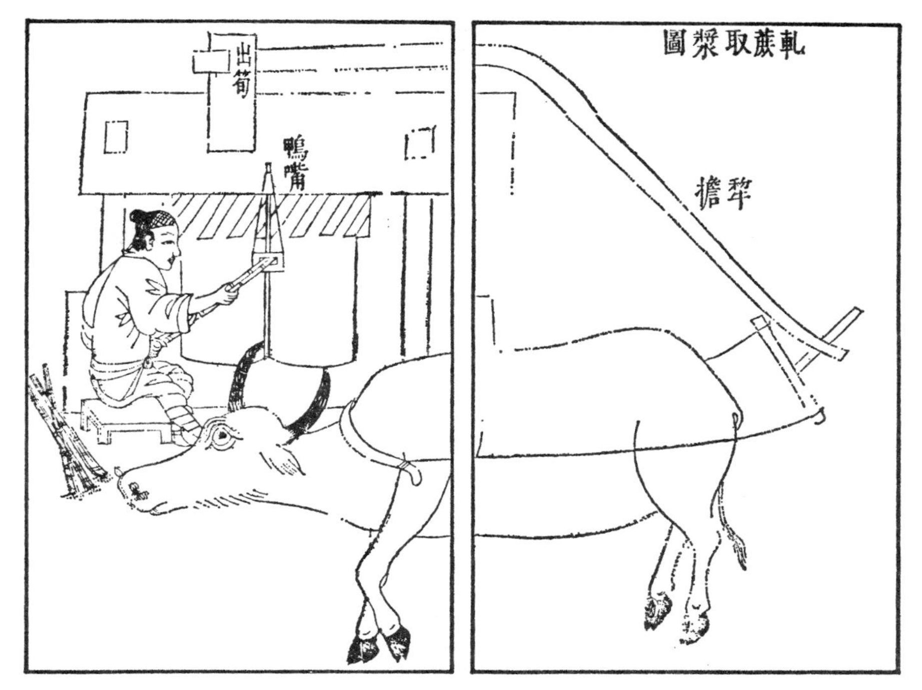

Figure 6–1. Crushing cane with a vertical-toothed roll crusher.

To every *tan* of cane juice is added one-half a pint of lime.[8] In boiling the [clarified] juice for sugar, three cooking pots should be arranged to form a triangle [and used simultaneously]. First the thick syrup [that has been obtained after boiling in the other two pots] is transferred into one pot, then more thin [uncooked] juice is gradually added into the two other pots. If the fire lacks even one handful of fuel wood [from the proper amount], the sugar will become "irregular sugar"; it will be full of foam and not of much use.

MAKING WHITE SUGAR

In south China, in Fukien and Kwangtung, for example, old sugar canes that have stayed in the fields during the winter are pressed in roll crushers, as described above, and the resultant juice is collected in large jars. [While the juice is cooking], the strength of the fire should be determined by the degree of boiling shown by the juice. When the bubbles are very numerous and fine, like those in boiling soup, and when touched with fingers [the syrup] will stick to the hand, then the cooking is sufficient [for "graining"]. The cooked syrup [or *massecuite*], now a dark yellowish color, is kept in barrels, where [approximately 65 per cent of the total sucrose] will crystallize into dark granules, [and 35 per cent remains in solution as molasses].

Next, a [conical] earthenware funnel (potters can be taught to make these), having a wide top and pointed base, is placed over a large jar [Figure 6.2]. The small hole at the base of the funnel is plugged with straw, and the dark sugar granules are transferred from the barrels into the funnel. When these granules are consolidated, the straw is removed from the hole, and yellow-earth water is poured over the contents. The dark impurities [incuding some molasses] in the sugar are thus run off into the jar, while only white sugar is left inside the funnel.[9] The uppermost layer, about five *ts'un* thick, is extremely white, and is called "Western sugar" (because sugar from Western countries is most white and excellent). Below this upper layer the sugar is somewhat yellow brown in color.

To make rock sugar, the "Western sugar" is [dissolved in water] and boiled with the addition of albumen, followed by skimming off the floating scum. When the correct temperature is reached, split pieces of new green bamboo, cut into one-*ts'un* bits, are thrown into the syrup and left over night.[10] The sugar will naturally crystallize into "rocks." There is no definitely fixed quality of sugar used to make lions, elephants, dolls, etc.

Figure 6–2. Removing dark impurities from crystallized raw sugar.

There are five grades of white sugar. The best is "rock mountain," next followed by "clustered branch," next "glossy jar," and next "small grain." The lowest grade is "sandy bottom."

MAKING OF ANIMAL-SHAPED CANDY

To make animal-shaped candy, a large pot is filled with fifty catties of sugar and heated gradually from below. The sugar syrup will boil gently if the fire is placed to one side of the pot; however, it will boil over and overflow to the ground if the fire is placed directly under the center of the pot. A mixture of the whites of three eggs and half a pint of water is added slowly to each pot of sugar, with the aid of a spoon. The sugar syrup and the added mixture in the pot are slowly heated so that the impurities and scum will float to the surface and then are skimmed off with a rattan strainer. The clarified syrup, being clear and clean, is transferred into a copper pot and warmed over a low fire of coal briquettes. After being cooked for a certain time, the syrup attains a correct temperature [i.e. chemical and physical properties] and is poured into the molds. Each mold, in the shape of a lion or an elephant, etc., is composed of two pieces of pottery that come together [to form an entire mold]. As the liquid sugar is ladled into the mold, the latter is immediately turned around, allowing the syrup to flow off. Since the mold is cold and the syrup hot, a layer of sugar will naturally adhere to the [inside of the] mold and become solidfied.[11] This is called feast candy and is used at banquets.

HONEY

In all parts of the country there are honey bees, but their numbers are naturally smaller in areas where large quantities of sugar cane are produced. Of the honey made by bees, 80 per cent come from crags and caves [i.e. wild bees], while 20 per cent come from domestically kept bees. There is no fixed quality for honey: it can be green or white, yellow or brown, all depending on the local conditions and the nature of the flowers. There are for instance, rape-flower honey, grain-flower honey, and hundreds of other varieties.

Whether wild or domesticated, the bees all have their bee kings,[12] who live in specially built chambers the size of a peach. The king's position is inherited by his son. The king never hunts among the flowers; he is waited on everyday by many other bees, who take turns in supplying him with nectar taken from flowers. Twice a day (in Spring and Summer, the season of honey-making) the king travels abroad, accompanied by eight other bees that are on duty. The king goes by himself to the opening [of the beehive], from where he flies off, with four attendant bees supporting his belly with their heads and four others flying beside him. They return after a few hours, keeping to the formation as before. The keepers of domestic bees either hang barrels below the eaves or place wooden boxes outside the windows as hives, on which scores of round holes are drilled to allow passage of the bees. Nothing would happen if one or two bees are killed by the keepers, but if the number reaches three the entire group would swarm to attack the people. This is known as a "bee rebellion."

Bees are a favorite food of bats. When one of the latter gets into a beehive through a crack it will devour large numbers of them. But if a bat is killed [by the beekeeper] and hung in front of the hives, no other bat will dare eat the bees again. This is commonly known as "warning through execution."

If a new colony of domestic bees is to be started by a keeper with a group transferred from his neighbor's hives, a son of the bee king must also go with the group to serve as its king. Upon leaving, [the young king] is surrounded during the flight by an entourage resembling honor guards. Sometimes the country people entice [a new colony of bees to their houses] by spreading fragrant wine-mash [outside the houses].

In making honey the bees build up honeycombs, which have a spongy appearance. [Honey is] made with the substance emitted by the bees after digesting the nectar gathered from flowers. The fragrance [of the honey] is achieved by the admixture of some human urine, such being the magical powers of that which is itself deteriorated or odorous. To get the honey, the comb is cut, which will kill most of the young bees [i.e. larvae] in it. Yellow wax is found beneath the honeycomb. Among crags in the mountains, where the [wild] honey has not been gathered for years and is fully ripened for harvesting, the local people usually pierce the combs with long poles, and the honey will flow out freely. Those honeycombs that are not so old and full can be gathered [individually] by climbing onto the rocks, and are cut and treated in the same way as domestic honey. Most of the wild honey found inside earthen caves is produced in North China. The damp climate of South China is favorable for the production of honey on mountain rocks but not for that of cave honey. Out of each catty of honeycombs twelve ounces of honey can be extracted. Honey is widely used in northwestern China, being of equal importance with cane sugar as a sweet condiment.

MALTOSE

Maltose can be made from rice, wheat, sorghum, or millet. According to the "Grand Regulations" chapter [in the *Book of History*], "this [i.e. making maltose] is the ultimate development in making sweet condiments out of grains." [13] The method of making maltose consists of soaking the grains of wheat or rice and the like [in water] and allowing them to sprout. They are then dried in the sun, cooked and treated, and maltose is obtained. The best quality is white. A reddish variety, resembling amber, is called jellied maltose; it melts in the mouth, and is a favorite in the palaces. The confectioners of the south refer to maltose as malt sugar in order to distinguish it from cane sugar.

There are hundreds of ways in which maltose is prepared by candymakers to please the human palate, which can not be enumerated here. There is a variety called "nest of silken threads" which is used [exclusively] by the

Imperial household. It is possible that the method of preparing this particular sweetmeat will be passed down to the future generations, [thus making it available to the common people].

NOTES

1. Fruit cane and sugar cane have long been a major cash crop for the farmers of south China. The fruit cane is over one inch in diameter with more juice and softer fibers. It is a perennial plant, and can be cultivated up to latitude 30 north in central China and 31 north in Szechuan. In contrast, the sugar cane is slender, less than one inch in diameter, with tougher fibers and less juice, but it has higher sugar content. It is an annual plant, and is not cultivated north of latitude 28 north in central China and 30 north in Szechuan. Probably it was also eaten fresh occasionally by the early Chinese, as indicated in *Materia Medica.*

The sugar beet is not mentioned in the present book since the possibility of extracting sugar from beets was discovered only in 1747.

2. The ancient Chinese cane sugars, such as *che-chiang* in the Ch'in dynasty and *che-t'ang* in the Han dynasty, were used in the form of syrup. Genuine crystallized cane sugar was first sent to China from Indo-China in the Han dynasty as tribute, and its manufacture in China was started in the T'ang dynasty. It is related in the *Hsi-yü chuan* of *The History of T'ang* that Emperor T'ai-tsung (627–649) had sent emissaries to India to learn the method of sugar making. He subsequently ordered the officers at Yang-chou to offer as tribute the sugar cane from which sugar was made. Up to then, however, sugar seemed to have remained a luxury item, its use limited to the imperial palaces. According to *T'ang-shuang p'u,* written by Wang Shao of the Sung dynasty, and as related in the present book, sugar making was popularized by the monk Tsou in Szechuan in the Ta-li period (766–788) of T'ang. The manufacture and use of sugar soon became widespread in China, although the techniques continued to undergo improvements. The manufacture of "rock sugar" in China was started at least in the Sung dynasty (960–1279) if not earlier, as pointed out by T'an Tan-chiung.

3. This protects the bud of the "seed" stock from damage by frost and snow. Almost to the end of the nineteenth century sugar cane was propagated entirely by planting the stalks, the bud at each joint germinating and producing a new cane of the same character as the parent stalk. Not until about seventy years ago was it discovered that some of the seeds could be made to germinate under favorable conditions. This discovery resulted in the creation of many new varieties.

4. Rain Water (yü-shui) is one of the twenty-four solar periods in the agricultural year. It occurs sometime in late February or early March of the modern solar calendar.

5. Cane juice contains, besides sucrose, a small amount of organic acids, and some protein. A high percentage of protein hinders the crystallization of sugar, hence in order to

obtain white sugar it is necessary to harvest the cane after it has fully matured at the end of autumn, when the sugar content is at the highest point. In some parts of China as well as in Louisiana, where the cane must be harvested in November before the arrival of frost, complete maturity may never be reached. In Hawaii, where more favorable climatic conditions exist, twenty-two months are sometimes allowed to elapse between the germination and harvest of the cane.

6. The Five Ranges are mountains separating Kiangsi and Hunan from Kwangtung and Kwangsi, hence Kwangtung is often referred to as Ling-nan or "south of the Ranges." The climate south of the Ranges is warm and approaches the subtropical.

7. "Iron pellet" is used here to translate *t'ieh-ting* or "iron ingot"; apparently it serves as a bearing.

8. Lime is used to neutralize the organic acids of the juice and to coagulate a number of undesirable impurities. The resultant aggregates rise to the surface or settle to the bottom. *T'ang-shuang p'u* indicates that the Chinese in the Sung dynasty were sometimes unable to determine the suitable amount of lime for making crystal sugar, and for this reason they produced a sort of molasses-sugar.

9. White sugar is referred to as *T'ang-shuang,* i.e. "sugar frost," in premodern Chinese writings. The early white sugar had the texture of coarse flour rather than the loose granulated appearance of the modern product. It is now known that certain earthen materials are good adsorbents and/or ion exchangers. Therefore, the slurry of yellow mud, on passing through the dark raw-sugar granules, removed the color and odor.

10. Small sugar crystals attached to the bamboo strips act as seeds or reactive centers to promote the deposition of additional sugar.

11. The animal candy would therefore be thin, crisp, semi-transparent, and hollow-centered confections.

12. Bee king (*feng-wang*) should, of course, be "queen bee." Although it is possible to interpret the word "wang" as "nü-wang," i.e. female king, this text clearly does not warrant such a translation.

13. Maltose was known to the ancient Chinese as *t'ang, i, fu,* and *san,* and was used in China long before cane sugar. References to it are found in writings dating from the Chou dynasty (1122–256 B.C.), and possibly it was made at a much earlier date.

PART II

7

CERAMICS

Master Sung has observed that, with earthenware brought about through the use of water and fire on clay, even the daily labor of a thousand workers is not enough to fill the demands of a large country, so numerous are the people's needs for pottery. As shelter against wind and rain, houses are built, and so there is need for tiles. In order to defend the country the rulers must construct strategic defense works, hence city walls and ramparts are built of bricks, and invaders are kept out. Sturdy earthen crocks preserve wine to a good vintage, while clean pottery vessels are instruments for containing the sacrificial offerings of wines and bean sauces. The sacrificial dishes of Shang and Chou times were made of wood; was it not because the people then wanted to show great respect [towards the spirits]? In later times, however, ingenious designs began to appear in various localities, human craftsmanship exerted its specialities, and superior ceramic wares were produced, beautiful as a woman endowed with fair complexion and delicate bones. [These wares] sparkle in quiet retreats or at festive boards, a concrete sign of civilized life. It is hardly necessary to adhere [to the ways of Chou and Shang] forever.[1]

TILES

The material for making tiles is a sandless clayey earth, which is obtained by digging about two *ch'ih* under the earth's surface. Within a hundred *li* [of any place] there is sure to be a suitable clay for the construction of houses. The tiles used for the houses of the ordinary people are all made from the quarter sections [of a cylinder]. The process of making such tiles involves the preparation of a cylindrical core mold, on the surface of which are four ribbed demarcation lines. The clay is mixed with water and made into a high rectangular pile. Then the moist clay pile is sliced by the iron wire of a bow, with the wire located at 0.3 inch away from the back of the bow. The sliced clay layer is lifted like a piece of paper, and wrapped around the cylindrical core mold. When slightly dried, the clay is removed from the mold, and naturally falls apart into four pieces [Figure 7.1]. There is no standard size for tiles, which range from a measurement of eight to nine inches on each side for the large ones to seven-tenths of the above size for the smaller ones. Those used to catch rainwater along house roofs, called drain tiles, must be the largest in size, so that they can sustain a continued rainfall without overflowing.

When the green tiles are made and dried, they are piled in the kilns and fired. The firing may last either one day and night or two days and nights, depending upon the number of tiles placed in the kiln. The methods used for the water-tempering and glazing of tiles are the same as those for bricks. There are "water drip tiles," to be placed at the eaves; "cloud tiles," below the roof ridge; "curved cover tiles," over the roof ridge; and tiles variously shaped like birds or beasts, to be affixed at the two ends of the ridge. These latter tiles have to be handmade, piece by piece; but in so far as they are prepared through the use of water and fire and firing in kilns, they are made by the same method [used for making other kinds of tiles].

Figure 7–1. Making tiles.

Tiles used by the Imperial Household, however, are quite different from the above. The *lazuli* tiles are made piece by piece, either flat or rounded in shape, with the aid of round bamboo or polished wood molds. Only clay from T'ai-ping prefecture [in Anhui province] is used for [making imperial tiles]. (This clay must be transported 3,000 *li* to Peking, in the course of which indescribable damage is caused by adulteration with sand, and by the lawless behavior of the transport laborers and boat crews. Even the imperial mausolea are built with such tiles, and no one has dared give advice to the contrary.) When the tiles have been shaped, they are first placed in the "glazed tile" kilns, in which [they are fired] to the proportion of 5,000 catties of fuel wood per 100 pieces of tiles. The fired tiles are painted with a mixture of pyrolusite [MnO_2] and palm hairs for green color, and with [a paste of] ochre, rosin and rushes [juice] for yellow. The painted tiles are then placed in another kiln to be fired again at a lower temperature. The gemlike brilliance of *lazuli* colors will then be achieved. Occasionally this kind of tiles is used in the palaces of princes (living) in the provinces as well as in temples and monasteries, although the materials and methods of glazing may be somewhat different from one place to another. Ordinary people, on the other hand, are forbidden to use [*lazuli* tiles] for their houses.

BRICKS

Clay for making bricks is also obtained from the ground, in color either blue, white, red, or yellow (that found in Fukien and Kwangtung is mostly red; blue clay is produced chiefly in Kiangsu and Chekiang and is called "good clay"), the best being the kind that is adhesive, fine-textured and sandless. The clay is mixed with water and trampled over by several oxen driven by a man, so that it becomes a thick paste. After that, wooden frames are filled with this clay paste, the surface is smoothed with wire-strung bows, and so the green or unbaked bricks are formed [Figure 7.2]. For the construction of city walls and house walls, there are two kinds of bricks known as "recumbent brick" and "side brick." The former are oblong pieces, used in building city walls and the houses of well-to-do people; they are placed solidly one upon another. Those people who wish to economize, however, usually put down alternate layers of "recumbent bricks" and "side bricks," and using stone fragments and bits of earth they fill the inner spaces; this is done to reduce the cost. Aside from building walls, the bricks used for making foundations are called "square-frame bricks"; for supporting tiles at corners of eaves, "*Huang-pan* bricks"; for constructing curved arches of small bridges, rounded doorways, or graves, "knife bricks" or "curved bricks." One side of the "knife brick" is cut off wedgewise; when several of these are packed closely together and constructed into a dome, the structure will not collapse even when trodden and weighed down by horses and carriages.

To make "square-frame brick," the clay paste is first put into the square frame, and then a smooth board is placed over it; two persons step on the board in order to pack the mass and ensure the solidity of the green brick. [After being air dried,] the green brick is then fired to harden it. The bricks used for laying foundations are first trimmed around the edges by stone workers. The knife bricks are larger than the wall bricks by one-tenth of an inch; the *huang-pan* brick is one-tenth the size of the wall brick, while the square-frame brick is ten times as large as the wall brick. After the green bricks are made and [air-dried], they are placed inside a kiln and are fired for one complete day and night for 3,000 catties [of bricks], and twice that length of time for 6,000 catties. The fuel for firing bricks is either wood or coal; when the former is used the brick produced will be bluish gray in color, while the latter will make white bricks.[2] At the top of a wood-fuel kiln three holes are opened on one side for the emission of smoke. When the process of firing is completed and the fire is withdrawn, these holes are sealed with mud, and then water is employed for the superficial glazing on quenching [of the bricks].

[As to the amount of heat applied in the firing process], if it is one ounce [i.e. degree] less than the proper temperature, the glaze will have no luster; if three ounces less, the product is known as "low temperature brick," with the clay colors plainly showing. This brick will disintegrate immediately after exposure to frost and snow and change back to clay. On the other hand, if the heat applied is one ounce more than the proper temperature, cracks will appear on the brick; if three ounces more, the size of the bricks will shrink and they will be full of crevices. The whole piece is warped, breaks like iron fragments when struck, and is altogether unsuitable for use. Skillful [builders], however, utilize these pieces by burying them underground as foundations for walls, thus putting them to the same kind of use as good bricks. The way to check the temperature in the kilns is to watch the clay through the kiln doors; under the attack of fire the clay will manifest an attitude of uncertainty, and appear similar to gold and silver at their melting point. The pottery works foremen will recognize [the proper temperature].

The superficial glazing or quenching of [bricks and tiles] is done by pouring water onto the level space atop the kiln which is surrounded by a raised wall on all four sides. Forty *tan* of water are needed per 3,000 catties of bricks or tiles in the kiln [Figure 7.3]. The water, once having permeated the earthen covering [of the kiln], will react with the heat [to form superficial glaze on the brick surfaces]. When the right proportion of water and heat is achieved, then the quality and durability of the [bricks] are assured.[3]

The coal-fuel kilns are twice as high as the wood-fuel ones, having tops that are domed but not sealed. Inside the kilns are laid round cakes [briquettes] of coal one and one-half *ch'ih* in diameter, and each layer of coal is alternated with a layer of bricks [Figure 7.4]. The bottom layer consists of reeds that serve as kindling.

A large factory for the making of bricks for Imperial Palaces is situated at Lin-ch'ing [in Shantung], and is operated under the direction of an official of the Board of Works. Previously many different kinds of bricks were produced here, such as "secondary brick," "tally brick," "flat-bodied brick," "look-board brick," "axe-blade brick," "square brick," and so forth, but half of

Figure 7–2. Making bricks.

Figure 7–3. Superficial glazing of bricks and tiles by water-quenching.

Figure 7–4. Coal-fired brick kiln.

Figure 7–5. Making bottles or small jars.

Figure 7–6. Making large jars.

these varieties were eliminated in later times. The bricks are transported to Peking by loading forty pieces per rice-tribute boat, and half as many pieces per ordinary private boat. The fine-quality square bricks, used in building the central halls [in the Palace], are made in Soochow and transported [to the capital]. As to the making of *lazuli* bricks, we have already mentioned it in the section on tiles. The fuel [for these bricks] is taken from the *T'ai-chi-ch'ang,* and the firing is done at the Black kiln.[4]

WATER JARS AND THE LIKE

Of pottery jars there are hundreds of kinds, from the large-sized heavy water jars to the medium-sized bowls, down to small vases and cannisters. Their shapes and styles differ according to the locality; they cannot be enumerated here. All the articles are round ware, without square angles. After the right kind of clay has been obtained, the making of these wares is performed on a potter's wheel. The experienced artisan is able to measure with his eye the amount of clay needed for an item, take nearly exactly that amount into his hands, and with the help of another person place it on the revolving potter's wheel; then with one single molding the article is done. [See the upper part of Figure 7.5.] However, the large "dragon and phoenix jars" used in the Court (produced by kilns at Ch'ü-yang in Chen-ting prefecture, and I-chen in Yang-chou prefecture), and the large figured jars of South Chihli province [modern Kiangsu and part of Anhui] are made by an entirely different method: the clay is allowed to form very thickly over the body of the jar, in order that figures and designs may be carved on the wares. This is why the price of such jars is fifty or one hundred times [that of ordinary jars]. The ears and spout on some of the large and small earthen jars are made separately and then affixed to the body with the aid of a liquid glaze [see the lower part of Figure 7.5]. All pottery wares have [solid] bottoms except in Shensi and further west, where the steaming pots [5] used for cooking purposes are bottomless [i.e. having perforated bottoms], and are made from clay instead of wood.

Glaze is applied to all the interior and exterior surfaces of fine pottery wares and usually to one-half of the body of a coarser ware. No glaze, however, is used on the interior of such objects as crockery mortars, because their rough surfaces should be retained for the easier braying and pounding of things; "sandy" pots and pans [for cooking] are also left unglazed in order to allow easier conductivity of heat and better cooking. Glaze material is locally obtainable. In Kiangsu, Chekiang, Fukien, and Kwangtung a special kind of fern, called "chueh-lan-ts'ao" is used, a plant that [commonly] serves as kitchen fuel for the local people. It is not over three *ch'ih* in height; the leaves and branches, somewhat resembling those of fir trees, are fibrous yet not prickly (there are dozens of names for this plant, depending upon the locality). The potters, after gathering these ferns, burn them, and the ashes are then placed in a cloth bag and washed [or classified] with water. The coarse particles are discarded as refuse, while the finest particles are [saved as concentrate and are] mixed with a pulp of red earth and water at a ratio of two parts fern ash to one part pulp. After the mixture is thoroughly stirred it is painted on the unbaked ware. A shiny glaze will naturally result after firing. I do not know what material [for glazing] is used in north China; special materials are also used in glazing the yellow pottery of Soochow. Only rosin and pyrolusite are employed, however, in the glazing of the "dragon-and-phoenix wares" used in the Imperial Palaces.[6]

Small ceramic articles are fired in "bottle kilns," and large ones in "jar kilns." These kilns are operated separately only in Shansi and Chekiang; in other provinces both the small and large articles are produced from the same kiln. Wide-mouthed large water jars are made by first fashioning each one in two cross sections; then the two pieces are joined together and held in place by wooden pegs both inside and outside. Small-mouthed jars are also made in two sections; in piecing them together, however, wooden pegs cannot be used [on the interior], but instead a ring-shaped circular tile, which has been prefired in another kiln, is placed inside the jar [where the seam is]. The joint is then reinforced on the outside of the jar with wooden pegs. The earthen quality [of the ring and jar] will naturally make them adhere to each other. [Figure 7.6.]

The bottle and jar kilns are not built on level ground but along hilly slopes, each kiln works occupying a tract of land between one hundred to two

Figure 7–7. Bottle kilns inter-connected to jar kilns.

or three hundred *ch'ih.* Several dozen kilns are built together, one above the other up the slope, in order to avoid trouble caused by the lack of drainage, while at the same time the heat will be able to ascend gradually from the lowest to the topmost kiln [7] [Figure 7.7]. In cases where it is necessary to have several dozen [kilns] for the proper firing of the wares, the potters suffer from [the considerable amount of capital required for the production of such cheap commodities,] and the enterprise must be started with the joint effort and resources of many persons. When the construction of the [inter-connected] brick kilns is completed, their tops are covered with a three-inch layer of extremely fine earth. The walls of each individual kiln are equipped not only with "smoke windows" at intervals of five *ch'ih* but also with a double door that opens from the middle. When charging the kilns, the smallest articles are placed in the one on the first or lowest level, and the largest water jars are put in the last or highest one. Fire is started in the lowest kiln, while two persons carefully watch the temperature. The amount of fuel needed is generally 100 catties of wood per 130 catties of pottery. When the correct temperature is achieved the door [of the fired kiln] is closed, after which the fire is lighted in the second kiln, and so on, until the last kiln is fired.

WHITE PORCELAIN (SUPPLEMENT: BLUE PORCELAIN)

The raw material used by porcelain makers for their finest wares is a white soil known as white clay, which is produced at only half a dozen places in China. They are: in north China, Ting-chou in Chen-ting prefecture [in Ho-pei], Hua-t'ing district in P'ing-liang prefecture [in Kansu], P'ing-ting in T'ai-yuan prefecture [in Shansi], and Yü-chou in K'ai-feng prefecture [in Ho-nan]; in south China, Te-hua in Ch'üan-chou [in Fukien] (the clay is actually obtained from Yung-ting district, but the kilns are in Te-hua), and Wu-yuan and Ch'i-men in Anhui (white clay produced in other places than these does not stick together when molded into pottery ware, and so is often used as plaster for walls). The Te-hua kilns specialize in making porcelain buddhas and delicate figurines, things that are of no great practical value. The products of Chen-ting and K'ai-feng kilns, on the other hand, have an earthy color and lack lustre. The porcelain of all these places together cannot compare with that produced in the Yao prefecture of Kiangsi province. In the districts of Lung-ch'üan and Li-shui in Ch'u-chou prefecture, Chekiang province, however, there is manufactured a kind of glazed bowls and cups that are blue-black like lacquer and are called "Ch'u-kilns ware." At the foot of Hua-liu Mountain in Lung-ch'üan district there were, during [late] Sung and Yüan dynasties, kilns operated by two Chang brothers, whose wares were highly prized. These are the pieces called in curio shops the "Brothers-kiln ware."

But the porcelain which has enjoyed far-flung fame, and which all China's neighboring countries are eagerly seeking, is that produced from the kilns of

Ching-te-chen in Fu-liang district of Yao prefecture. From ancient times to the present day this place has been a center of porcelain manufactures. White clay, however, is not locally available, but must be obtained from two [near-by] mountains in Wu-yuan and Ch'i-men. One [mountain] is called Kao-liang Mountain, where "nonglutinous rice" clay [China-clay] of a hard character, is found; the other is called K'ai-hua Mountain, and here "glutinous rice" clay [China-stone] is found, which is soft and pliable in quality. Porcelain wares can be fashioned only out of a mixture of these two clays.

These run-of-mine clays are molded into square blocks, and transported to Ching-te-chen on small boats. The porcelain makers then put an equal amount of each kind in a bowl, and pound the mixture for an entire day, after which [the powdered clay] is placed in a large water jar to be classified [by means of decantation]. The fine particles suspended in the upper part of the water column are poured into a second jar, whereas the coarse particles settle on the bottom of the first jar. The particles in the second jar are further classified to result in a suspension of superfine material, which is poured into a third jar. The particles that sink to the bottom [of the second jar] are called medium-sized material. After decantation, the fine-sized clay pulp is poured into an oblong ditch built of bricks beside the kilns, so that the clay pulp can be dried with the help of the [waste] heat. The dried clay is again mixed with clear water to form a paste, which is used for making the body of porcelain ware.

There are two kinds of porcelain bodies. The first is called "pressed ware." This consists of such articles as vases, jugs, incense burners, boxes and the like, which are not uniform in shape; also included are imperial articles like porcelain screens, candlesticks, etc. The way to make this kind of wares is first to fashion yellow-clay molds, whether whole or in two sections cut vertically or across; white clay paste is then cast into the mould, the seams are painted over with a liquid glaze, and the resulting product, after firing, will be an entire unbroken piece. The other kind of porcelain body is called "round ware." This consists of the tens of thousands of dishes and cups and the like that are needed in everyday life: these actually constitute nine-tenths of all porcelain made, with the "pressed ware" taking up the remaining one-tenth. To make "round ware" a potter's wheel must first be set up. The wheel revolves around a wooden axle [or spindle] which has been planted in the earth to a depth of three *ch'ih* to insure stability, with the exposed portion about two *ch'ih* in height. The vertical axle is mounted with two horizontal disks, one on top and one at the bottom, which are turned from the rims by means of short bamboo sticks. A mandrel made of sandalwood protrudes from the centre of the upper disk. In the making of cups and dishes, which need not conform to specific molded styles, the clay paste is placed over the mandrel that is made to revolve by turning the disks. The worker then presses his first finger with the nail clipped off, over the bottom of the clay, while the thumb lightly shapes the clay body as it revolves [Figure 7.8]. A cup or a bowl will thus be formed (new apprentices are allowed to discard their spoiled pieces and renew their effort with new clay). After long experience, when the artisan's skill is highly developed, all items so made will be identical as though they had come out of the same mold. If small cups are intended, the mandrel is used as it is. If medium-sized dishes or large bowls are made, the mandrel is first enlarged with clay and used after it is dried.

After a clayware article has been made by hand on the wheel, it is pressed once with the mandrel, [which acts as a finishing mold]; this is followed by slight exposure to the air and another pressing while the clayware is still damp. When the ware is thoroughly dried, as indicated by its white color, it is dipped once in water [Figure 7.9] and mounted on a mandrel again to be polished twice with a sharp knife [Figure 7.8] (while the ware is being polished, the slightest tremor of the hand will result in jagged crackles in the final fired product). Thereafter the ware is inspected and repaired for blemishes and turned on the wheel for drawing a line along the rim [Figure 7.10]. The next step is the decoration of the ware by either writing words or painting [designs or figures] on it; after that is done, the ware is sprayed with a little water and then is ready to be glazed. No blue coloring material is used for the [glazing and painting] of "crackled" ware, "thousand-grains" ware, or "brown-colored" cups.

To make a crackled porcelain ware, the clay body is exposed directly to the sun after polishing with the knife until it is very hot and then is dipped once quickly in clear water; the resulting ware, after firing, will show crackled marks.[8] The "thousand-grains" ware is fashioned by very swift dotting of the surface with a liquid glaze, and the "brown-colored" ware is prepared by brushing the clay surface with the boiled juice of old tea leaves. (The Japanese, who highly value antique crackled ware, are willing to pay thousands of taels of silver for a genuine piece. There is an antique crackled-ware incense burner whose date is unknown; an iron nail whose gleam is not yet tarnished by rust is embedded in its bottom).

The liquid glaze used at Ching-te-chen for coating white porcelain ware is composed of water, the clay from Hsiao-kang-tsui, and the ash of the leaves of peach-bamboo,[9] and it has the appearance of a clear rice broth (the porcelain figurines produced in Ch'üan-chou are glazed with a mixture of pine-needle juice, water, and clay; it is not clear what materials are used for glazing the dark-blue colored porcelain of Ch'u-chou). This is placed in a large water jar [Figure 7.11]. The interior of the wares to be glazed is first rinsed with this liquid glaze, then a certain amount of the glaze is applied with fingers to the rim of each article, and it will naturally flow and spread to cover the entire [outer] surface.[10]

The indispensable ingredient for blue color in porcelain decoration is pyrolusite [MnO_2] (it is also used as a coloring material by varnish makers in the process of heating their oils). This mineral is obtainable near the earth's surface, not more than three *ch'ih* underground at the deepest, and is found in all provinces. It comes in top-, medium-, or low-grade varieties. Before use it should be roasted in a red-hot charcoal fire, the top-grade material will then turn peacock blue; the medium, light blue; and the low-grade will have a color

Figure 7–8. Shaping (lower) and polishing (upper) clayware with potter's wheel.

Figure 7–9. Dipping thoroughly dried wares in water.

Figure 7–10. Decorating porcelain ware.

Figure 7–11. Dipping clayware in liquid glaze.

Figure 7–12. Porcelain kiln.

close to that of brown earth. Each catty of untreated ore will produce only seven ounces of top-grade pyrolusite, while the amounts of the medium- and low-grade variety are proportionately even less. The [coloring materials used] for decorating the finest pieces of porcelain as well as the dragon-and-phoenix ware for imperial use are top grade. The price of such a high-grade pyrolusite is twenty-four taels [of silver] per *tan,* while the medium grade costs half that price and the low grade, three-tenths. Among the blue coloring material used in Yao prefecture [i.e. Ching-te-chen], that which is produced in the mountains of Ch'ü and Hsin prefectures [both in Chekiang] is the best and is known as "Chekiang stock." The kind produced in Shang-kao, etc. is medium grade, and that from such places as Feng-ch'eng is low grade. After it is roasted, the blue coloring material is ground into a very fine powder with pestle and mortar (the inside surface of the mortar is unglazed and rough for better abrasion) and is then mixed with water for painting. Its color is black while being ground and suspended in water, but after [painting on porcelain and] firing the color turns into greenish blue.

If a "purple-cloud" color crackled porcelain cup is desired, the ware is first wetted with an aqueous solution of rouge, and is then placed in a wire net for convenience in heating over a charcoal fire. After that, the ware is brushed once with a pad of cotton moistened with rouge, and the coloring effect is achieved. The kind of colored porcelain called "Hsüan-red" ware is made by the extremely skillful coloring in conjunction with the slight roasting of a special kind of prefired ware; it is erroneous to think that the red color of cinnabar can remain fast on the ware even after firing. (The process for making the "Hsüan-red" ware was lost at the end of the Yuan dynasty, but it was rediscovered after many experiments during the Cheng-te reign [1506–21].)

After being decorated and glazed, the procelain wares are packed into box frames or saggers (a slight heavy-handedness in packing them will result in warped shapes after firing). The saggers are made of coarse clay; each packed [piece of] porcelain ware is supported by a clay disk; and the empty space at the bottom of each sagger is filled with sand. One sagger will have room for one large porcelain article, or about one dozen small pieces. Good saggers can be used repeatedly for more than ten firings, but poorly made ones will disintegrate after one or two.[11] After the kiln is completely loaded with packed saggers, a fire is kindled. Toward the top of the kiln there are twelve round holes, called skylights. The firing should continue for twenty-four hours, of which the first twenty hours consist of firing from the kiln door, allowing the heat to rise from the bottom upward; then lighted wood is thrown in through the skylights and burns for four more hours, allowing the heat to travel from the top downward [Figure 7.12]. While they are in the burning kiln the porcelain pieces are soft like cotton wool. To test the extent of firing, one article is taken out of the kiln with a pair of iron tongs, and the fire is stopped when the specimen is determined to be sufficiently heated. In sum, without counting the minute details, a portion of clay must pass through seventy-two different processes before it is finally made into a cup.

SUPPLEMENTS: TRANSMUTATION IN KILNS, AND MOHAMMEDAN BLUE

In the Cheng-te reign [A.D. 1506–21] a eunuch was appointed to supervise the manufacture of porcelain for imperial use. That was the time when the process of making Hsüan-red was still lost, and [failing to produce it], the eunuch paid the penalty with his and his relatives' lives. One of the potters killed himself by jumping into the burning kiln. Later he appeared in another person's dream [and imparted the secret of the red color], who was able to produce the desired red. News of this event immediately became widespread and was known as transmutation or chanciness in the kilns. Subsequently the story was enlarged by miracle-fanciers to include such points as the kiln producing strange objects such as deer, elephants, etc.

Mohammedan blue is the "deep blue" glaze material of the Western Region [Persia], the best quality of which is also called "Buddha's head blue." The top grade pyrolusite, after firing, is similar to the "deep blue" in color. It is not true that the "deep blue" can retain its original color after being exposed to the intense heat of the porcelain kilns. [For this reason, the "deep blue" painted on porcelain must be covered by a layer of colorless glaze prior to firing.]

NOTES

1. Black earthenware of the Lung-shan civilization and painted pottery of the Yang-shao civilization have been found in different parts of China, and the An-yang excavation sites have yielded pottery of the Shang dynasty (1783–1122 B.C.). The earliest bricks and tiles yet discovered (in 1955 in Cheng-chou, Honan) date from the Warring States period (403–221 B.C.). According to a journalistic account these are burnt hollow bricks engraved with hunting scenes.

The specimens of the bricks and tiles of the succeeding Han dynasty (206 B.C.–220 A.D.) can be found in many Western museums. They include narrow rectangular bricks used as headers and binders for the construction of walls and tombs, square tiles to cover surfaces, inscribed bricks to record burials and other personal affairs, miscellaneous bricks used as supports and bases for pillars, and hollow tomb bricks with holes in both ends. Some of the large hollow Han tiles from Honan province measure almost six feet in length and two feet in width. Dies were pressed into the clay before firing, leaving intaglio designs. These illustrate games, feasts, hunting, and fishing and also give some idea of commerce and industry.

Bricks and tiles of the Han and Wei periods are all unglazed, although glazed pottery first appeared in China during the Han dynasty. Glazed tiles and bricks were probably first used during the T'ang dynasty (A.D. 618–916). The History of Yüan records that four kilns were established in 1276 at Ta-tu for the making of plain, white-glazed bricks and tiles; three hundred workmen were employed there. The technique of making glazed tiles and bricks was further improved and manufacture flourished in the Ming and Ch'ing dynasties. Today glazed bricks and tiles are little used in China, though the unglazed ones are still used as building materials.

2. The bluish-gray color is apparently caused by the deposition of fine smoke particles from the wood fuel in the pores of the bricks. These smoke particles are subsequently fixed by the fusible matter on the surface of the brick. The blueing of bricks and tiles by smoke is practically eliminated when the wood fuel is replaced by coal, particularly anthracite, since a coal fire gives a cleaner atmosphere and a higher temperature. The term "white" is often used very loosely in clay working and is used to include all shades from a true white to a distinct cream or even a slight reddish or pale yellow color.

3. The water-quenching (i.e. with steam) of red-hot bricks and tiles requires great skill and care. An improper cooling of kilns is the source of many defects, especially crazing, cracks, dunts, and feathering or crystallization.

4. *T'ai-chi-ch'ang* and Black kiln were establishments administered by the Board of Works during the Ming dynasty. The former was a fuel dump, and the latter a place where bricks and tiles for palace use were made.

5. It can be deduced from the Chinese word *tseng* that these bottomless steaming pots are similar to the currently used double boiler for cooking. The upper half of the pot has holes on its bottom to permit the passage of steam, and is then said to be bottomless.

6. Chinese ceramic glazes: The first people who used glaze were the Egyptians of the Badarian and predynastic periods; they applied it to objects of stone and quartz; modern excavations at Mohenjo-Daro in Pakistan have uncovered glazed materials made in approximately 3000–2750 B.C.; and in Assyria and Babylonia colored glazes were used at about 1100 B.C. In China, the excavations at An-yang reveal that glaze of a sort was used by the Chinese in the Shang dynasty (1760–1120 B.C.). There is some probability that it was also used in the succeeding Chou and Ch'in periods, but the dating of the specimens in question is still uncertain. In the Han dynasty (206 B.C.–220 A.D.), the Chinese had a considerable trade with the Romans by the overland route through Turkestan and by sea via India and Persia; the silk of China was exchanged for western goods, among which Syrian and Egyptian glaze was almost certainly included. The first use of lead glaze in China at that time must have been learned from the West, although the Chinese soon developed notable techniques of their own.

7. Lo Tsung-shan found that the ancient kilns in I-hsing of Kiangsu province were interconnected to save fuel and were elevated like a stairway to facilitate draught. Wang Chin, a modern Chinese scientist, believes that such interconnecting kilns have been used in China for over one thousand years. Waste fuel gases from the first kiln fired preheat the green goods in the rest of the kilns before entering the stack. Furthermore, the waste heat in the cooling kiln can be partly extracted as hot air and passed into the next kiln for combustion.

8. In porcelain of the Sung and Ming dynasties, crackle varies from a "fishing net" pattern to what has been called "truite" or fish roe. Production involves the selection of a suitable combination of body and glaze and control of the heating and cooling rate. A delayed crazing, which may take months or years to complete, can give rise to a secondary crackle over and above the original one. Examples of this can be found in the lines of crackle of some Manchu wares that show two different colors, one generally black and the other light brown. It is assumed that the original crackle of the initially fired ware was stained with one kind of coloring material, and that the secondary crackle was stained with a different color.

9. Peach-bamboo is a large bamboo grown chiefly in Szechuan province and has no relation to the peach tree.

10. Prior to the Ch'ing dynasty, both large and small round articles were dipped in a large jar filled with liquid glaze. Irregularly shaped wares were painted with a goat-hair brush. With these processes, the pieces were often too thinly or too thickly coated, and some of the large items were broken during the dipping. In the Ch'ing dynasty, small round pieces were still glazed by dipping, but vases and large round vessels were glazed by a primitive spraying process. A bamboo tube, approximately one inch in diameter and seven to ten inches long, with one end covered with fine gauze, was dipped into the liquid glaze and then blown through from the other end. This process had to be repeated from three to eighteen times, depending upon the article and the glaze.

11. A passage in *Ching-te chen t'ao-lu* states that the characteristic yellow foot of porcelain wares produced by the *Ch'ai yao* of the Later Chou dynasty (951–960) was caused by the yellow earth placed on the inner bottom of saggers. Saggers were known to the Chinese long before this time and were used to protect porcelain wares from uneven temperature and kiln gases, as described in the present book.

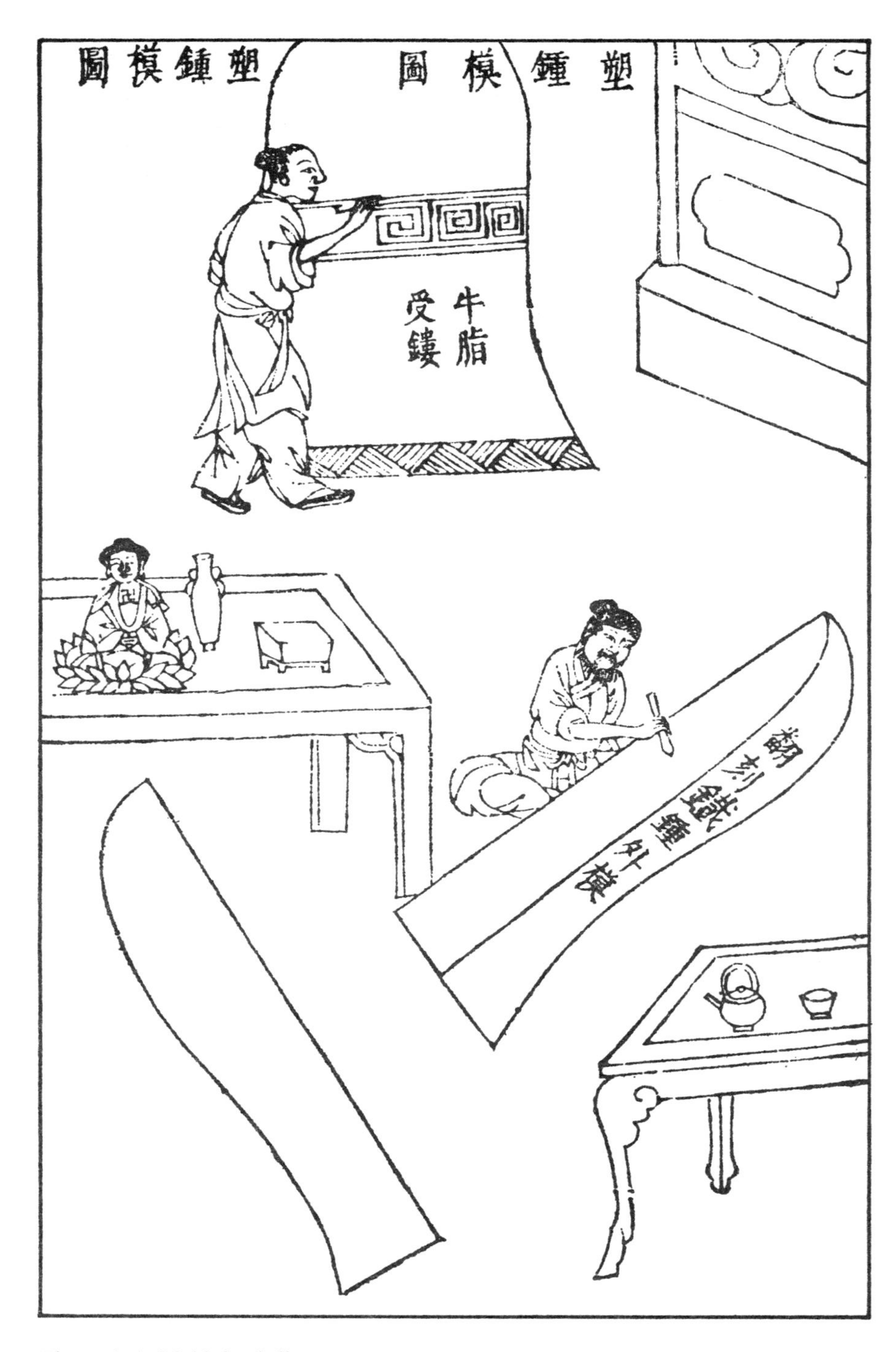

Figure 8–1. Mold for bell.

8

CASTING

Master Sung observes that ever since [the copper of] Shou-shan [1] was mined by the Yellow Emperor, mining has had a long history in China. In the time of the Emperor Yü (2198 B.C.?) the chiefs of the nine provinces [2] presented metals to the Emperor for the casting of his *ting*-tripods,[3] and thenceforth the art of smelting and casting metals was increasingly improved with time.

In its natural state metal is enclosed in earth; likewise, when it is being made into useful implements it is also enclosed in earthen molds. The implements may be fine or coarse, large or small. That which is blunt is used as a mortar [for pounding grain], that which is sharp is used for the ploughing the earth; make it thin [into pots] to stand between water and fire, and the people will have nourishment; make it ir to hollow [bells] and harmonious sounds will fill the air. The devout fashion metal into images, so that divine likenesses are seen in this mortal world—the more skillfully made ones borrow even the spirit of the Heavens. [Metal] coins are minted to circulate even to the remote corners of the empire—so numerous that it is impossible to count them. Such profusion can not have been achieved by human effort alone.

TING-TRIPODS

No information is available regarding the casting of the *ting*-tripods before the time of the emperors Yao and Shun. Then emperor Yü cast his nine *ting*. At that time the nine provinces had been demarcated, the annual tributes from each region had been fixed, the rivers had been dredged and brought under control, and the *Tributes of Yü* [4] had been written. Fearful, however, that the rulers of later ages might increase the taxes by great amounts, and that later kingdoms might become oppressive, or that those in charge of irrigation and flood control might not follow the correct methods, Yü had all these matters inscribed and cast into the *ting,* which would be more durable than books, so that posterity could act according to the rules laid down. This was the reason for casting the nine *ting.* As the centuries passed people's knowledge (about antiquity) became dim. It is said that such phrases as "pearls and fish" and "foxes and woven leather" were inscribed on the *ting,* but the words might have been mutilated and worn by time. The ignorant, however, took these to be sentences denoting frightful phenomena,[5] giving rise to the statement in the *Spring and Autumn Chronicles* that "if one knows the designs of the spirits then ghosts and monsters can be avoided."

By the Ch'in dynasty [221–206 B.C.] the *ting*-tripods of Yü had been lost. Although the Great *ting* of the state of Kao, and the two Fang *ting* of the state of Lü, were cast in the Spring and Autumn period [770–403 B.C.], their inscriptions by no means had the same intent as those of Yü; that is, these things were antiques in name only. In later ages there has been a multitude of documents and books, but *ting*-tripods were never made again. I have therefore noted their history here.

BELLS

Foremost among the metal musical instruments is the bell. When it is rung a large bell can be heard at a distance of ten *li,* and a small one at more than one *li.* Bells therefore are sounded for assembling people at the sovereign's audience, or when the officials go out, and for the accompaniment to singing at local festivities. In temples and shrines the ringing of bells is done to indicate the devotion of the worshippers as well as to rouse the serious attention of the gods.

Bells of high quality are made of copper [alloys]; those of lower quality, of iron. The Audience Bell now used in the North-Star Pavilion [6] is made entirely of bell metal. For casting one such bell 47,000 catties of copper are used, together with 4,000 catties of tin, fifty ounces of gold, and 120 ounces of silver. The finished bell weighs 20,000 catties, and measures 1.15 *chang* high, with a diameter of 8 *ch'ih* at the mouth. The double-dragon neck of the bell is 2.7 *ch'ih* high. Such are the specifications of an Audience Bell.

The method of making a bell of the 10,000-catty [class] is the same as that for casting *ting*-tripods. A pit one *chang* and several *ch'ih* deep is first dug in the ground. In the pit a houselike dry-mold foundation is constructed, the core of which is built of lime and mortar, and no crack is to be allowed to appear. After the core is dried it is covered with ox fat and yellow beeswax several inches thick, the proportion being eight parts of ox fat to two parts of beeswax. The foundation and core of the mold should be protected against the sun and rain (the casting must not be undertaken in the summer months, because then the fat and wax will not congeal). When the fat and wax have been applied securely on the core, they may be engraved with decorative designs, texts, or

Figure 8–2. Casting a ten-thousand-catty *ting*-tripod.

Figure 8–3. Casting a thousand-catty bell and a statue of Buddha.

pictures, down to the last detail [see upper part of Figure 8.1]. Next, very fine earth and charcoal powder are pounded, screened, and mixed into a mud paste, which is gradually spread on [the surface of the engraved wax] until it is several inches thick. When it is thoroughly [air] dried, heat is applied from the outside so that the fat and wax will melt, and flow out entirely through apertures at the base. The bell or the *ting*-tripod will then be cast in the cavity thus vacated between the core and mold.

Ten catties of copper are needed to fill the cavity previously occupied by one catty of fat and wax, therefore for every ten catties of fat and wax used, 100 catties of copper [alloy] should be prepared for use in casting. The copper is melted when the cavity [of the mold] has been thoroughly drained. But as it is not possible to cast such 10,000-catty quantities of molten copper by hand, [melting] furnaces are built all around [the molding pit], and earthen trenches are made which decline toward the pit and connect the furnaces with the "copper hole" [i.e. pouring gate] on the bell mold. The trenches are surrounded with red-hot carcoal. When the metal is melted in the furnaces, the plugs in the trenches are removed (the trenches are at first closed by earthen plugs), and the molten metal [is tapped from the furnace and] flows like water down the trenches and eventually gets into the mold. And so a bell or a *ting*-tripod is cast [Figure 8.2].

The same method is used in the casting of 10,000-catty iron bells as well as cooking pots, except that the molds used for these iron castings tend to be less expensive.[7]

Not so much work is involved in casting articles weighing less than 1,000 catties. The process requires about a dozen kettle furnaces, made of earth and reinforced with iron bars, and shaped like winnowing baskets. At the lower part of each furnace two iron tubes are inserted horizontally, so that two carrying poles can be put through, and the furnace is placed on an earthen mound. All furnaces are fired at the same time. When the metal is melted, the carrying poles are inserted, and each kettle is carried to the mold by two or more men, depending upon the weight of the load. The molten metal in the furnaces is poured into the mold in rapid succession, and it will naturally coalesce inside the mold [Figure 8.3]. If the pouring is not done rapidly, however, the metal that is poured in first will begin to solidify before the next lot comes in, thus making it impossible for the different parts to adhere and giving rise to defects [cold shuts].

For casting iron bells without the expenditure of too much fat and wax, an outer layer [or cope] of the mold is first made from earth, which is then cut into two sections either longitudinally or across. Dowel pins are used to insure the positive alignment of these two sections when the mold is closed. The reversed version of words are inscribed on the inside of this cope [see lower right of Figure 8.1]. The inner layer [or core] of the mold is smaller in size. On the basis of accurate calculations, a predetermined space can be established between the inner and outer layers of the mold. After the words are engraved, the surface is spread over with a thin layer of ox fat, so that later the bell casting will not stick to the mold. The cope is then laid over [the core], and the separate sections are sealed together with mud. The mold is then ready for casting.

The above method is also employed in making big chimes and musical boards.

COOKING POTS

The cooking pot contains water and withstands fire; it is an indispensable article of daily use. It is made of either cast iron or [remelted] junk consisting of old cast iron objects. The size varies, the commonest being two *ch'ih* in diameter at the mouth and the metal is about two-tenths of a Chinese inch in thickness. The smaller pots have a diameter half as big, but are of the same thickness.

There are two layers to the mold, an inner and an outer. The inner layer or core is made first and allowed to dry thoroughly, after which the size of the pot is calculated and used as a basis for making the outer layer or cope of the mold. The latter requires great skill, for the slighest inaccuracy will render the mold useless. The mold is left to dry completely.

Pig iron is placed in an earthen melting furnace with the interior shaped like a cooking pot. The rear of the furnace is connected to a bellows box by a blast pipe, while the front is equipped with a spout for discharging the molten iron. Each furnace contains enough iron for casting ten to twenty pots. The fluid molten iron [in the furnace] is tapped from the spout and poured into the molds with ladles, which are made of pure iron coated with earthen materials, each ladle holding approximately the amount needed for casting one pot [Figure 8.4]. Before the cast pot becomes cold the cope of the mold should be lifted and the pot inspected for defects. This is done at this stage because, while the iron is still red-hot, the cracks, blow holes, and other defects of the casting can be mended by pouring a little bit of molten iron onto the spot, which is then smoothed over with wet grass to leave no trace. Much mending is required if the pig iron is being cast for the first time. However, mending is not necessary if old cast-iron junk is re-used for casting. (In Korea it is the custom to discard old pots in the mountains, and never use them for re-casting.)

To test the quality of a pot after it is finished, one should strike it with a light stick; a sound resembling that produced by a wooden musical instrument indicates that the pot is good, but an inferior noise is the sign that the iron had not been refined enough and that the pot will be easily damaged.

In large temples and monasteries there are huge "thousand-monk pots," which are said to have the capacity of cooking porridge from two *tan* of rice. What an imbecility it is!

STATUES

The method used for casting bronze images of Buddhist or other gods is the same as that employed in the making of Audience Bells. While the bells and *ting*-tripods are cast in one piece, the statues, however, can be cast in sections; [8] therefore, the engraving of designs is much easier. The piecing together of the mold, on the other hand, must be executed with the greatest skill.

Figure 8–4. Casting cooking pots.

CANNON

Refined copper is used in the casting of Western-ocean cannon, the Red-hair Barbarian cannon and French cannon. Equal amounts of refined and raw [or blister] copper are used in making such arms as signal guns and muskets. For making guns like Hsiang-yang, Chan-k'ou, First General and Second General, iron is used.[9]

MIRRORS

The mold used for casting [bronze] mirrors is made of ash and sand, and the bronze is an alloy of copper and tin (zinc is not used). It is stated in the *K'ao-kung-chi* [10] that "the mixture of the same amounts of copper and tin is called mirror grade [of bronze]." Its surface shines when a coating of quicksilver is applied; bronze itself cannot be so bright.

The palace mirrors used during the K'ai-yuan period [713–41] in the T'ang dynasty were made of equal parts of silver and copper, and that is why each such mirror cost several taels [of silver] in price. The vermillion spots on these mirrors are caused by the essence of silver and gold (gold was sometimes included in the alloy in the old days). In the present dynasty the products of the Hsüan furnace [are well known]. These had their accidental origin in a fire that razed a treasury, in which gold and silver were fused with copper and tin into one mass, thus giving rise to this special product of the furnace (a genuine Hsüan furnace article will have a golden sheen). The mirrors of T'ang and of Hsüan furnace are the products of prosperous times.

COINS

For the convenience of the people copper is minted into coins and circulated. On one side [of the coin] are inscribed characters indicating it is "currency" of a given dynasty, and the affairs concerning copper money are administered by a bureau in the Board of Works. Coins in circulation are usually computed at ten coins to 0.1 tael of silver, but there are large coins that are equal in value to five or ten [ordinary] coins. The disadvantage of coins is that they can be easily minted illegally, which is harmful to the people. That is why in all parts of the empire circulation of coins is frequently stopped.

The approximate proportions of the materials used in minting coins is about six or seven parts of copper to three or four parts of zinc (called "water tin" in Peking). About one-fourth of the latter is usually lost through heat in the process. In the present dynasty only the Yellow Coins minted by Pao-yüan chü [11] in Peking and the Dark Furnace Coins of Kao-chou in Kwangtung are of high quality (the Kao-chou coins circulate widely in the Chang-chou and Ch'üan-chou areas,[12] each one of these being worth two coins of such localities as Nanking, Kiangsu, and Chekiang. There are again two varieties of Yellow

Figure 8–5. Casting coins.

Coins: the "gold back" coins made of four-fire brass, and the "sealing wax" coins made of two-fire brass.[13]

The crucible in which copper is melted for minting coins is made of extremely fine powdered earth (crushed dry earth and brick fragments) mixed with charcoal powder (ox hooves are used in Peking, the reason for which is not clear); the proportion being seven parts earth to three parts charcoal. The latter is used because its heat-inducing properties can help the earth in melting the metals. The crucible measures eight Chinese inches in height and two and one-half inches in diameter at the mouth, each having the capacity of about ten catties of copper and zinc. Copper is melted first, then zinc is added and melted, and the molten alloy is poured into the coin molds.

The coin mold is made as follows: a [square] frame is constructed with four pieces of wood (each measuring 1.1 *ch'ih* long and 1.2 Chinese inches wide), which is then filled and tightly packed with finely screened earth and charcoal powder, on which some powdered charcoal of fir or of willow is sprinkled; however, resin and clear oil are applied to the mold if it is to be baked.

Next, one hundred specimen coins (struck of tin) are placed on the mold, with either the head or the tail uppermost. Then another frame, constructed and filled also as above, is placed on top of the first mold. There are now two

Figure 8–6. Filing coins.

Figure 8–7. Casting silver coins in Japan.

molds, the one with impressions of one side of the coins, the second with those of the obverse side. Turn the molds over, so that the specimen coins will drop onto the second mold. [The first mold is then removed.] Now a third mold, constructed in the same manner, is placed over the second one, and turned over. The process is repeated until about a dozen pairs [i.e. cope and drag] of [impressed] molds have been prepared. These are tied securely together with ropes [to form stacked molds]. A pouring hole [or sprue] originally had been left on the top side of the frame. The minter now lifts the crucible of molten metal with a pair of forceps, helped by another worker who holds the crucible from the bottom with another pair of forceps, and the metal is poured into the molds through the pouring holes [Figure 8.5]. When the metal has cooled the molds are untied, and there they are—one hundred shining coins, looking like so many flowers and fruit hanging from a tree. This is because hollow passage ways had been left in the mold [between the impressions of the coins], so that when the molten brass flows through the sprues and runners a treelike pattern is formed. The coins are taken out and cut [from the in-gate] and are ready for filing and polishing.

The edges of the coins are first filed. For this purpose several hundred coins are strung on bamboo or wooden sticks [14] and filed together [Figure 8.6]. The surfaces of the coins are then polished, which is done individually.

The proportion of zinc in a coin determines its quality, and it is easily detected by observing whether a coin is heavy and thick or light and thin. Since zinc is cheaper than copper, the illegal minters sometimes put in as much as fifty per cent of zinc in their products. If a coin gives off a woody noise when thrown against a stone, it is of low quality; the high grade coins made with nine parts of copper to one part of zinc will produce a metallic sound when thrown on the ground.

If old brass articles are remelted for minting, there will be a ten per cent loss in the smelting process, for part of the zinc will be lost, and the resultant product will contain a higher percentage of copper; therefore it is better than coins struck with new copper.

The silver coins of such countries as Liu-ch'iu [Islands] are made directly by holding the molten silver in a die or metallic mold which is attached to the ends of a pair of iron tongs and immersing it immediately in cold water. Each action produces a coin, as shown in the accompanying illustration [Figure 8.7].

SUPPLEMENT: IRON COINS

Iron is a cheap metal and no coins were made of it in ancient times. Minting of iron coins began in the T'ang dynasty when copper coins were not available in certain border marches like Wei and Po, and it was done as a temporary measure. In the heyday of a dynasty silver would be cast into sacrificial vessels; when the territorial lords are on the wane then even iron is made into money. I record it here for the melancholic pondering of the learned.

NOTES

1. Shou-shan is a mountain in modern Shansi province where, according to the *Historical Records,* the Yellow Emperor, the first historical ruler of China (ca. 2700 B.C.), had copper mined.

2. In antiquity China was divided into nine major parts or provinces.

3. *Ting,* a particular kind of three-legged vessel.

4. The *Tributes of Yü* is a chapter in the Chou-dynasty *Book of History* dealing with the geography, products, and taxes of the different parts of China.

5. In Chinese folklore the pearl-bearing oyster and the fox were both said to possess magical powers.

6. North-Star Pavilion (*Pei-chi Ko*) is a hall in the Imperial Palace.

7. Modern scholarship indicates that iron castings were produced in China soon after bronze was first cast there, whereas the Western world did not achieve this technique until about nineteen centuries later. Several reasons have been given for the earlier Chinese invention of this technique: The Chinese had a box-bellows superior to the devices used elsewhere. Too, bronze was expensive in early China, because copper and tin had to be obtained from the southwestern frontier, then largely under the domination of warlike tribes. Iron ore and coal, however, were abundant in northern China. Lastly, the ancient Chinese foundrymen were apparently able to pour iron at a relatively low temperature through the addition of carbon and phosphorous.

8. Apparently the Shang and Chou foundrymen usually cast the parts separately and assembled them afterwards, although they also turned out some intricate one-piece castings.

9. See Chapter 15, "Weapons." "Red-hair Barbarian" usually referred to the Dutch, while "French," or *Fo-lang-chi,* designated people from Portugal and Spain rather than from France.

10. A part of the *Rites of Chou.*

11. *Pao-yüan chü* was the official minting establishment under the jurisdiction of the Board of Works in the central government.

12. Chang-chou and Ch'üan-chou are in modern Fukien province.

13. Four-fire brass contains approximately seventy per cent copper and thirty per cent zinc, while *two-fire brass* contains these two metals in equal proportion.

14. Copper coins were round in shape, made with a square hole in the middle so they could be strung together when handled in large quantities. "One string" was a unit statutorily denoting 1,000 cash.

9

BOATS AND CARTS

Master Sung observes that the world is composed of people and goods from different localities, yet among them there is constantly a process of journeying back and forth and mutual exchange. If all things remain forever in the same spot, how is each to find its proper place in the world? There are persons who must embark on travels under difficult conditions even though they are men of high rank; and there are things, cheap and common though they be, that are needed by people who must obtain them through commerce.

In our empire, the southerners rely on the boat and the northerners the cart as the chief means of transportation. By [the subjects'] journeying and sailing across great distances our imperial capital is kept brimful of power. How is it then that the innovators of boats and carts are not accorded reverence and worshipped at proper shrines? The sailors live afloat on the ocean year after year, regarding the expanse of tossing waves as no more dangerous than level ground. Certainly they differ not from Lieh-tzu, who "lightly rode on a breeze as he walked." And is it not right to call such persons as Hsi-chung,[1] as described in the historical records, divinely inspired men?

BOATS

Nowadays, as in ancient times, boats are known by hundreds, even thousands, of names. Some are named according to their shapes (such as "sea eel," "river flounder," "mountain shuttle," and so on), some according to capacity (the amount of cargo that can be carried in a boat), and some according to quality (the various kinds of wood used in building the boats)—all too numerous to be described. People along the seacoast can observe the ocean-going ships, those who live along rivers will see the grain tribute boats. But those that spend their whole lives in mountainous country or on the plains will see no more than a simple scull or two, or some rafts tumbling down the stream. The following are brief descriptions of a few types of boats that will serve as examples for all the others.

Grain Tribute Boats

The capital being an area where large numbers of people and soldiers congregate, [grain] is sent thither from all over the empire in order that there may be an adequate food reserve, and thus arises the need for the grain transport system.

When the Yuan dynasty (1279–1368) assumed rule over the entire country, the capital was settled at Yen-ching [Peking] which was called Ta-tu [or Great Capital]. The route of grain transport from the South began at Liu-chia Harbor at Soochow, and at Huang-lien-sha at Hai-men, whence the boats headed out to sea, sailing directly to Tientsin. The boats employed were sea-going vessels.[2] [In the Ming dynasty] the above procedure was followed until the reign of the Yung-lo emperor (1403–1424), when the system was changed to river transport because of the many dangers of the stormy sea route.[3] The grain transport boats of today are modelled on the shallow flat-bottom craft built under the direction of Ch'en [Hsüan], the Earl Pacifier of Rivers.

The various parts of a grain boat [Figure 9.1] may be described as follows: The keel, or bottom, is the floor; the hull is the wall; the bamboo sections [gunwale?] are the tiles [atop the wall?]; the bow cross-beam [*lit.* "crouching lion"] is the front gate, and the stern cross-beam forms the sleeping chamber. The mast is the arc of the bow [for shooting arrows], and the sails are wings. [To a boat] the scull is as the horse to a cart, the rigging is as shoes [to a man's feet], the ropes and chains are as muscles and bones to a hawk. The steering paddle is the vanguard, the rudder is the commanding general, while the anchor is the stronghold where the army camps.

The bottom of the early grain boat was fifty-two *ch'ih* in length, made of planks two *ts'un* thick. These were fashioned from huge timbers, the preferred wood being first cedar and secondly chestnut. The fore-deck was nine *ch'ih* five *ts'un* in length and the aft-deck was also nine *ch'ih* five *ts'un.* The bottom [at the waist of the ship] was nine *ch'ih* five *ts'un* in width, while it measured six *ch'ih* wide at the bow and five *ch'ih* at the stern. The bow cross-beam measured eight *ch'ih* and the stern cross-beam seven *ch'ih.* There were fourteen beams: [4] the "dragon's mouth beam" is ten *ch'ih* wide and (placed) four *ch'ih* above the bottom; the "using-the-wind beam" was fourteen *ch'ih* wide and three *ch'ih* eight *ts'un* above the bottom; the "cut water beam" near the stern of the boat was nine *ch'ih* wide and four *ch'ih* five *ts'un* above the bottom. The boards forming the foot passageways[?] on both sides of the boat were seven *ch'ih* six *ts'un* wide altogether. These were the early measurements and each boat was

Figure 9–1. A grain tribute boat.

capable of carrying nearly 2,000 *tan* of grain (although [in practice] only 500 *tan* were delivered to each boat). Later, when the boats were constructed by the transport corps, the length of the boat was increased by twenty *ch'ih,* and the bow and stern were widened by over two *ch'ih,* thus augmenting the capacity to 3,000 *tan* of grain. The locks on the Canal being only twelve *ch'ih* wide, (these new boats) were barely able to pass through them. Nowadays, the boats used by officials in their travels are constructed in the same way; the only difference being the addition of larger windows and doorways which are elaborately and colorfully decorated with paint.

In constructing a boat, the bottom or keel must be laid first. Along the sides of the upper surface of the bottom are erected the "bottom walls," which touch the bottom at the lower end and support the planking of the hull above. The beams are placed at regular intervals, while the part of the boat that rises vertically on the sides is the hull. The large timber covering the hull is the "principal planking," on the top of which is the gunwale. The spot in front of a beam for stepping the mast is called the "anchor altar," under it the cross-planks used to secure the below-deck portion of the mast are called "ground dragons." The outermost beams ahead and astern are called bow and stern cross-beams [or "crouching lions"], the timbers supporting them are called "lion holders"; under the [bow] cross-beam the wood that closes the front of the boat is called the "breakwater plank."

Toward the middle of the foredeck there is a square opening called "the well" (in it are stored ropes and such things). Still further forward stand two wooden stakes for winding up the ropes; these are the capstans. At the stern of the boat, the portion that rises slanting [from the bottom] upward is called "straw-sandal sole." The wood that closes the stern end of the boat is the "short plank," and beneath that is the beam called "stop-foot beam." The aft end of the boat contains the helmsman's quarters, the top of which forms the poop [*lit.* "pheasant's roost"]. Here a man sits aloft to manipulate cordage from the sails when the boat is catching the wind.

A boat that approaches 100 *ch'ih* in length must carry two masts. The main mast is located just two beams forward of amidships, and the foremast stands some 10 *ch'ih* further ahead. On grain boats the tallest mainmast measures about 80 *ch'ih,* with the shorter ones reduced by 10 or 20 per cent. The below-deck portion of the mast measures somewhat over 10 *ch'ih,* and the portion used for hanging the sail is about 50 or 60 *ch'ih* long. The foremast, however, is less than half the length of the mainmast, and its sail is not quite one-third the size [of the mainsail].

In the six prefectures of Soochow, Huchow, etc., not only must the grain boats pass under arched stone bridges along their transport routes, but also the waterways are without the stormy dangers of the great rivers, therefore the sizes of the masts and sails have all been reduced. For boats used in Hukuang and Kiangsi provinces, however, which ply across great lakes and the Yangtse River and often encounter unexpected storms, the proper measurements of the anchors, chains, sails, and masts must be strictly adhered to before a safe voyage can be undertaken.

The size of the sail is determined by the width of the boat itself: if the sail is wider than the boat dangers may ensue, and if narrower, will not be effective. The sails, made of woven rattan matting interlaced with bamboo strips, are folded into layers to await hoisting and unfurling. On a grain boat ten men are needed to hoist a mainsail to the top of the mainmast, but two men are quite enough for hoisting a foresail.

To connect the sail cordage, a wooden ring measuring one *ts'un* across is first affixed to the masthead. [A sailor] then climbs up the mast, carrying the cordage around his waist, and passes it in three strands [through the ring, thence to connect with the battens on the sail [5]].

One section at the top of the sail will furnish power equivalent to that which can be obtained from three lower sections; [the important thing is] to regulate them so that a proper balance can be struck. In a favoring breeze the entire sail is extended right up to the top, and the boat will speed ahead like a galloping horse. If the wind becomes gusty, then the sail can be taken in section by section. (In case the sail is too full in a strong wind and cannot be taken in with speed, hooks are used to pull it down). Should the wind become extremely strong, only one or two sections of the sail need be extended.

Making use of the breeze that comes from the side, or beam wind, is known as tacking. When the boat is going downstream, the sail hangs on the mast and [the boat] takes a zig-zag course. First the boat may tack to the east, gaining only by inches, or is even possibly set back a few hundred *ch'ih*. Before it touches shore, however, the rudder is turned, the sail dipped, and the boat is now tacking to the west; taking advantage of the direction of the river current together with the force of the wind, the craft can now speed over a dozen *li* in a matter of moments. When sailing on the calm water of a lake, tacking can also be used, with slower results. Going upstream, however, a boat cannot tack and gain any distance at all.

It is the nature of boats to follow the current of the water, just as it is the nature of grass to bend to the wind. Therefore, the rudder is used in order to control the water force and regulate the direction of the boat. A turn of the rudder will create an eddy in the wake of the boat. In construction, the base of the rudder should be even with the keel of the boat. If it extends as little as one *ts'un* longer [than the keel], then there is the danger of the rudder sticking fast to the bottom of shallows, while the boat itself is able to glide over them; and should the wind be strong and contrary, the boat would be in extreme difficulty. If the rudder is shorter by even one *ts'un,* on the other hand, it will be weak in manipulation and cannot effectively turn the boat around. The stretch of water responsive to the rudder extends (from the rudder) to the bow of the boat: directly under the boat bottom the water assumes the nature of a fast headlong current, thus automatically setting the direction of the boat. All this is marvelous beyond words!

The handlebar used to steer the rudder is called the tiller. When you wish the boat to head south, the tiller is turned to the north; and when you wish to head north, it is turned south. If the boat is too long and faces a strong beam wind, so that steering becomes difficult, an additional plank [leeboard?] is immediately lowered into the water to counter the force [of the wind]. The main body of the rudder is fashioned out of a straight length of wood. (On grain boats it is three *ch'ih* in circumference and over ten *ch'ih* long.) The tiller is set across the top of the rudder, while at the lower end an opening is made and an axe-shaped wooden board is inserted through it and secured with iron nails, thus fending against the water. The high-rising part of the stern of the boat is called the aftercastle [lit. rudder-house].

Iron anchors are used to moor the boats, and five or six anchors are needed for each grain-boat. The heaviest, weighing about 500 catties, is known as the "watchdog anchor"; in addition, two anchors are used at the bow of the boat, and two at the stern. In case contrary winds should confront a boat in midstream, so that she could neither proceed nor draw close to the shore (or if, having come near the river bank, it is found that the bottom of the water is filled with shoals instead of sand, and it is impossible for her to be moored to shore, and only the anchors could be used where the water is deep), then the anchors should be dropped into the water. The anchor chains are wound around the capstan. When the claws of the anchor hit the mud and sand in the bottom, they will sink in and hold fast. In extremely dangerous situations the "watchdog anchor" is lowered. [The capstan] to which this anchor is secured is named the "main stake," an indication of its importance. Or, if in a flotilla the boat ahead has had to slow down, and there is danger of a collision should the second boat dart forward with the current, then the stern anchors ought to be lowered at once to reduce the speed of the second boat. When the wind subsides, the anchors are lifted by means of winches, and the boats can continue on their voyage.

To caulk the plankings of a boat, finely chopped flax is first stuffed into the space between the boards; next a mixture of screened fine lime powder and *t'ung* oil is applied to the crevices to complete the caulking. In Wen-chou, Ta'i-chou [both in Chekiang], Fukien, and Kwangtung, however, pulverized oyster shells are substituted for lime.

The cordage for sails on a boat is made of hemp stalks twisted roughly together; such a rope of one *ts'un* in diameter is capable of sustaining a weight of 10,000 catties without breaking. The rope for tying the anchor is made of split green bamboo bark, which is first boiled in a pot and then twisted together. Similarly, the tracking ropes [for pulling the boat against the current] are also made of boiled bamboo fibres. About every 100 *ch'ih* the rope is looped onto [the next section], and when obstacles are encountered the two sections can be disconnected. The nature of bamboo itself is "straight" [i.e. high in tensile strength], that of its split bark is capable of sustaining 1,000 catties on one [strand of fibres]. However, for boats going upstream through the Yangtse Gorges [the tracking ropes] are not made of twisted fibre strands. Instead they are fashioned out of lengths of bamboo about one *ts'un* wide, with each length attached to the end of the next until a long pole, called the "fire stick," is made. This is because the rocks along the river here are sharp as knives, and it is feared that the bamboo-fibre ropes may be easily cut and damaged.

As to the types of wood [used in constructing a boat], the mast is made from the straight trunks of fir, using one on top of another to achieve the desired height; all along the entire length of the mast it is encircled securely with iron hoops. Spaces are left in the foredeck of the boat for stepping the masts. When the mainmast is to be erected, it is brought to the boat on several large vessels, [and after the mast is placed in position] it is pulled upright by means of a long rope attached to the masthead. For the boat beams and the ribs, the wood of cedar, oak, camphor, elm, or locust is used. (Camphor wood felled in the spring or summer invites termites after a certain period of time.) For the planking of the hull, any wood will do. The rudder-post is made of elm or oak; the tiller, of an evergreen or elm; and the sculls, of fir, Chinese juniper, or catalpa. These are the major varieties of wood used.

Sea-going Vessels

Of the sea-going vessels, the ships used in transporting rice during the Yuan dynasty and at the beginning of the present dynasty were known as the "over-ocean shallow ships"; next [in importance] were those called "wind-penetrating ships" (also called "sea eels"). The voyages undertaken by these vessels were all within a distance of some 10,000 *li,* going past such places as Ch'ang-t'an (Long Beach), Hei-shui-yang (Black Water Ocean), Sha-men-tao (Sandgate Island), etc., where there are fortunately few hazards, and the costs of construction of these boats are not one-tenth as much as those for the ships sailing on missions to the Liu-ch'iu Islands and Japan or trading with Java and Borneo[?].

The sea-going rice transport boats are longer than the [inland] grain boats by sixteen *ch'ih,* and wider by two and half *ch'ih.* All other equipment is the same, except that the rudder-post of a sea-going boat must be made of an extremely strong timber, called *T'ieh-li-mu;* also, for some unknown reason, fish oil is mixed with the *t'ung* oil for caulking.

Ships plying [between China and] foreign countries generally are constructed along the same patterns. The sea-going vessels of Fukien and Kwangtung (ships sail from Hai-ch'eng in Fukien, and from Macao in Kwangtung) have attached to both sides [of the bulwark] fencelike posts made of split bamboo which help to fend off the ocean waves; the ships from Teng-chou and Lai-chou [in Shantung], however, are different. The ships of Japan have sailors stationed along the sides to maneuver their vessel by means of wooden planks or paddles, but Korean ships differ from these. However, the compasses placed fore and aft on board a ship, in order to determine the course; the great beam amidship, protruding a few *ch'ih* beyond the hull on either side; and the location of the centerboard—these are common to all ships. The centerboard [lit. "waist rudder"] is not shaped like a stern rudder. It is instead a broad plank cut into the form of a wide knife, which is lowered into the water for the purpose of keeping the ship's balance; it cannot be turned or steered. The top of the centerboard is attached to a crossbar, which is tied to a beam. When the ship encounters shallow water, the centerboard is lifted. This resembles [the handling of] a rudder, therefore it is called "waist rudder."

On board the sea-going ships a few *tan* of water is stored in bamboo barrels, enough for two days for the entire ship's company. Fresh water is taken aboard en route when the ship touches at islands. As to the destination of the ships, to which country or island each one should go, and what course to set, these are clearly indicated on the compass [6] and are probably not to be determined by man's will alone. The helmsman rules over the entire crew. He is a man who possesses thorough knowledge and a steadfast sense of duty, not one endowed with a mere dash of bravery.

Miscellaneous Boats

The tax boats of the Yangtse and Han Rivers: The hulls of these tax boats are very long and slender, with a row of about ten cabins on board, each cabin large enough only for one person to sit or sleep in. These boats [Figure 9.2], are equipped with a total of six oars located fore and astern, and a small mast with sail. The oars help to balance the boats in stormy weather. When sailing with the current, when there are no contrary winds, these boats can cover a distance of over 400 *li* in a day and a night, and even going upstream they are capable of covering over 100 *li.* As a large portion of the government's salt revenue comes from the Huai-an and Yang-chou [northern Kiangsu] regions, these boats are especially constructed to transport the silver

Figure 9–2. A six-oar tax boat on the Yangtze and Han Rivers.

and consequently are known as "tax boats." Travelers who wish to make good time also hire them for their journeys. This type of boat is seen in the region bounded by Chang-chou and Kung-chou [in Kiangsi] in the south, Ching-chou and Hsiang-Yang [in Hupeh] in the west, and Kua-chou and I-chen [in Kiangsu] in the east.

The wave-riding boats of Kiangsu: The area in Western Chekiang,[7] covering some 700 square *li* and centering around P'ing-chiang [i.e. Soochow], is laced by numerous winding streams and channels. On these waterways the "wave-riding boats" (the smallest of these are called "pond boats") number in the thousands. They are used by travelers, rich and poor alike, as one would use horses, carts, or footwear. And even the smaller boats feature windows, passage-ways, and cabins. They are built mostly of fir wood. With passengers and goods on board, such a boat must be kept in perfect balance, otherwise it will overturn with the least unevenness of weight. Hence it is commonly called the "balance-boat."

These boats usually ply within the 700 square *li* area. Occasionally, however, a traveler might wish for the sake of convenience to hire one for a voyage all the way up to T'ung-chou or Tientsin in the north [near Peking]. On such a journey the boat only has to cross (the Yangtse) at Chinkiang, which is done when the winds are calm. Next it passes the intersection of Ch'ing-chiang-p'u, and goes upstream on the Yellow River for 200 *li* in shallow water, after which it will enter the safe channels and locks [of the Grand Canal]. If (a wave-riding boat) sails on the [more turbulent waters of] the upper Yangtse River, it will capsize and sink at the first encounter with a storm; hence the use of this boat is avoided there.

The forward power of the wave-riding boat is provided mainly by a great scull, which is located at the stern and worked by two or three men at a time. Sometimes the boat is pulled along by trackers with ropes. As to the sail, it consists of a small mat hardly larger than a man's palm, and cannot be depended on.

The Hsi-an boats of Chekiang: In Eastern Chekiang,[8] between Ch'ang-shan and Ch'ien-tang, for a distance of 800 *li* the rivers all run into the sea without connecting with other inland systems. The Hsi-an boats therefore sail only between the small rivers of Ch'ang-shan, K'ai-hua, Sui-an, etc., and the Ch'ien-t'ang River, and are not found elsewhere. In form the boat has a cabin, which is constructed of woven straw matting, and shaped like an inverted water jar. It carries a cloth sail about 20 *ch'ih* high, which is secured with cotton ropes. It was thought at first that cloth sails were used because they could be taken down more easily when the powerful tides of the Ch'ien-t'ang estuary were encountered; but this does not appear to be true. Moreover, cloth sails probably cost more than those made of woven bamboo bark. It is not clear to me why cloth sails are used.

The "clear-stream" and "mizzen-sail" boats of Fukien: These boats ply between the small streams of Kuang-tse and Ch'ung-an, and the Hung-t'ang seawall of Foochow; beyond this point the route takes to the sea. The "clear-stream" boats are used to transport merchants and their goods, while the "mizzen-sail" boats, being larger, contain living quarters and are used by the families of officials on their travels. Both types of boats are constructed with fir bottoms, which are frequently damaged by rocks and shoals. When this happens, the boat is immediately steered toward shore where it is unloaded and the damages repaired.

Instead of the stern rudder, these boats are equipped with a huge steering-paddle at the bow, and the boat is thus steered by turning at the head.

These boats sail in flotillas of five. When passing dangerous shoal waters, the speed of the boats is reduced by having the four others line up behind the lead boat and pulling each other back with ropes. The boats sail all year round, not lying up even in deep winter, thus giving the utmost convenience to travelers on waterways. The sail, however, is not actually used.

The "eight-scull" and other boats of Szechuan: The rivers of Szechuan are the headwaters of the Yangtse and Han Rivers; but the Szechuan boats only go as far as Ching-chou [in Hupeh], where one must change boats for voyaging further downstream.

Going upstream [on the Yangtse], the Gorges are entered at I-ling. The boat trackers, using "fire sticks" fashioned out of giant bamboos split vertically into four or six sections and tied together with hempen ropes, make their way among the rocks along the shore and pull the boat forward to the beat of drums, which are sounded by men in the boat as though for a race. Between mid-summer and the Moon Festival [the 15th day of the 8th lunar month, approximately in mid-September], however, high waters in the river seal off the Gorges, and for several months all boat traffic is suspended until the water level has fallen; then communications can be resumed.

At the extremely dangerous spots [in the Gorges], such as Hsin-t'an, the men and goods are all removed from the boat and a portage is used for about half a *li,* the empty boat being left to ride out the turbulent waters. These boats are built with a rounded bottom and pointed, narrow bow and stern, designed to cope with the shoal waters.

The "full-sail" boats of the Yellow River: These craft are used to go from the Yellow River to the Huai River, and from the Huai upstream into the Pien River. They are expensively built of cedar wood, and are of varying sizes, the larger ones having a capacity of 3,000 *tan,* the smaller 500 *tan.* When sailing downstream a beam is laid across the foredeck, while the boat is propelled along by two large sculls located one on each side. The anchor chains, rigging, and sails carried by these boats are similar to those of the boats of the Yangtse and Han Rivers.

The "black castle" and "salt" boats of Kwangtung: These boats are found between Nan-hsiung in northern Kwangtung and the provincial capital [Canton]. Beyond this point the route from Hui-chou and Ch'ao-chou to Ch'üan-chou and Chang-chou lies along the coast, where sea-going craft are used. The black castle boats are used by officials and high-ranking personages, while the salt boats are employed to transport merchandise. There are catwalks or passageways built onto both sides of the bulwark. The sails are made of woven reed mats, which are hung on a double mast instead of a single one. With such an arrangement, [the sails] are less maneuverable than those of the central China boats. When traveling upstream the boat relies on tracking, which is similar to the practice of other provinces.

The Shensi boats of the Yellow River: Constructed mainly at Han-ch'eng [in Shensi], the larger of these boats are capable of carrying several tens of thousands of catties of stones down the Yellow River to supply the needs of the Huai and Hsü-chou region [in northern Kiangsu]. The boat is built with a bow and stern that are equal in width, and low-lying cabins that have rather flat roofs. When going downstream in a swift current the boat is helped along by a pair of large sculls worked from either side, and it does not depend on the wind for progress. [When traveling upstream] on the return voyage trackers are used, sometimes as many as twenty-odd men pulling the boat at a time. There are even occasions when the boats are abandoned and the men return without them.

CARTS

Carts or carriages are used to the best advantage on level terrain. In ancient times in the area covered by the states of Ch'in, Chin, Yen, and Ch'i,[9] chariots were always used when the states engaged in warfare, hence such designations as a "thousand chariot state" or "ten thousand chariot state"

Figure 9–3. A four-wheeled cart drawn by eight mules or horses.

originated during the Warring States period. After the bloody struggle between Ch'u and Han, use of other devices was greatly increased. In the south, more and more naval engagements were fought between marine craft, while infantry and horse contended in land battles. In the north, barbarian tribes warred with armoured cavalry. Thus the war chariot fell into disuse. Nowadays horse- [or mule-] drawn carts are used for transporting freight, and we may regard the mule carts of today as having evolved from the chariots of old.

The mule carts are built with either two or four wheels, and the axletree is the base on which the superstructure of the cart is constructed. On a four-wheeled cart there is an axletree both fore and aft, on which rest short posts that support straight beams, and on the beams rests the body of the cart. When the cart is not in motion and the horses [10] are unharnessed, the cart will be as level and secure as one's own living-room. In contrast, the body of a two-wheeled cart will remain level only while it is drawn by the horses; when the horses are unhitched, the cart must be propped up with posts, otherwise it will overturn.

The wheels of a cart are also termed [*lun*] *yuan* (commonly called *ch'e-t'o*).[11] On a large cart the hub (commonly called "cart brain") in the center of a wheel measures one and a half *ch'ih* in diameter (see the Chu commentaries in the *Hsiao-jung* poem, [*Book of Odes*]) and supports the spokes all round itself while allowing the axle to come through the center.

There are thirty spokes to each wheel. The inner end of each spoke is attached to the hub, and the outer end to the interior rim. In a wheel, that part which inwardly holds the wheel together and connects outwardly to the felloe, making a complete circle, is known as the interior rim. Where the felloe or exterior rim terminates it is called *lun yuan* or the shaft of the wheel.

When a mule cart is unhitched, its composite parts are taken apart for safekeeping. To assemble a cart, first the two axletrees are mounted [though the wheels], next the other parts are put together step by step. Such items as the front stretcher, the front crossboard, the yokes for horses or oxen, and the rear crossboard of the cart body, rise from the axletrees that serve as the bases.

A four-wheeled mule cart [Figure 9.3] has a capacity of fifty *tan*. The team for such a cart may consist of as many as twelve or ten mules or horses, or eight at the least. The driver stands on an elevated spot in the middle of the cart. The horses are harnessed to form two rows.[12] (The ancient war chariots were drawn by teams of four in a row, and the four animals are each designated as the *ch'an* [outer] and the *fu* [inner] horses.) A long hempen rope is attached to the neck of each horse, then tied together and passed inside the front crossboard and placed along the sides of the cart. The driver wields a long hempen whip measuring seven *ch'ih* with a handle of equal length, which he administers to the back of any animal that is not pulling its full weight. Two other men in the cart have the job of [controlling the speed by] stepping on the ropes. These men must be familiar with the nature both of the horses and of the ropes. When the horses are trotting too fast, the men will immediately step on the ropes [thus causing the team to slow down], otherwise the cart may upset and crash.

If while the cart is going along a road it becomes necessary to avoid running into pedestrians walking ahead, the driver will immediately call out a command, which halts the entire team of horses. At the place where the reins of the horses are passed through the front crossboard into the body of the cart, they are wrapped with ox-hide for reinforcement; this is what was termed a "driving aid" in the *Book of Poetry*.

The team of a cart is not fed within houses or yards. Instead, trays of woven willow are brought along in the cart, and the animals are untied and given their feed [on these trays] out in the open air. Persons embark or disembark from the cart by means of a small ladder. Crossing a highly arched bridge, the strongest horse is chosen from the entire team of ten and tied to the rear of the cart; when the cart goes down the decline of the bridge, the nine front horses are made to proceed slowly, while the rear horse is to pull backward, thus braking the forward speed of the vehicle. Otherwise, a dangerous upset may ensue.

The journey of a cart is stopped short when it encounters either a river, a mountain, or a narrow winding path. The route for carts sometimes may reach an extent of some 300 *li*—it is only a means of compensating for the want of boats in a region that lacks waterways.

In constructing carts, first attention should be given to choosing long pieces of wood for the axles, and short ones for the hub. The best types of wood

Figure 9–4. Single-wheeled cart drawn by two mules or horses.

Figure 9–5. A single-wheeled cart pushed by one person, south China.

for these purposes are locust, jujube, sandalwood, or elm. However, some people hesitate to use sandalwood, which tends to become considerably hot after long and continuous use in motion. The choicest material is either jujube or locust wood cut from trees whose trunks are large enough for a man to put his arms around. The other parts of the cart—the front and rear crossboards, the body of the cart, and the yoke—may be made of any kind of wood.

Another type of cart is the ox cart, which is widely used in Shansi for transporting fodder and army provisions. Where the roads are narrow the oxen are equipped with large bells that hang from their necks. These are called "herald bells," and serve the same purpose as the bells placed on the teams of the mule carts.

In north China there is also a single-wheeled cart, which is pushed by a man from the rear and at the same time drawn by a donkey in front [Figure 9.4]. Travelers who do not wish to ride [horseback] often hire such a cart for their journeys. An awning is erected overhead to serve as cover against the elements. The passengers must sit facing each other, one on each side [of the wheel, which is placed in the center], otherwise the cart will tip over. This cart is used northward to Ch'ang-an [in Shensi], Chi-ning, [in Shantung], and up to Peking. When not carrying passengers, these carts may transport merchandise. Each cart has a capacity of four or five hundred *tan*.

Oxen-drawn covered wagons are much used in central China. There are two wheels connected to the cart by an axletree, which is level and keeps the front and rear sections of the cart in perfect balance. Short boards are laid over it, and, on the boards the body of the wagon is mounted, in which a man can safely sit. The [empty] wagon will not tilt even when the animals are unharnessed.

As to the single-wheeled push-cart of the southern provinces [Figure 9.5], it depends solely on the strength of a single person. Its load limit is about two *tan,* and it must stop when it encounters any unevenness in the terrain. It does not range beyond 100 *li* at the most.

There are many other kinds of carts which cannot all be described here. Since the natives of the south never see a mule cart, and those who spend their whole lives in the north never see a large boat, I have recorded the above general accounts for their information.

NOTES

1. Hsi-chung was traditionally the official in charge of the construction of chariots under the Emperor Yü (2205–2198 B.C.) of the Hsia dynasty. Ancient Chinese books, such as *Mo-tzu, Hsün-tzu, Lü-shih ch'un-ch'iu,* and *Tso chuan,* designate Hsi-chung as the first Chinese to build carts from wood.

2. More than 46,000 *tan* of grain were shipped in 1283, 295,000 *tan* in 1284, 578,520 *tan* in 1286, and 3,522,163 *tan* in 1329. The large seagoing ships could each carry 8,000–9,000 *tan* of grain. Eunuch Cheng Ho made seven expeditions to Southeast Asia during the period 1405–30. He employed as many as sixty-two ships on one voyage and sailed as far as the east coast of Africa. They were all the same size: 44 *chang* long and 18 *chang* wide; each one could carry 3,000 to 4,000 *tan* of grain.

3. The grain tribute boats used on canals and rivers were introduced in the T'ang dynasty, and became common in the Sung. The construction of the Grand Canal in the Sui dynasty (A.D. 589–618) led to large-scale inland shipping between the south and north. On his visit to Yang-Chou (in Kiangsu province) in 605, the Emperor Yang of Sui went by way of this canal, and his ships extended over a distance of 200 *li.*

Largely replaced by seagoing ships in the Yüan period, river boats became important again during the fifteenth century. *The History of Ming* and the "Book of Money and Commodities" of the *Ming History* relate how boat construction was supervised first by Sung Li and then by Ch'en Hsüan. The various designs of grain transport boats are schematically shown in *Lung-chiang ch'uan-ch'ang chih,* written in the early 1550's by Li Chao-hsiang, who served for a number of years as a supervising official of Lung-chiang Shipyard near Nanking. The materials used in building grain boats are listed in *Ta Ming hui-tien* (The Institutes of the Ming dynasty).

From the middle to the end of the Ming dynasty, an average of about 4,000,000 *tan* of grain was transported each year over canals and rivers, totaling some 1,000 miles, with differences in water levels of up to 116 *ch'ih.*

Piracy was also a problem. Yabuuchi Kiyoshi points out, *Tenkō Kaibutsu no Kenkyū* (Tokyo, 1953), that the system of grain transportation was changed several times in the Ming dynasty. From 1415 on, the burden passed gradually from the peasants to the army.

4. These beams are the cross-beams in the hull of the boat exclusive of the bow and stern *fu-shih.* A diagram showing the position of these beams may be found in Li Chao-hsiang, *Lung-chiang Ch'uan-ch'ang chih.*

5. Handling silks on a Chinese junk has been described fully by Alan Villiers, in *The Way of A Ship* (New York, 1953), pp. 98–99. Although the sails depicted in Villier's book are those of modern junks, the principal features appear to fit those described in the present chapter. Particularly notable is the way individual sections of a sail are manipulated with battens, so that each section can be easily extended or taken in as the need may be.

6. Crude compasses in the form of the south-pointing ladle (*Ssu-nan shao*) and the diviner's board (*shih* or *lo-pan*) were constructed from lodestone in the later Han dynasty (A.D. 25–220).

7. Western Chekiang was a term used during the Ming period to denote the area later known as Kiangsu.

8. Eastern Chekiang corresponds to the province later called Chekiang.

9. Ch'in, Chin, Yen and Ch'i are respectively the modern provinces of Shensi, Shansi, Hopei, and eastern Shantung.

10. The author is not consistent in his use of the words "horses" and "mules," even though both kinds of animal were used for pulling carts.

11. *Yuan* is not a wheel, but rather the shaft of a carriage or cart. This word is very likely a misprint in the original text.

12. Presumably in column rather than line; the team shown in the illustration would require a modern football field to maneuver.

Figure 10–1. Preliminary steps in needle making.

10

HAMMER FORGING

Master Sung observes that it is by working on metals and wood that artifacts are made. How could the skills of even Kung-shu Pan and Ch'ui [1] be manifested if the world were devoid of tools? Without forceps, hammers, and the like, the "five weapons" and "six musical instruments" would be incapable of fulfilling their respective functions. Out of the blazing fire of the same furnace will emerge a host of objects of different sizes: one may weigh 30,000 catties and be capable of anchoring a battleship in a raging sea, while another may be as light as a feather and fashion embroideries on a ceremonial robe. Yet if we assign all the credit of superb smelting and casting to supernatural forces, it seems that proof of this can be found in the story of the Mo-hsieh and Kan-chiang swords, and in the rise heavenward of two other famous swords which turned into two dragons [after being discovered].[2]

MAKING IRON ARTICLES

For making iron articles wrought iron is used. First an anvil is cast to provide the space for hammering; it is truly said that "all things stem from a pair of forceps" [for holding the red-hot iron while it is being hammered]. The wrought iron that has just emerged from the forge is termed "unfashioned iron," and in the course of heating and hammering it will lose three-tenths of its volume in the form of sparks and droppings. If, however, unrusted old iron articles are remelted to be made into new or the original objects, then this is called "used iron" and the loss amounts to only one-tenth of the volume.

Coal accounts for seventy per cent, and charcoal thirty per cent of the fuel used for heating the iron in the forge. In places where no coal is produced, the ironsmith would first select very hard wood and make hardwood charcoal (it is commonly called "fire arrow" and can be burned in a furnace without closing its door) which gives even greater heat than coal. In using coal for fuel [one should select] a kind known as iron-coal [anthracite] which contains intense internal heat but does not waste it in high flames. It is different from the ordinary charcoal though resembling the latter in appearance.

Pieces of [wrought] iron can be joined together by covering the juncture with yellow mud and hammering it after heating in fire. The waste mud will then be gotten rid of, while its essence will have bound the two parts together so securely that they can never be separated except by cutting with an axe at a red hot temperature.

The hammer forged articles of wrought iron and steel are not very hard before the correct balance of water and fire has been achieved, and therefore should be quenched in clear water immediately after being taken from the forge. This is called the hardening of steel and iron, meaning that before this the soft properties of the iron and steel were not yet entirely removed.

The countries of the West [i.e. Europe] have a remarkable chemical [solder] which they use for soldering. In China, however, we use powdered white brass for soldering small articles.[3] For large objects we can only bind the pieces together by exhaustive hammering, but the result of this process will not hold up over the years. That is why though the Western countries can produce large cannon by forging and hammering, in China they are made only by casting.

Knife and Axe

Among the iron weapons,[4] the thin ones are knives and swords, and the ones with thick backs and thin blades are axes. The best swords are coated with steel that has been obtained after a hundred smeltings, but the core of the sword is still made of [wrought] iron; otherwise [that is, if the entire blade is made of steel] the sword will break off when it is used with vigor. The ordinary knives and axes are inlaid with steel on the cutting edge only. But the steel coating on even the costliest knife, which can chop nails and cut through ordinary iron, will disappear and the iron core will show up after the blade has been ground and sharpened a few thousand times. In Japan, the back of the knife blade is less than 0.2 inch wide, yet will stand on one's finger without toppling. The process for fashioning this kind of knife is not known in China. The hardening of knives, swords, and axes by quenching in water is done after they have been coated or inlaid with steel. But their sharpness will depend on grinding.

The hollow space in axes and mallets, into which the handle is fitted, is made in the following way: first, a "bone," called a "sheep's head," is prepared by hammering a piece of cold iron; second, a piece of hot iron is placed around the cold iron bone [and is hammered to form either an axe or a mallet]. The cold and hot iron do not adhere to each other, and [when the "bone" is withdrawn] a hollow space is left [in the axe or mallet that has been made].

After long use, the mallet used for breaking stones becomes hollowed on all sides owing to the constant hammering. It can be repaired by filling the cavities with molten iron, and then can be reused without difficulty.

Hoe

The hoes and other farm implements are made of wrought iron, but the sharp edges are coated with cast iron and quenched in water to achieve the desired hardness. Generally 0.3 ounce of cast iron is used for a shovel weighing one catty; if less than this amount [of cast iron] is used the implement will not be sturdy; and if more, too brittle.

Files

Files are made of pure steel. While the steel body of the file is still soft before quenching, its surface is grooved by means of steel chisels. The strokes are made obliquely against the surface [of the file] and the pattern is that of criss-crossed slanting grooves. After the file is grooved, it is heated to red-hot, followed by slight cooling, and then quenching. To re-groove a file after long use, heat it first to get rid of the hardness, and then make the lines with a chisel as before.

Figure 10–2. Making an anchor.

To make saw-teeth, the "straw-leaf" file should first be used, following with the "arc-sharpening" file; to polish copper coins, the "square-long-pull"; to polish locks, keys, and the like, the "square strip"; to cut and fashion ivory and horns, the "sword-face" (called *tin-file* in the Chu commentary); and to work on wood the file is marked with round holes instead of criss-crossed lines, and is called "incense" file. (The powder of ram's horn, mixed with salt and vinegar, is applied to the file before the lines are marked on it.)

Awl

Awls are hammered out of pure wrought iron. The "round" awl is used for binding books and the like; the "flat" awl for leatherwork; the "snake-head" awl is used by carpenters when small holes are needed for the passage of strings, nails, etc. About 0.2 inch above the sharp end, this awl assumes a shape that is rounded on one side and concave on the other side, with two protruding edges to facilitate drilling. For work on copper plate, the "chicken-heart" awl is used. A three-ridged awl is called the "revolving" awl; and a square one with a sharp end, the "boring" awl.

Saw

Saws are made by cold hammering of thin strips of wrought iron that have been cooled after leaving the forge; no steel is added, and no quenching applied. The teeth of the saw are made by means of a file. Each end of the saw is attached to a piece of wood. By fastening these two pieces of wood to the two ends of a bamboo stick, the saw is stretched tightly to become straight. The long saws are used to split lumber lengthwise, the short ones to cut wood across, and the finest-toothed are used for cutting bamboo. When the teeth become blunted, re-filing will make the saw usable again.

Plane

Planes, which in ancient times were called *chun,* are employed to smooth off the surface of wood. The plane is made by grinding to a high sharpness a piece of steel-inlaid iron about an inch long, with the blade slightly showing at an angle out of a wooden receptacle. The large kind, used by barrel-makers, is operated by placing the plane on its back with the blade showing; then the wood is pulled over it. This is called "push-plane." The type intended for ordinary use has a handle protruding on each side, which enables the worker to push forward. For very fine work the carpenters use a variety called "pick-thread plane," which has a blade about 0.2 inch wide. There is also a kind known as "centipede plane" that is used to make the surface of wood extremely smooth; on the wooden body of the plane are laid about a dozen small blades which resemble the centipede's feet.

Figure 10–3. Final steps in needle making.

Figure 10–4. First step in making the copper gong and copper drum.

Figure 10–5. Second step in making the copper gong and copper drum.

Wood-Chisel

The wood-chisel is forged out of wrought iron and inlaid with steel at the blade. The upper part of the chisel is a hollow cylinder for holding the wooden handle (an iron bone, generally known as "sheep's head," is first prepared; it is used as a mold for [making] both the sharp part and the handle-holding part of the chisel). The lower end of the wooden handle is knocked into the hollow cylinder of the chisel. The [sharp] end of the large chisel is about an inch wide, that of the small one only 0.3 inch. If the work demands the making of round holes, scoop-chisels are fashioned for the purpose.

Anchor

In a storm the fate of a ship depends on the anchor. Those used on battleships and sea-going vessels weigh up to 30,000 [*sic.,* probably 1,000, because the word *chün,* meaning "30 catties," apparently is used interchangeably with *chin,* "a catty"] catties. The four "claws" of the anchor are forged first, and then attached to the trunk one by one. For an anchor weighing less than 300 catties an anvil one foot in diameter is placed beside the forge; when the iron is thoroughly red-hot, the coals are removed and the hot metal is lifted onto the anvil with wooden staffs coated with iron [and hammered]. For those weighing around 1,000 catties a wooden scaffold is erected, on which stand many men holding an iron chain that is attached to the anchor [below]. Huge iron rings are placed at the ends of the chain, so that [the anchor] can be lifted and turned as it is being hammered jointly [by a number of workers. See Figure 10.2]. The drugs [fluxes] used [in joining the different parts] are not yellow mud, but consist of old wall plaster screened to a fine powder. One man continually scatters it onto the places of jointure to prevent cracks. Of all the things forged and hammered, the anchor is the largest in size.

Needle

To make needles, iron is first hammered into a slender strip, then it is pulled through the small holes of a flat iron rod, thus becoming thin wires. These are then cut into needles of one inch in length [Figure 10.1]. One end of the needle is filed to form a sharp point; the other end is hammered flat with a small mallet, the hole is drilled with a steel awl, and the exterior is also filed smooth. The needles are then placed in a pot which is heated over a slow fire. The needles [still in the pot] are covered with earthen materials mixed with pine wood, hardwood charcoal, and bean jam, and heated from beneath.[5] A few needles are stuck into the surface of the mixture for testing the proper heating time and temperature: when these outside needles crumble as they are pinched with fingers, it is time to uncover the needles underneath and to quench them in water [see Figure 10.3].

All needles used for sewing and embroidery are of a hard quality. The best embroiderers of Ma-wei, however, use the "willow-branch soft needle." The difference [between hard and pliant needles] depends upon whether the red-hot needles are quenched in water [or cooled slowly in a dying] fire.

COPPERWORKS

In order to make "copper" articles, red copper is converted into yellow brass [by the addition of zinc or native zinc carbonate]. White brass is prepared by mixing red copper with arsenic and other drugs [see Chapter 14], but its cost of production is twice that of other [copper alloys] and so only the extravagant undertake to manufacture it. Yellow brass prepared with zinc carbonate as an ingredient should be hot hammered before cooling, whereas that made with zinc is to be cold hammered after cooling. Musical bronze is made with a mixture of tin [see Chapter 14 for the method].

In the manufacture of musical instruments, the metal must be of one piece without soldering. All other articles of round or square shape can be soldered over a flame: when powdered tin is used [as solder], the process is called "minor soldering," and when powdered bronze is used, it is called "major soldering." (Bronze fragments are first ground into powder, and held together with cooked rice. Later the rice is washed off with water and the bronze powder remains in place. Otherwise the particles of the latter would be scattered.) Red copper powder is used, however, to solder silver articles.

Among the musical instruments, the *cheng* (commonly called gong) is hammered [6] directly from the heated metal without casting; the *cho* (commonly called copper-drum) and the *ting-ning* [small bell], however, are made by first casting [the metal] into round pieces and then hammering them. For hammering the gong or the copper-drum the metal is placed on the ground, and the combined labor of many men is required for hammering a large instrument. [As the instrument takes shape] its size is gradually enlarged with the progress of hammering, resulting in the resonant sound of the instrument [Figure 10.4]. The raised part in the middle of the copper-drum is made first, and then the article is cold hammered to produce the [proper] sound. The slightest difference in the strokes will determine whether the sound will be male or female; the former is achieved with many repeated strokes of the hammer [Figure 10.5].

The color of copper ware becomes dull white after hammering, but the yellow gleam will reappear when it is polished with a file. Under hammering, the loss of copper is only one-tenth that of iron. Copper not only has an odor, but also is more valuable [than iron]. The copper artisans, therefore, are more highly ranked than iron artisans.

NOTES

1. Kung-shu Pan of the state of Lu was a famous artisan of the Spring and Autumn period (770–403 B.C.). Ch'ui was a skilled artisan of legend who lived in the days of the Emperor Yao (ca. twenty-second century B.C.). Some say he belonged to the time of the Yellow Emperor, ca. 2700 B.C.

2. *Mo-hsieh* and *Kan-chiang* swords were named after the famous metallurgist Kan-chiang, who made the swords, and his wife Mo-hsieh. The dragon swords were said to have been discovered by Lei Huan of the Tsin dynasty (A.D. 265–419), who dug them out of their underground hiding place.

3. Wang Chin believes that the Chinese term *la* in the *History of Sui Dynasty* (A.D. 581–617) and in the *New History of T'ang Dynasty* (A.D. 618–906) represents a lead-tin alloy which was employed in large quantities for minting coins. It is possible that this alloy was also used by the Chinese at that time as a soft solder. Hard solders of copper base and silver base were definitely used in China by at least the early seventeenth century, as described not only by the present book but also by Fang I-Chin in *Wu-li hsiao-shih* (A minor Treatise on Natural Phenomena).

4. The use of copper and bronze for making knives, axes, hoes, needles, and awls in the Shang dynasty (ca. 1783–1122 B.C.), is evidenced by the articles found in the An-yang excavations. It was probably not until the seventh century B.C. that iron began to replace bronze in the making of some of these implements. In the Freer Gallery there are specimens thought to date from this period. One of these, a long hafted pickaxe (dagger axe), has a core of iron (possibly meteoric) covered with bronze. Similar objects have been described by Chang Hung-Chao in his *Lapidarium Sinicum*. What may be an echo of this period is contained in a passage in the *Kuo-yü* (Discourses of the States), which purports to record a conversation between Kuan-tzu and Duke Huan of Ch'i (reigned 683–641 B.C.): "The lovely metal (bronze) is used for casting of swords and pikes. The ugly metal (iron) is used for casting of hoes and axes." The increasing use of iron at a later date is noted in the *Kuan-tzu* book, written probably by the Chi-hsia academicians in the late fourth century B.C. A passage in this book enumerates the following items as within the jurisdiction of iron administrators: "each woman certainly has a needle and a knife . . . each tiller of the soil certainly has a pointed plough, a forked plough and a large hoe . . . each cart builder certainly has an axe, a saw, an awl, and a chisel."

5. Wrought iron, being malleable and ductile, was particularly suitable for the primitive method of wiredrawing. The wrought-iron needles were apparently converted into steel by heating the iron needle in a pot together with charcoal combined with, or instead of, other carbonaceous materials. The top of the pot was covered with clay or other inert material and every effort was made to render the pot as airtight as possible. The carbon was slowly adsorbed by the iron needles in solid solution up to its limit of solubility. The amount of carbon adsorbed by the red-hot iron needles increased directly with the temperature and length of time of the carbon-needle contact.

6. Hammering not only shapes the instrument but also increases the hardness of the copper or bronze.

Figure 11–1. Calcination of stone into lime.

11

CALCINATION OF STONES

Master Sung observes that among the five elements [1] earth is the "mother" of everything, but of her "offspring" the metals are not the only valuable members. When metal and fire are brought together, the former flows in a molten state, and its uses cannot be excelled. But with calcining or heating, stones can be made to perform wondrous things. Water, for example, will soak and spoil things, and will seep into every crack. It is said that water does not neglect the most minute bit of space, yet [we can] make a certain mixture to defend ourselves from it. [This substance] if used on a ship at sea, can fend off the ocean waves or it can strengthen city walls when applied to bricks or mortar. This most precious matter is obtainable without our having to seek it in distant places: such are the benefits derived from the calcination of stones. Furthermore, the alum [and vitriol] of five colors, and the masterly qualities of sulphur, all result from the application of intense heat. Such [alchemist's] skills climax in the distilling of litharge, but how can the artful arguments of the necromancers compare [in their marvels] even to one ten-thousandth of the creations of nature?

LIME

Lime can be used after being burned, and is never corroded by water once the correct properties are achieved. It is used to fill crevices as a water repellent in myriads of ships and myriads of walls.[2] Within one hundred *li* [of any place] there are always some calcinable stones, the black-colored stones being the best and the yellow second. They are usually buried two or three *ch'ih* deep beneath the surface of the earth and must be dug out before calcination. Stones that have been eroded by wind on the earth surface can not be used to produce lime. Nine-tenths of the fuel used for calcination consists of coal and one-tenth charcoal. The coal is first made into briquettes with the admixture of mud; then the briquettes are placed with the stones in alternating layers, with wood lining the bottom of the pile. A fire is lighted, and the stones are burned [Figure 11.1]. The best lime [thus obtained] is called "mine lime" and the lowest grade "kiln refuse lime." After the stones have been burned to a state of crumbling, they are placed [in open air] to be pulverized by wind. For quick results, however, water is poured over them, whereupon they will disintegrate.

When lime is used for caulking ships, it is mixed with *t'ung* oil and fish oil, folded into thick silk pongee and thin silk gauze, and pounded into the space to be caulked. When lime is used for building a wall, the lumps are first screened off, and the remainder is mixed with water and applied to bind the wall stones together. When used to line the inside of a well, it is again mixed with oils. When used to plaster walls, it is first decanted and then mixed with the fibres of paper and then applied. When used to line tombs or reservoirs, one part of lime together with two parts of river sand and mud are finely stirred into glutinous rice and the juice of the *yang-t'ao* vine. The resultant material, known as [Chinese] concrete, is extremely durable and never corrodes. Other uses of lime, such as the making of indigo and paper, cannot all be enumerated here. Along the coasts of Wen-chou, T'ai-chou [both in Chekiang], Fukien, and Kwangtung, where the stones cannot be calcined for lime, Nature has provided the oyster [shells] as a substitute.

OYSTER LIME

Oyster beds [*li-fang*] are found along the sea coast where rocky mountains, rising out of the sea, are constantly pounded by the salt waves. The Fukienese call them *hao-fang*. After many years these [beds] become several dozen *ch'ih* [in thickness] and cover several *mou* in area. They have a craggy appearance resembling garden rocks and animal shapes, and are gradually pressed into the shore rock. After a long period of time the inside melts to form a fleshy lump, called oyster meat, which is most delicious to the palate.[3]

In order to get the oyster shells for calcination, the worker wades into the water and collects them with mallet and chisel [Figure 11.2]. (The "oysters" sold in pharmacies are the small odd pieces gotten this way.) Then the coal is piled up, the fire is made, and the method of burning the shells is the same as the calcination of limestone described above. Oyster lime can be used in the same way as stone lime: as mortar for walls and bridges and mixed with *t'ung* oil as caulking material for ships. Some people mistake clam shell powder (i.e. clam powder) for oyster lime. That is because they have never troubled themselves to learn the nature of things.

Figure 11–2. Removing oyster shells from rocks with mallet and chisel.

Figure 11–3. Coal mining.

COAL

Coal is obtainable everywhere,[4] and is used for the smelting and calcination of metals and stones. South of the Yangtse River, coal is found in mountains that are bare of trees. We need not discuss the situation in North China.[5] There are three kinds of coal: anthracite, bituminous, and powdered. The large pieces of anthracite coal are about the size of a bushel. The coal is produced in places like Yen, Ch'i, Ch'in, and Chin [that is, the provinces of Hopei, Shantung, Shensi, and Shansi in North China]. It is kindled with a little charcoal, and can burn for a whole day without the use of bellows. The fragments beside the large pieces can be used as fuel after being mixed with clean yellow mud and made into cake-shaped briquettes.

Of the bituminous coal, which is mostly produced in Wu and Ch'u [i.e. in the middle and lower Yangtse region], there are two types: the high volatile type is known as "rice coal" and is used in cooking, while the low volatile is called "iron coal" and is used in smelting and forging metals. The coal is first dampened with water before being placed in the furnace, and the bellows must be used to bring it up to red heat. While the fire is going, coal should be added repeatedly. The small fragments of this coal, fine as flour, are called "automatic wind." These fragments, when made into briquettes with the addition of mud and water, will burn constantly throughout the day and night similar to anthracite coal. Half of the briquettes are used for cooking and half for smelting, calcination, and the manufacture of cinnabar. As for burning lime, alum, or making sulphur, all three kinds of coal can be employed.

Experienced coal miners are able to find underground coal by the color of the earth. Coal is reached in pits about fifty *ch'ih* deep. The first appearance of the coal seam is accompanied by strong poisonous gas; some people therefore erect a thick hollow bamboo pipe on the coal, thus drawing up the poisonous gas and enabling the men to shovel and pick the coal underneath. Sometimes the coal seams extend in several directions from a shaft; in which case the miners simply follow the seams, with timber built overhead [in the mine] to prevent collapse [Figure 11.3]. After a coal mine has been depleted, the shafts are then filled with earth, and more coal will appear there again after twenty or thirty years—it is inexhaustible [*sic*].

The round stones at the bottom of and around the coal seams, called "copper coal" [i.e. carbonaceous shale containing cupriferrous pyritic minerals] by the local people, are mined as raw material for making black vitriol and sulphur (see below). This kind of round stone, used only for making sulphur, has a partly oxidized, penetrating odor and is called "smelly coal" [i.e. carbonaceous shale containing pyrites and native sulphur]. It is found occasionally in Fang-shan and Ku-an in Yen-ching province [modern Hopei], and in Ching-chou in Hu-kuang province [modern Hunan and Hupei].

The essence of coal disappears with the element of fire after burning, and leaves no residue or ash. This means that it is a special manifestation of Nature placed between the species of metal and that of earth and stone. The marvel of the Heavenly Design is also shown by the fact that coal is not produced in regions where grass and wood abound. As a fuel for cooking, coal is surpassed [by wood or grass] only in the making of bean curds (the bean curds are bitter if made over a coal fire).

ALUM STONE AND WHITE ALUM

Alum is obtained from the treating of [alum] stones with heat. White alum is available everywhere, and is produced mostly in Chin-chou in Shansi, and Wu-wei in South Chihli [modern Kiangsu].[6] Alum resembles gypsum and costs little. When alum is dissolved in boiling water, however, and the liquid is used [as a mordant] for dyeing, the colors will stay fast like a layer of skin, never to fade on account of moisture. It is therefore needed by the candied-fruit makers and by the manufacturers of figured paper and red paper. When made into dry powder, white alum is effective in removing impurities and watery infections, and is therefore much needed by the curers of eczema.

To make white alum, dig from the earth stone lumps that have formed an underground layer, and apply heat to the pile together with coal, as if lime were being made. When heating has been sufficient, [the calcined stones] are allowed to cool and then put into water [for lixiviation and decantation]. The [supernatant] liquid is then heated, and alum is obtained when a substance, commonly called butterfly alum, flies out of the kettle of boiling liquid. Heating continues until the liquid is thick. Then it is purified [by fractional crystallization] in a large water jar. The substance that crystallizes on top is called "hanging alum" and is extremely white. That which sinks to the bottom [of the jar] is called "jar alum," and the light and fluffy kind, resembling cotton wool, is known as "willow-catkins alum." When the liquid has been entirely evaporated by heating, the snowy-white residue is called "Szechuan stone," and that which has been burned by the alchemists, "dry alum."

BLACK VITRIOL, RED VITRIOL, YELLOW VITRIOL, AND GALL VITRIOL

Black, red, and yellow vitriols all come from the same source—are variations of the same material. [For making vitriols] take the associated gangue stones of coal (commonly called "copper coal") and put them into a furnace at the ratio of 500 catties of stones to over 1,000 catties of coal briquettes (the kind known as "automatic wind" that needs no bellowing), which surround the stones on every side. A mud-brick wall is built to encircle the furnace [Figure 11.4]. On top of the furnace is a round hole the size of the rim of a tea cup, through which flames can shoot, and around the opening of which a thick layer of vitriol residue is applied. (It is not known how far back [the use of] this residue dates, but without application of this old residue, no new furnace can be operated successfully.) A fire is then started at the bottom of the furnace and lasts for ten days before burning out, and golden flames

Figure 11–4. Burning copper coal for black vitriol.

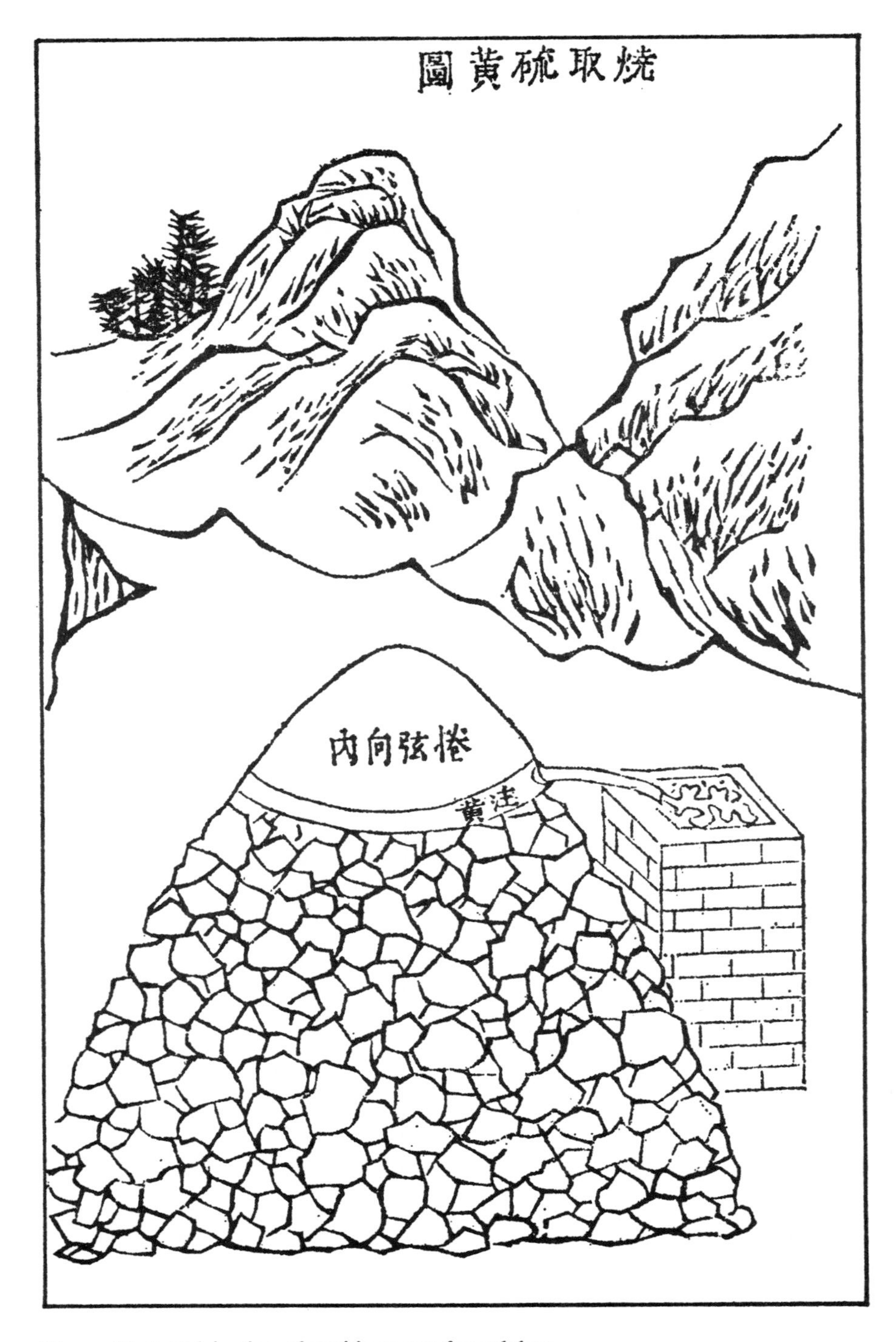

Figure 11–5. Calcination of pyritic stones for sulphur.

often shoot up through the opening. (For making sulphur, see below.) After ten days of burning, the calcined products are taken out after they have thoroughly cooled. The half-crumbling fragments, called "time vitriol," are selected and put aside for the making of red vitriol. The fine, lime-like powder, on the other hand, is immersed in water for six hours in a large jar, and then boiled in a pot. When the volume of every ten *tan* of the liquid is reduced to one *tan* by boiling, heating can be considered sufficient. After the material is dried in the pot, the upper layer will be high-quality black vitriol, and the bottom will consist of vitriol residues (to be used later for covering the opening of a new furnace). Black vitriol is needed by dyers, and is manufactured in only five or six places in all China. Generally speaking, about 200 catties of black vitriol can be obtained from 500 catties of the original stone.[7]

The "time vitriol" (popularly called "chicken-droppings vitriol") that has been put aside is now roasted in a crucible, after four ounces of yellow earth have been added to every catty [i.e. a ratio of 1:4] of the material, and red vitriol is obtained. It is used by plasterers and varnish-painters.[8]

Yellow vitriol [9] is produced in a very peculiar fashion. Between late autumn and early winter when the weather has turned cold, the mud-brick wall around the furnace wherein black vitriol is made will be covered with a layer of this substance, similar to the saltpeter that appears on brick walls north of the Huai River.[10] This matter is scraped off the wall and is known as yellow vitriol. It is used by dyers. When it is applied to articles of a light golden color and held over a fire, the color will immediately turn into the deep hues of true gold. Among imported yellow vitriol there is a special variety called "Persian vitriol," which shows gold-colored threads inside when broken into pieces.

In Shansi and Shensi provinces, where sulphur is made on the mountains, [calcined pyritic] residues are left on the ground. In two or three years these will have been washed by rain into gullies, and the essence naturally becomes black vitriol. It can be sold and used without calcination. The best of the black vitriol is sometimes falsely labeled "Stone gall." Stone gall is also known as gall vitriol [or blue vitriol] and is found in such places as Chin-chou and Shih-chou. Gall vitriol is a natural formation found in mountain caves, green in color with a gemlike lustre. A heated iron article quenched in an aqueous solution of gall vitriol becomes the color of copper.[11]

The *Materia Medica* [12] mentions five kinds of alum and vitriol, but does not discuss their distinctive qualities. Both the "Kun-lun vitriol," which looks like black mud, and the "iron vitriol" resembling red Kaolin, are products of the Western Regions.[13]

SULPHUR

Sulphur is produced by condensation of vapors arising from the heat treatment of rocks. In some books it is called "alum liquid," because burning [or sulphur] stones were mistaken for alum stones. The confusion arises from the fact that the stones out of which sulphur is made are partly a special white variety, and partly those from coal mines used for making alum; hence the term "alum liquid." It has also been maintained that sulphur is always present where there are warm springs in China, yet along the eastern sea coast and in Kwangtung in the south, sulphur is produced in places where there are no warm springs. The reason [for this fallacy] is that the similarity between the vapors of the warm springs and sulphur has caused people to speculate on their connection.

The shape of the sulphur-producing stones is the same as those obtained in coal mines. After the stones are dug out, they are packed with coal briquettes into a pile, around which an earthen furnace is built [Figure 11.5]. The pile contains 1,000 catties each of stones and coal, and the top of the furnace is covered with the old residue of previous sulphur calcinations. The center of the furnace top is dome-shaped with a round hole in it, through which gold-colored flames will appear when the heating is sufficient. A porcelain bowl has previously been prepared. This rises in the center and has a rim that is turned over like a bag. This bowl is now [inverted and] placed on the hole [of the furnace]. When the essence of the [sulphur] stone is heated, it seeks to escape in the form of a yellow vapor. This being stopped by the bowl, the yellow vapor then turns into liquid form, adheres to the bottom of the bowl, and flows into the baglike rim from which it flows through a small aperture into a clay pipe and thence into a small tank where it crystallizes into sulphur.

The same method of covering with an inverted bowl is followed as the yellow flame shoots up when coal mine stones are being heated to make black vitriol, in order to obtain sulphur. With each catty of sulphur realized by this method there would be a reduction of 30 catties in the output of black vitriol. When the essence of vitriol has been converted into sulphur, the rest is only waste material. Of the components of gunpowder, sulphur is pure positive [*yang*] and saltpeter is pure negative [*yin*]. When these two elements come together, the result is noise and change—this being a wondrous thing brought about by the celestial forces.[14]

Sulphur is not produced in the northern barbarians' country,[15] possibly because the people there do not know how to roast and collect it. Further, amazing cannon have been manufactured in the Western Ocean and by the Red [Headed] Barbarians, which shows that sulphur is produced everywhere in East and West, over a distance of tens of thousands of *li*. As to the "native sulphur" of the Liuchiu [i.e., Ryukyu] Islands, and the "water sulphur" of Kwangtung, they are only erroneous records in books [*sic*].

ARSENIC STONES

The material which can be calcined to produce arsenic oxide looks like earth, but is firmer; and resembles rocks, but is finer in size. It is obtained by digging a few *ch'ih* beneath the earth and is termed "Hsin-stone" because it is mined both in Hsin district [i.e. Kuang-hsin], Kiangsi, and in Hsin-yang, Honan. Recently, however, Heng-yang [in Hunan] is the place where the manufacture of arsenic [oxide] flourishes, and where the output of one firm can be as much as 300,000 catties. There is often turbid green water lying over the

Figure 11–6. Production of arsenic through calcination of minerals.

spot where arsenic stones are located, and this water should be pumped off before digging operations begin.

There are two kinds of arsenic [compounds], red [AsS] and white [As_2O_3], each resulting from the color of the stone from which it is sublimed and condensed. For the calcination of [arsenic stones to make] arsenic, an earthen kiln is built which is faced with rocks on the outside [Figure 11.6]. A chimney is built into the upper part [of this furnace], and the opening of the chimney is covered by an iron pot turned upside down. When fire is lighted beneath, the [arsenic] fumes will ascend through the chimney to become condensed in the pot. When the condensed layer is estimated to be about one inch thick, the fire is extinguished. After this layer is thoroughly cold, the fire is started again and the process is repeated until several layers have accumulated on the pot. The latter is then taken down, broken apart, and arsenic is obtained. The little bits of iron we see in arsenic are the fragments from the broken pot. This is the only way to make white arsenic. Red arsenic can be found in the flue gases and dusts from the parting or refining furnaces of silver, copper, or gold.

During the process of calcination, the operator must stand some 100 *ch'ih* to the windward of the kiln; all grass and trees close to the leeward side of the kiln will die. Arsenic workers must be transferred [to other work] after two years, otherwise all their hair will fall off. Arsenic is a substance of which an overdose of but one-hundredth or even one-thousandth of an ounce is fatal to man, yet its sale annually amounts to tens of thousands of taels of silver; this is because in Shensi it is mixed with wheat and legume seeds before planting. Furthermore, it is used against weasels, and people in Chekiang sprinkle it at the roots of young rice shoots to insure a good harvest. Otherwise, how much can the demand for this commodity amount to, if it is used only in gunpowder and the dyeing of copper [to a whitish color]?

NOTES

1. The five elements in old Chinese cosmology were metal, wood, water, fire, and earth.

2. The "white lime surfaces" found in the ruins of Hou-kang during the Anyang excavations indicate that lime and gypsum plaster may have been known to the Chinese of the Yang-shao civilization (ca. 2700 B.C.). *K'ao-kung chi,* a section of the *Rites of Chou,* states that shell lime was used to ornament a temple of the Hsia dynasty (2197–1766 B.C.).

3. The author was apparently not aware that the oyster is a living creature.

4. Historical evidence points to the use of coal as a fuel during the late Han period (A.D. 25–220), or earlier. For example, several hundred thousand catties of coal were stored by Ts'ao Ts'ao, and the term *shih-t'an* (stone charcoal) appeared for the first time in the *History of Later Han*. In the Sung dynasty (960–1279), the coal tax was a source of important national revenue, and the coal industry maintained many people. The "Book on Money and Commodities" in the *History of Sung* notes more than twenty government-operated coal depots, and relates that the marketing of coal was a government monoply during the reign of Ch'ung-ning (1102–10). The extensive use of coal in China during the Yüan dynasty (1279–1368) is described in *The Travels of Marco Polo* and later in the *Journals of Matthew Ricci:* "Nature is more propitious to the man in the north, where this coal is found in plenty and of better quality. It is mined from the earth and widely distributed throughout the country at a low price, indicating its abundance and making it possible for even the poorest to use it in their kitchens and in heating the bath."

5. The implication is that coal is used universally in North China because of widespread deforestation.

6. The technique of making white alum was definitely known to the Chinese in the sixth century A.D. In the T'ang dynasty (618–906), the tax collected from the manufacture of white alum and vitriol at Yang-ping-hsien of Shansi province was considered a regular national revenue. In the reign of Hui-tsung (1101–25), the alum and vitriol industry was tightly controlled by the government. This control resulted in an annual revenue of as much as 290 million copper cash.

7. Black vitriol is a term used in Chinese to denote ferrous sulfate, now known as green vitriol. This Chinese misnomer may be caused by the fact that, from ancient times to the present day, green vitriol has been used in China chiefly for blacking textile fabrics and leather, though it has also been used in medicine, as a disinfectant, and as a wood preservative. When green vitriol is used in black coloring materials, tannin is added. Tannin reacts with ferrous sulfate to form the ferrous salt of tannic acid. This salt is oxidized in air to the black ferric tannate. The tanning agents used by the early Chinese dyers were extracted from "wu-pei-tzu" (Chinese gall or gall nut), *"tsao-chiao"* (Gleditschia Chinensis), *"Kan-kuo"* (valonia), myrobalon bark, sumach bark, birch bark, oak bark, elm bark, and many other tannin-containing materials.

8. Red vitriol was probably red ochre, also called hematite though it is an amorphous earthy variety of hematite. The reddish colors on the Yang-shao pottery (ca. 2700 B.C.) excavated from the ruins at Honan and Kansu provinces came from natural red ochre or red ochreous clays, or a mixture of both. It is no exaggeration to say that red ochre was the most commonly used red pigment of ancient China.

9. Yellow vitriol was probably fibroferrite, $Fe_2O_3.2SO_3.10H_2O$, and utahite, $3Fe_2O_3.3SO_3.4H_2O$.

10. Huai-pei, or the north side of the Huai River, is a region of East China where the soil is highly alkaline.

11. The third century book *Po-wu chih* (*A Record of Nature*) relates that, in the third year of Hunag-ch'u (A.D. 222), Wang Pao presented thirty catties of gall vitriol to emperor Wen-ti. This indicates that the manufacture of gall vitriol was still in its infant stage and the high-grade product was reserved for imperial use. In *Pao-p'u-tzu* (ca. early fourth century) it is stated that verdigris, known in China as copper blue or copper green, was manufactured by applying vinegar to the surface of copper, and was used as a wood preservative. The Chinese production of gall vitriol was continually increased in the succeeding Liang and T'ang dynasties.

12. *Materia Medica,* the comprehensive botanical and mineralogical encyclopedia, known in China as *Pen-ts'ao kang mu,* was written by Li Shih-chen in A.D. 1596.

13. Western Regions is a general term used traditionally to indicate the countries to the west of China—from modern Sinkiang westward to central Asia.

14. Sulphur was used by Ma Chün as early as the Three Kingdoms period (220–64) in making gunpowder for firecrackers.

15. A term usually designating the territories north of the Great Wall.

Figure 12–1. Roasting (right) and steaming (left) oil seeds.

12

VEGETABLE OILS AND FATS

Master Sung observes that although Nature has divided time into day and night, man is able to prolong the day artificially in order to carry out his tasks. The reason is not due to his being fond of toil or disliking leisure. If, for example, the weaver must burn wood for light and the student rely on the glow of snow to read by, how little work could be accomplished in this world!

Stored in the seeds of grasses and trees there is oil which, however, does not flow by itself, but needs the aid of the forces of water and fire and the pressure of wooden and stone [utensils] before it comes pouring out in liquid form. [Obtaining the hidden oil] is an ingenuity of man that is impossible to measure.

For the transportation of goods, and travel to distant places, men must depend on boats and carts. One drop of oil [in the axle] enables a cart to roll and one *tan* of oil used in caulking a ship [1] makes it ready for the voyage. Thus, neither cart nor boat can move without oil. Furthermore, cooking vegetables without oil is like letting a crying infant go without milk. The uses of oil are indeed varied and numerous.

GRADATION OF VEGETABLE OILS

For eating, the oils of sesame seeds, turnip seeds, yellow soy beans, and cabbage (also called "white cabbage" [i.e. celery cabbage]) seeds are the best. Next in quality come *Perilla ocymoides* oil ([the plant] resembles *Perilla nankinenis;* the seed is larger than that of sesame) and rape-seed oil (in the

South it is called "vegetable seed"); next, camellia or tea-seed oil [2] (this tree is over ten feet high, the seed resembling poppy seed, which should be shelled [and the oil is] obtained from the kernel); next, *Amarantus mangostanus* oil; the last in quality is hemp-seed oil (the seed resembles caraway seeds; the skin [of the hemp plant] is used to make rope).

The best lamp oil is that made from the kernels of the vegetable tallow-tree seeds, followed by rape-seed oil, linseed oil (planted in Shensi province, this plant is colloquially called "louse sesame"; the oil has a bad odor and is inedible), cotton-seed oil, and sesame oil (the latter burns off too quickly in the lamp), while Chinese-wood oil [i.e. *t'ung* oil] and "mixed" vegetable tallow-seed oil are the worst. (The poisonous odor of the Chinese-wood oil is sickening, and the "mixed" vegetable tallow-seed oil, prepared from both the tallow hulls of seeds and the seeds themselves of the tallow tree, tends to solidify and appear turbid.)

For making candles, the best material is the "unmixed" vegetable-tallow fat [derived solely from the white tallow layer on the outside surface of the tallow or *Stillingia sebifera* tree seeds]; next, castor oil; next, "mixed" vegetable tallow-seed oil solidified by the addition of [a certain amount of] white wax;[3] next, the various clear oils solidified with the admixture of white wax; next, camphor-seed oil [4] (it gives no less light than the others, but some people are allergic to its scent); next, holly-seed oil [5] (used exclusively in Shao-chou [in Kwangtung]; the latter is rated low on this list because the oil yield of the holly seeds is small); the lowest in quality is beef fat, which is used widely in North China.

The yield of oil per *tan* of seeds is: sesame, castor, and camphor, 40 catties; turnip, 27 catties (this oil is particularly palatable, and is good for the internal organs); rape, [usually] 30 catties but 40 catties can be realized if the crop is planted in a rich field, well cultivated, and if the pressing is skillfully done (after one year's storage the seeds will be completely dry and oilless). The yield of tea or Camellia seeds is 15 catties of oil per *tan* (the oil tastes like lard and is quite delicious; the residual meal can be used only as kindling or fish poison); [whereas] one *tan* of *t'ung* seeds yields 33 catties of wood oil. When the tallow hulls and kernels of the Chinese vegetable-tallow tree seeds are pressed separately, the yield per *tan* of seeds is 20 catties of "unmixed" tallow fat and 15 catties of kernel oil; but if the seeds are pressed without such a separation, then the yield is 33 catties of "mixed" tallow-seed oil. (These amounts of tallow fat and oils are obtainable only when the tallow hulls or seeds are thoroughly pressed to result in the "cleanest" cakes or meal.) From each *tan* of holly seeds, 12 catties of oil are obtained; from yellow soy beans, 9 catties (in Kiangsu the bean oil is used as food for humans, and the meal cakes are fed to pigs); from cabbage seeds, 30 catties (the oil appears limpid as green water). One hundred catties of cotton seeds yield 7 catties of oil (the oil is dark and unclear when first pressed, but becomes extremely clear after standing for half a month). Thirty catties of oil are obtained from one *tan* of *Amarantus* seeds (it tastes delicious, but is rather cold and laxative in nature), and more than 20 catties from one *tan* of linseeds or hemp seeds.

The above is a brief summary [of the well known oils]. I have not touched on other kinds,[6] of which the properties have not yet been completely tested, or have only been tried out locally and are not generally known.

METHODS AND IMPLEMENTS [FOR OIL-EXTRACTION]

Aside from pressing, oils can be extracted from seeds by other processes. Castor oil and *Perilla ocymoides* oil are prepared by the boiling process. Sesame oil is produced in Peking by grinding the seeds, and in Korea by using mortar and pestle in a pounding process. The rest [of the oils], however, are obtained by the pressing process.

The large timber used for making an oil press [Figure 12.2] must be an armful in diameter[7] and hollowed in the center. Camphor wood is best suited for this purpose, followed by sandalwood and alder wood. (A press made of alder wood will rot rapidly if the moisture from the ground is not eliminated.) The grains of these three kinds of wood are circular and spiral, and do not run up and down vertically. The wood, therefore, will not split at the ends even though the center post is subjected to heavy hammering with a pointed wedge. The press should not be made of wood that has longitudinal grains. North of

Figure 12–2. Press for making vegetable oil in south China.

the Yangtse River, where timber of such large dimension is seldom obtained, four pieces of wood can be joined together by iron hoops. The center portion is then hollowed out to serve as the pressing area; thus small wood is made to function as large timber.

The size of the hollow in the press varies with the dimension of the wood. Its capacity ranges from less than five pecks to over one *tan*. This hollow is fashioned by scooping out the wood with a curved chisel, so that the hollowed space becomes a flat trench with rounded ends. On the bottom of the trench a small hole is drilled [through the wood], and then connected to a trough [or pipe], thus permitting the flow of oil from the trench into a receptacle. The flat trench is about three or four *ch'ih* long and three or four inches wide, depending on the general shape of the wood. There are no standard dimensions for it. The pressing wedge and ram are made of sandalwood or oak, for no other wood is suitable. The wedge is cut and shaped with an axe but is not planed, since it is desirable that the surface should be rough; a smooth-surfaced wedge may [glance off and] swing backward [after the impact of the blow]. The ramming wood and the wedge are both hooped with iron rings to prevent their breaking.

The press is now set up. Next, materials such as sesame seeds or rape seeds are roasted slowly in a pan over moderate heat (tree-grown seeds, such as those of the Chinese vegetable-tallow tree and the *t'ung* tree, are rolled and steamed without preliminary roasting) until a fragrant scent is given off. Then they are rolled into fine fragments after which they are steamed. For roasting sesame and rape seeds, a flat-bottomed pan of no more than six inches depth should be used. The seeds in the pan must be stirred and turned rapidly. If the pan is too deep or the stirring too slow, the seeds will be parched and the result will be a reduced oil yield. The roasting pan is fixed on the stove at an angle very different from the steaming pot [Figure 12.1]. The trough of a [hand-operated] rolling mill is anchored by burying its bottom in the ground. (The surface of a trough made from wood has to be covered with iron sheets.) Over the trough is an iron roller suspended from a wooden pole and pushed [along the trough] to and fro by two workers standing facing each other. A man who has large capital can build an ox-pulled rolling mill with stones, the work of one ox being equal to that of ten men. Some seeds, such as cotton seeds and the like, are ground [i.e. using two millstones] instead of rolled.

The rolled seeds are screened, the coarser fragments being returned for a second rolling. The fine meal is placed in a double boiler and steamed. When it has been steamed sufficiently, it is taken out and wrapped in rice or wheat stalks to form cake-shaped parcels, which are tied either with iron hoops or split rattan cords. The size of the parcels should fit the hollow of the press.

Since oil is derived from the vapor [of steamed meal] and is a substance that is created from the void, inefficient handling of the steamed meal will allow the hot vapor to evaporate, thus diminishing the yield of oil. The skillful worker therefore always pours the meal out quickly, wraps it quickly, and just as quickly ties it up. This is the secret of obtaining a high yield of oil from seeds, but there are pressers who have worked a lifetime at the job and still are ignorant of this fact. The wrapped and tied meal parcels are then put into the press until it is packed full. The ram is applied and the wedge is inserted and struck, forcing out the oil like spring water. The remnants left in the parcels are called pressed cakes. The pressed cakes of sesame, turnip, or rape seeds are all rolled into a fine powder, and then screened to remove the stalks and husks. This cake powder is further treated by steaming, wrapping, and pressing [for a secondary recovery of oil]. This secondary pressing will yield half as much oil as that obtained from the primary pressing. In the case of Chinese vegetable-tallow and *t'ung* seeds, however, the oil is entirely squeezed out by the primary pressing, therefore the pressed cakes need not be re-treated.

The boiling process [for extracting oil from seeds] requires the use of two pots. The castor seeds or *Perilla ocymoides* seeds are rolled into meal, put into a pot, and boiled in water. When foamy oil rises to the top, it is skimmed and poured into the other pot, which is dry, and placed over a low fire to evaporate the moisture [contained in the foamy oil], thus resulting in water-free oil. However, the yield of oil from this method is low [as compared with that of the pressing method].

In North China, sesame oil is made by a grinding process that involves squeezing and twisting the ground meal in a coarse hemp sack. The details of this method shall be described at a later time.

VEGETABLE TALLOW FAT

The making of vegetable-tallow candles originated in Kuang-hsin prefecture [in Kiangsi]. [Vegetable tallow is prepared] by first steaming the tallow-tree seeds together with their hulls in a double boiler, and then pounding them with mortar and [foot-operated] pestle.[8] The mortar is about 1.5 *ch'ih* deep, and the pestle is made of stone, with no iron cap. The stone used is a fine-textured variety obtained from deep in the mountains [in Kuang-hsin], and the pestle being made to measure, weighing forty catties, is attached to a wooden lever for pounding [Figure 12.3]. The tallow hulls adhering to the seed-husks fall off during the process of pounding and are collected in a tray. These hulls are again steamed, wrapped, and pressed [to produce the "unmixed" vegetable tallow fat] as described above.

When the tallow hulls have been entirely removed, the black seeds will emerge. A small grinding mill, which is made of fine smooth stone that cannot be damaged by fire (this mill stone is also obtained from the mountains in Kuang-hsin prefecture), is heated by surrounding it with red-hot coals, while the black tallow seeds are fed into it and ground with quick motions [Figure 12.3]. When the husks or shells of the seeds are broken, they are winnowed

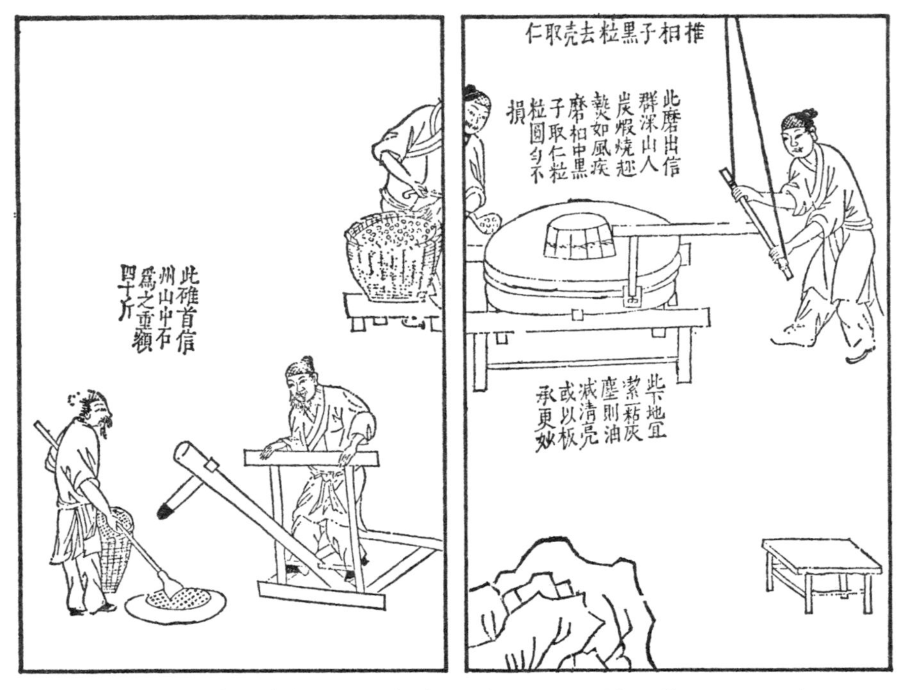

Figure 12–3. Pounding (left) and grinding (right) vegetable tallow tree seeds.

off, leaving behind white kernels that closely resemble the *t'ung* tree seeds. These kernels are then rolled, steamed, wrapped, and pressed (to produce the vegetable-tallow tree kernel oil) as described previously. The liquid oil [known in China as *ch'ing yu*] thus obtained is sparkling clean. If some of it is placed in a small dish and a grass wick is put in it and lighted, it will burn until dawn. No other oil can compare with such performance. It is usually not used in cooking, for although it is not harmful, some persons may be allergic to it. Hence it is better not to use it.

To make vegetable-tallow candles, a piece of bitter bamboo [9] is split vertically into two halves which are boiled in water until they are swollen. (Otherwise, the molten vegetable tallow will stick to the bamboo.) These two halves of bamboo are then held together with rattan strips to form a pipe, into which the molten tallow is poured from a pointed iron ladle. A wick is inserted, and when the tallow is solidified the bamboo pipe is unbound and removed and the candle is lifted out. Another method of making candles consists of forming the mold by wrapping a piece of paper around a wooden stick, thus creating a paper cylinder. The molten vegetable tallow is poured into it to form a candle. This kind of candle is highly durable, not deteriorating in quality through long storage or exposure to dusty wind.

NOTES

1. The use of *t'ung* oil and lime for caulking boats in the Yuan dynasty (1270–1368) and possibly at a much earlier date is mentioned in the *Travels of Marco Polo.*

2. Tea-seed oil is produced from the seeds of *Thea sasanqua,* which has long been cultivated in China for its seed, and is to be distinguished from *Thea sinensis,* which is grown for its leaves.

3. White (Chinese-insect) wax is made from the secretion of a small, winged insect, *Coccus pela* or *Coccus sinensis,* found chiefly in western China, where the insect is raised for the wax. Watson's translation of the description of Chinese-insect wax in *Materia Medica* is reproduced with slight modifications as follows:

> The insect is about the size of a louse, and after it has been propagated it remains upon the green branches of the tree, eating the sap and giving off from its body a secretion which adheres to the fresh stalks, this gradually changes into a white cere, which congeals to form the wax, appearing like frost upon the branches. The crude wax is melted and purified or steamed in a retort, in order to get rid of the impurities, and is then poured into moulds to cool. This forms the white wax of commerce. The insects produce the wax while they are young and of white color. When they are old, they are reddish black in color, and form balls upon the branches of the tree, at first of the size of a grain of millet, but in the second spring they grow to the size of a cock's head, are purplish red in color, and closely encircle the branches, appearing as if fruits borne upon the tree. The insect deposits its eggs, making a cell that much resembles a chrysalis, which is called "wax seed" (*la-chung*) or "wax egg" (*la-tzu*). The eggs within this cocoon are like small silk-worm eggs, and there are several hundreds to a cocoon. At the beginning of spring they are taken down, wrapped in bamboo leaves, and hung upon the tree. After hatching out the insects will adhere to the under side of the leaves and the other parts of the plant, where they begin the manufacture of wax. The ground beneath the tree must be kept very clean lest the ants eat the young insects.

4. Camphor-seed oil is obtained from the seeds of the *Cinnamomum camphora* tree belonging to the Lauraceae family.

5. Holly-seed oil is obtained from the seeds of the *Ilex pendunculosa* plant.

6. Master Sung neglects the renowned Chinese lacquer or *ch'i,* which was prepared from the sap of the varnish tree (*Rhus verniciflua*) by a traditional process that remained a carefully guarded secret until the middle of the eighteenth century.

7. "An armful in diameter" is from the expression *ho-pao,* meaning a circumference of the size made by one's two arms. In other words, the wood needed here would be a log with a diameter of approximately two feet.

8. Vegetable tallow is prepared from the hull of the seed of *Stillingia sebifera,* a tree found wild and also cultivated in central and eastern China. The seeds are approximately as large as coffee beans, black in color, and covered with a fairly hard layer of white tallow.

9. Bitter bamboo (*k'u-chu*) is a variety of large bamboo, *Phyllostachys bambusoides,* usually 50 or 60 *ch'ih* in height.

Figure 13–1. Steeping and washing cut bamboo.

13

PAPER

What, queries Master Sung, is the carrier by which knowledge of Nature's wonders and the mysteries of the Universe is transferred from ancients to moderns, and from Chinese to foreigners, so that those born in later times can learn it at a glance? If all communication between ruler and subject, teacher and pupil, were by word of mouth, could much be achieved? Yet with the use of a slip of paper or a slim volume of writing, teaching can be accomplished and [government] orders carried out, as easily as the breeze blowing or ice melting in the sun. It is indeed fortunate that in this world exists Old Sir Paper, from which both the sages and the ignorant have benefited.

Paper consists of fibres and tree barks. White color emerges as the "green" is killed, and thus was taken the first step toward [the existence of] myriads of books and hundreds of specialists and schools [of thought]. This is the refined use of paper. As to its ruder uses, such as shielding against the wind or protective wrapping of things, they were initiated in antiquity. How contemptible it is that people in the times of Han and Tsin [206 B.C. to 419 A.D.] should arbitrarily record their names [as the inventors of paper].[1]

RAW MATERIALS

Of the different kinds of paper, bark paper is made from the bark of the paper-mulberry tree [*Broussonetia papyrifera*] (also called "grain tree"),[2] silk-mulberry fibre, or hibiscus [*Hibiscus mutabilis*] skin. Bamboo paper is made from bamboo fibres. The best quality of this paper is extremely clean and white and is used for writing, printing, and as letter paper. The coarser grades are used to make articles for burnt offerings or wrapping paper.

The [ancient] *sha-ch'ing* or "killing the green" refers to chopping down the bamboo plants; while *han-ch'ing* or "sweating the green," to cooking and straining [the bamboo fibres]. The word "board" means the finished paper. "Boards" therefore were made by cooking the bamboo. In later ages, however, people mistakenly thought that the "boards" meant pieces of bamboo slats and arrived at the erroneous conclusion that the term "leather braiding" was a reference to stringing the bamboo boards together with strips of leather. In view of the fact that a great number of books were in existence in China before the burning of books was ordered by Ch'in [i.e. The First Emperor of Ch'in], would it have been possible to produce so many books if they had had to be made up with pieces of bamboo slats? [3]

Some foreigners in the west make paper out of palm leaves, but the Chinese, forgetting that leaves will wither after they are plucked from the tree, again took it to mean that it was on the leaves themselves that the scriptures were written.[4] This notion is as ridiculous as the idea of books made of bamboo slats!

THE MANUFACTURE OF BAMBOO PAPER

Bamboo paper is an industry of south China, especially in Fukien province. The topography of the mountain area having been surveyed after the bamboo shoots have started to grow, the young plants are cut down in early summer. The best in quality are those shoots that are about to put forth branches and leaves.[5] The bamboos are cut into lengths measuring about five or six *ch'ih*. A pit is dug right there in the mountain and filled with water in which the bamboo sticks are immersed. The water supply is constantly maintained by means of bamboo pipes so that the pit will not run dry. After soaking for more than one hundred days [Figure 13.1], the bamboos are carefully pounded and washed to remove the coarse husk and green bark or node (this is *sha-ch'ing* or "killing the green").

The mass of the pounded bamboo, having a hemp-like appearance, is mixed with a high-grade milk of lime and put into a cooking cask to be boiled over a fire for eight days and nights. In the boiling of bamboo, [the wooden cask is held on top of a metal] pot, which in turn is attached with the aid of mud and lime to a cone-shaped stove [Figure 13.2]. This pot, measuring fifteen *ch'ih* in circumference and over four *ch'ih* in diameter, and having a capacity of more than ten *tan* of water, is similar in dimensions to the salt-making pot of Kwangtung province. When the cooking cask has been placed on the pot, the boiling process begins. The bamboo pulp is cooked for eight days, and then the fire is stopped and the pulp allowed to cool for one day.

After cooling, the cask is removed and the bamboo fibres are taken out of the cask to be thoroughly washed in clear water. This is done in a washing pit of which the bottom and four sides are lined securely with wooden boards in order to keep out the dirt (this lining is not necessary in making coarse paper). When the fibres are washed clean, they are soaked in a solution of

Figure 13–2. Cooking the inner mass of bamboo in a pot.

Figure 13–3. Collecting bamboo pulp on top of a screen.

wood ash and transferred to [the cask of] a pot. The fibrous mass is pressed to form a flat top, covered with about an inch of rice stalk ash, and finally cooked by heating the water in the pot until the water in the cask is boiling. After this, it is transferred to another cask and strained with the passage of a hot solution of ash from top to bottom. [This is followed by washing it with a passage of hot water]; if the water cools off, it is reboiled to repeat the washing.[6] After some ten days [of repeated cooking, straining and washing], the bamboo pulp will naturally become odorous and decayed. It is then taken out to be pounded in a mortar (in hilly regions water-powered pestles are usually employed) until it has the appearance of kneaded dough.

[This pulp] is poured into a square-shaped tank [Figure 13.3]. The exact dimensions of a pulp tank, however, are governed by the size of the papermaking screen being used, which in turn is determined by the size of the paper sheets to be made. When the clear supernatant liquid in the pulp tank is three inches deep [as caused by the gradual settling of the fibrous mass], a chemical solution is added. (The material used for the preparation of this chemical solution has the form of the leaves of the "peach-bamboo," and is called locally by different names.) This chemical solution is capable of bleaching [the paper sheets] to a white color after drying.

The screen used for the manufacture of paper sheets is made of a rectangular frame covered with a mat woven of finely split and polished bamboo.[7] The worker holds the screen with both hands and submerges it in the fibre suspension of the pulp tank so that some of the latter remains on top of the screen. The thickness [of the paper] will depend on the way in which the screen is manipulated: shallow submerging results in thin paper, while a deeper dip produces a thick one. Water from the pulp will drain off around the screen's edges and back into the tank as the bamboo material is left on top. The screen is then inverted and the paper is dropped onto a wooden board until many such sheets have been piled together. When the number is sufficient, the sheets are covered with another board and [the upper and lower boards] are rope tied with the aid of a pole placed over the top board, as in a wine press, and all the water is squeezed from the sheets [Figure 13.4]. When this is done, each sheet is lifted up by means of a small brass tweezer and dried with heat.

For drying paper [Figure 13.5], a double wall of earthen bricks is erected, with the ground between the two rows covered by bricks. Holes in the bottom of the wall are left by the spaced omission of bricks. A fire is lighted at the first hole and the heat travels through the apertures and spreads to the wall surfaces where the bricks will become hot. The wet sheets of paper are spread onto the wall one by one, baked dry, and then taken off as finished sheets. In recent times the large-size sheets have become known as "Large Fourfold," which is highly prized as a writing paper.

After the ink and colors of waste paper are washed off, the paper is reduced to pulp by soaking in water and transferred to the pulp tank for making new paper. This process of papermaking not only eliminates the expensive operations of cooking and straining [the bamboo or wood], but also loses very

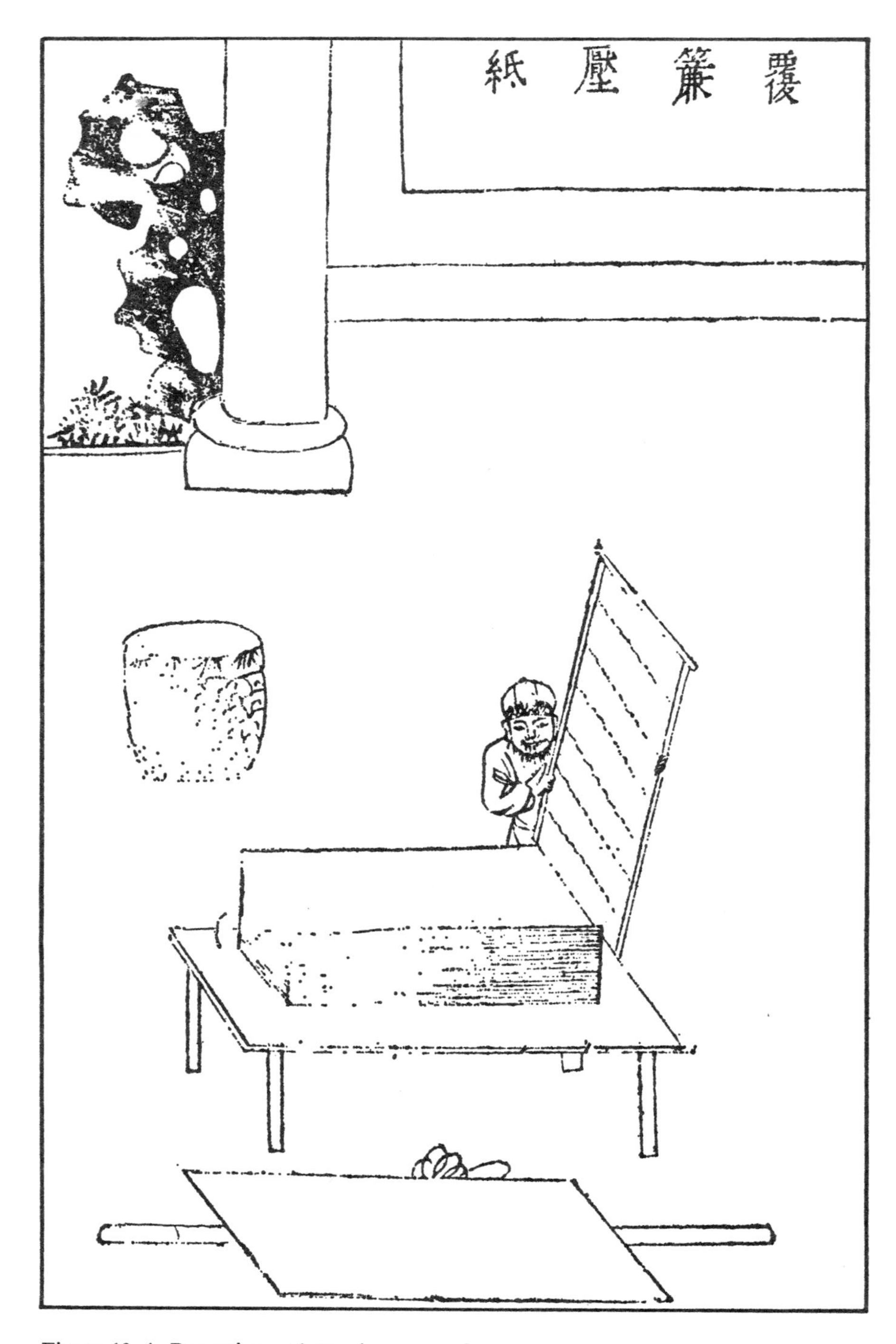

Figure 13–4. Removing and pressing paper sheets.

Figure 13–5. Drying paper sheets.

little of the waste paper so used. This procedure is not popular in south China where bamboo is abundant and cheap. In the north, however, even small strips or odd pieces [of waste paper] are saved for reprocessing, the product being known as "resurrected" paper. There is little difference whether the waste paper is made from bark or bamboo or is fine or coarse in quality.

In the manufacture of burnt-offering paper and coarse paper, the procedures of cutting the bamboo, cooking and straining the fibres with ash solutions, and washing the fibres with water, are the same as described above. These kinds of paper, however, are not baked dry after the sheets have been taken off the screen, but are first pressed to remove the excess water and then dried by the sun.

"Burnt-offering" paper is so called because during the T'ang dynasty [8] large amounts of sacrificial paper money were used instead of "burnt" silk for religious worship (in north China people make such money by cutting paper into strips; this is called "cut money"), and this particular kind of paper is given the present name. In accordance with recent customs in Hupei and Hunan, as much as 1,000 catties [of sacrificial money] are wastefully burned at a single occasion. [In reality] about seventy per cent of this paper [produced in the country] is used for burnt offerings, and thirty per cent for daily use. The coarsest and thickest variety of paper, called wrapping paper, is made of a mixture of bamboo fibres and rice stalks.

The stationery paper manufactured in such places as Ch'ien-shan in [Kiangsi] is made entirely of fine bamboo fibres and screened from a thick pulp. This paper fetches a high price on the market. The highest grade, known as "official stationery," is used as letter paper by high officials and wealthy persons. It is thick and smooth and without any fibrous ribs. When used for red ceremonial cards, it is first treated with alum water and then dyed red with safflower juice.

THE MANUFACTURE OF BARK PAPER

Paper-mulberry [*Broussonetia papyrifera*] bark is stripped from the tree in late spring and early summer.[9] Trees that are too old are felled at the root level and [the stump] covered with earth. In the following year new shoots will appear, of which the bark is of an even more excellent quality.

In making bark paper, forty catties of the most tender bamboo are added to sixty catties of paper-mulberry bark. Together, these are first steeped in a water pit, covered with lime juice, and finally boiled in a pot to become pulp. In recent times, a more economical method consists of mixing thirty per cent rice stalks with seventy per cent bark and bamboo. After suitable chemicals are added, however, [the resultant paper] can still be white.

Torn lengthwise, the strong and hard-sized bark paper will show ragged ends resembling cotton fibres, hence it is called "cotton paper." It is extremely difficult to tear this paper crosswise. The highest grade is used to paper the windows of the Imperial Palaces and is called "window gauze" paper. Produced in Kuang-hsin prefecture [in Kiangsi], each sheet is over seven *ch'ih* long and more than four *ch'ih* wide. It is dyed in various colors by the addition of coloring material to the pulp tank. Dyeing is not done after the sheets have been formed.

The second in quality is the "fourfold" paper, of which the whitest variety is called "superior" paper. That product which is bark paper in name, but is actually made from a mixture [of bark,] bamboo, and rice stalk, is called "document" paper. Another kind, made from the skin of hibiscus stems, is termed "small bark" paper, but in Kiangsi province it is known as "middle" paper. It is not known what grass or wood is used as raw material for the kind of paper manufactured in Honan. This paper is produced in large quantities and has a market in Peking. Silk-mulberry fibre paper, made from the bark of the silk-mulberry tree, is extremely thick, smooth and soft; the type produced in Chekiang is necessary to the silk producers in the lower Yangtse region as a repository of silkworm eggs. "Small bark" paper is used for the manufacture of [oil-paper] umbrellas and oil-paper fans.

The very long and wide bark paper is formed in an extremely large pulp tank. The screen used here is too big for one person to handle and is worked by two persons standing opposite each other. In the case of "window gauze" paper, several persons are needed. When used for painting, bark paper is first sized with alum water to eliminate fluff from the surface. That side of the sheet which adheres to the screen is considered the right side as the texture of the free side is inevitably rougher even though the pulp has been reduced to a state of subdivision as fine as clay.

It is not known what the "white hammered paper" of Korea is made of. In Japan, paper is sometimes made without use of the screen. [The method is as follows:] when the paper pulp is ready, a large slab of slate is placed on a brick platform below which a strong fire is burning to heat the stone. The worker then dips a plasterer's brush into the pulp and brushes the pulp thinly over the stone. In a short time, it dries and becomes a sheet of paper which can be easily lifted. I do not know if the same method is employed in Korea or China.

The paste-impregnated paper of Yung-chia [in Chekiang] is also made of silk-mulberry fibres while the Hsüeh T'ao note-paper of Szechuan is made from hibiscus skin. [For making the latter], the bark of hibiscus is cooked to a pulp, then the aqueous extract of powdered hibiscus flower petals is added. This process was probably first devised by Hsüeh T'ao [10] and has been known by that name down to the present day. This paper is famous for its beauty, not for the quality of its material.

NOTES

1. Chinese records, such as *Shuo-wen* (A.D. 100), *History of Former Han* (A.D. 100), and *History of Later Han* (A.D. 450), show clearly that a sort of paper or near-paper made of raw silk was used in China since the later part of the Former Han dynasty (206 B.C.–24 A.D.). In 1942, while working at a Han ruin in Tsakhortei, south of the Bayan Bogdo Mountains in modern Ninghsia province, Lao Kan and Shih Chang-ju discovered what is probably the oldest paper in the world. From the dated inscriptions on wood, found by the Swedish archaeologist Folke Bergman in 1931 in the same Han ruin, Lao Kan deduces that this piece of paper was made around A.D. 98. Other pieces of early paper discovered in Turkestan date from A.D. 250 to 300.

2. "Grain tree" or *ku-shu* is actually different from the paper-mulberry tree. In early times, however, these two were spoken of as one.

3. "Boards" and "leather braiding" were actually employed prior to the invention of paper. To the end of Chou dynasty (256 B.C.), writing was managed with a bamboo pen and ink of soot or lampblack on slips of bamboo or wood. The bamboo strips, being stronger, could be perforated at one end and strung together with either silken cords or

leather thongs to form books. The older sites excavated by Sven Hedin, Aurel Stein, and the Prussian expeditions in Chinese Turkestan yield both wooden and paper documents. At several places the gradual displacement of wood by paper in the third, fourth, and fifth centuries A.D. can be traced.

4. Palm leaves refer to the *pattra* of India, where people actually did write the Buddhist sutras on leaves. G. Sarton relates in his *History of Science* (p. 25) that palm leaves were used in Ceylon and India for writing. From the leaves of talipot (*Corypha umbraculifera*), which grows in Ceylon and Malabar, was produced a kind of papyrus, in narrow strips, called olla.

5. While not as long as spruce fibres, bamboo fibres are much longer than those from any of the deciduous trees. Bamboos run to the hundreds though not all are suited for paper making due to difficulties in pulping and bleaching. These difficulties are caused chiefly by the high silica content of certain bamboo stems.

6. This technique is still used by the modern paper-pulp industry, except that plant ash has been displaced by manufactured soda ash. Washing removes possible alkali and dissolves intercellular matter. Incomplete washing renders subsequent bleaching difficult and impairs the quality of the pulp.

7. *Tung-t'ien ch'ing-lu,* written by Chao Hsi-huo in the Sung dynasty, relates that in north China the screen (matting) was transversely laid, causing the northern paper to show a horizontal grain, while in south China the matting was vertically laid, so that southern paper showed a longitudinal grain. It is not known whether the same distinctions existed in the Ming period.

8. In the T'ang dynasty, Buddhism enjoyed great popularity and influence, and the country was dotted with monasteries and temples.

9. Paper-mulberry bark is used for making the fine paper called *Hsüan-chih.* It derives its name from being produced in Hsüan-chou of Anhui. In the Yung-hui era (A.D. 650–56) of the T'ang dynasty, a monk of that region used this kind of paper for writing Buddhist books. It can also be made from the *Dalbergia Hupeana* tree.

10. Hsüeh T'ao was a famous courtesan of the Late T'ang dynasty who was known for her calligraphy and poetry.

PART III

14

THE METALS

Master Sung observes that men are divided into ten classes and that every class, from prince to sedan chair bearer, is necessary to the social order. Similarly, the earth has produced the five metals [1] for the benefit of all the people and their posterity. [Of the metals] the precious are found once in 1,000 *li,* or 500 *li* at the closest, while the less valuable are always found in large quantities in places that are not easily accessible. The value of high grade gold is 16,000 times higher than that of black iron. But suppose that such [iron] articles as pots, food vessels, hatchets, and axes are not available for daily use, then even though gold is obtained and its price is high, there would not be people [to take advantage of it]. The livelihood of all depends on the trading of that which the people have for that which they have not, as provided in the chapter on goods and exchange in the *Rites of Chou.* Whoever set up the standards for differentiation of the good [metal] from the bad, and for their weights and measures, founded a system that will last forever.[2]

GOLD

Gold ranks highest among all the metals and will never change once smelted and cast into shape. In the case of silver, although none of it is lost in the smelting process, it nevertheless requires that the final blow of the bellows be struck at exactly the right moment, when the silver will flash its gleam once; thereafter it will not appear again [on the molten surface] despite continued use of the bellows. Gold, however, will continually flash its gleam with repeated blowing of the bellows. The stronger the fire, the more the gleam emerges. This accounts for its precious quality.

In China, gold is produced in more than one hundred places, which we can not enumerate here. Of the kind produced in mountains, the large pieces are called "horse-shoe gold," the medium "olive" or "pendant gold," and the small "melon-seed gold." Of the gold obtained from water and sand, the large pieces are called "dog-head gold," [3] and the small "wheat-husk gold" or "chaff gold." The gold obtained by underground mining is called "sand-powder gold," but the larger size is known as "bean gold." All these varieties of run-of-mine gold need to be washed and smelted before they can be made into nuggets.

Most gold is produced in the southwest. The miners dig tunnels of some dozen *chang* into the hillside. Gold is usually found by the detection of its particular gangue, which is brown in color, one end having a charred appearance. Placer gold is found mostly in the Gold Sand River (called the Li River in ancient times) in Yunnan. Having its source at T'u-fan [in modern Ch'ing-hai], this river flows through Li-chiang Prefecture [in northern Yunnan] and reaches Pei-sheng district after a meandering course of more than 500 *li*. Gold is taken from several parts of it. In addition, gold is also panned from the river sand in such places as T'ung-ch'uan in northern Szechuan and Yuan-ling and Hsü-p'u in Hukuang province. A piece of "dog-head gold," which is called "mother gold," is obtained perhaps once in a hundred or a thousand cases, and the rest [of the gold pannings] all look like wheat chaff. When smelted, the crude gold bullion will turn light yellow and will become reddish-golden after it is refined.[4] A variety is produced at the "gold fields" at Tan-yai [in northwestern Hainan Island], where it is mixed with sand and earth and is obtained without deep digging. Too frequent mining, however, will exhaust the stock, and the yield is likely to be limited when a place has been worked over a period of years. In the caves of the southern barbarians,[5] a kind of gold is produced that looks like black iron when first mined. It is obtained from under dark, carbonaceous stones in pits several dozen feet deep and at first is soft to the bite. Sometimes the miners steal this gold by swallowing it, which leaves no ill effects. In such places as Ts'ai and Kung districts in Honan, and Lo-p'ing and Hsin-chien in Kiangsi, gold is obtained through washing and smelting fine sand that has been mined from deep shafts dug into the ground. The returns for all this labor, however, are small. Gold deposits in China are generally spaced every 1,000 *li*. It is recorded in the *Ling-piao lu* [6] that the people [in Kwangtung] washed goose and duck droppings for gold flakes and bits, some getting one ounce in a day and others none. This is probably not true.

Gold is a heavy metal. If a square inch of copper weighs one ounce, a piece of silver of the same dimensions will weigh 1.3 [*sic*] ounces; and if a square inch of silver weighs one ounce a piece of gold of the same dimensions will weigh 1.2 [*sic*] ounces. Gold is also soft and can be bent like a willow twig. The grades of gold are classified in terms of "seventy per cent green," "eighty per cent yellow," " ninety per cent purple," and "pure red" gold, which can be easily determined by placing [the gold] upon the "gold-testing stone" [touchstone]. (This stone is found in abundance in the river at Kuang-hsin

Prefecture [in Kiangsi], ranging from the size of a bushel to that of a man's fist, and becomes shining black like lacquer when boiled in goose soup.) [7] The only element that can be mixed with pure gold for the purpose of adulteration is silver; nothing else will do.

To separate gold [from its alloy] the latter is first beaten into a thin sheet and cut into small pieces. Each piece is then wrapped in mud and placed into a crucible. When smelted together with borax, the silver is absorbed by the mud, letting the pure gold flow out. The residual mud is placed into another crucible, and a little lead is added so as to recover from the mud all the silver that has been absorbed therein.[8]

Since gold is the most splendid and precious decoration in the world, it is used only after it has been first made into gold leaves. A one inch square leaf is made from 0.007 ounce of gold and 1,000 gold leaves will cover an area of three square feet. The leaves are made by the vigorous hammering of thin gold sheets, which have been placed between layers of ink paper (the mallet used for this purpose is short-handled and weighs about eight catties). Produced in Soochow and Hangchow, this ink paper is made of the membrane in giant bamboos from Tung-hai and smoked with bean-oil lamps. Each piece of paper can be used fifty times for the hammering of gold leaves, and is then discarded. So ingeniously is this paper manufactured that it is not damaged or torn [by the hammering], but afterward can be used by medicine shops to wrap cinnabar.

When the gold leaves are done, they are taken out of the ink paper and placed upon a piece of cured cat skin stretched tightly over a square frame, over which the ash of slim incense sticks has been sprinkled. Using a blunt knife, one cuts [the gold leaf] into pieces one inch square; then, with a light stick moistened with saliva and while holding his breath, the worker picks up the pieces and stores them between small pieces of paper. For decoration, the gold leaves are applied to areas brushed with varnish (for the decoration of characters [such as written or carved in a sign] the sap of the paper-mulberry is most frequently used as a base). In Shensi the people make "leather gold" by affixing gold leaves to sheep skin cured to an extreme thinness, so that it can be cut and made into articles of clothing or ornament, all of which gleam with golden splendor.

When articles coated with gold leaves become worn, the gold should be scraped off and burnt, so that the gold is preserved in the ashes. A few drops of clear oil is added [to the ash] for settling the gold, which is then reclaimed through washing and smelting.

To make simulated gold, the makers of fans in Hangchow use silver foil as base, on which safflower seed oil is painted and then heated near a fire. In Kwangtung, however, the goods are painted with a liquid prepared by soaking the skins of cicadas in water, and then held briefly over a fire. But this process produces a color that is not truly golden.

For any golden article that is light-colored owing to the low gold content, the rich appearance of pure gold can be achieved by painting it with yellow vitriol [ferric sulfate] and holding it over a charcoal fire. The new coloring will fade after a certain amount of exposure, but reheating will bring it back. [For yellow vitriol, see Chapter 11.]

SILVER

In the past, silver mines were worked in Chekiang and Fukien. These were intermittently mined at the beginning of this dynasty, while the mines in Jao, Hsin, and Jui prefectures in Kiangsi have never been opened. The good [producing] silver mines are located at Ch'en-chou in Hukuang; T'ung-jen in Kweichow; Chao-pao-shan at I-yang, Ch'iu-shu-p'o at Yung-ning, Kao-tsui-erh at Lu-shih, and Ma-ts'ao-shan at Sung-hsien, all in Honan; Mi-le-shan at Hui-ch'uan in Szechuan; and Ta-huang-shan in Kansu, together with others that cannot all be enumerated. The yields, however, are limited. After a mine is opened, the deficit in the amount produced [as against the official estimates] is made up by quotas allotted [to the people in the locality], while thefts, fights, and riots would ensue if the mining laws are not strict. The unrelenting enforcement of prohibition [of mining at random] has therefore been necessary.[9] In the northern provinces, such as Chihli and Shantung, the earth is of a cold nature and the rocks are thin, so that no gold or silver is produced there. But the eight provinces mentioned above combined cannot produce half as much silver as Yunnan. The mining and refining of this metal, therefore, can be carried on continuously only in the latter province.

Most of the Yunnan silver is produced in Ch'u-hsiung, Yung-ch'ang, and Ta-li, with Ch'ü-ching and Yao-an ranking second, followed by Chen-yuan.[10] The presence of silver ore in rocks and caves is indicated by heaps of surface stones that are slightly brown in color and scattered in such a way as to present the appearance of forked paths. Much time is required to dig the mine shafts, which are from ten to twenty *chang* deep, and the ore can be reached when outcrops are found in the earth. The ore deposit is buried deep in the earth, branching off in different directions, so that each man should follow the outcrops individually and find the ore by tunnelling. The top [of the tunnel] is reinforced by wooden planks inserted horizontally to prevent collapse [Figure 14.1]. Following the vein and carrying a lamp, the miner digs with a shovel until the ore deposit is found. Its proximity is indicated by the presence of small yellow stones in the vein, or of thread-like marks on the stones.

[The following are the various kinds of silver ore:] That which contains native silver is called *chiao* [reef or lode]; that which exists in small fragments is called *sha* [gravel, i.e. alluvial or glacial deposit]; and that which has branch-like markings on its surface is called *k'uang* [mine] and is usually surrounded by valueless boulders. The size of boulders varies from that of a fist to a bushel-box. The lode and gravel ores have the appearance of the rock beneath a coal seam but are not very dark in color. These ores vary in grade. (The business people who strike silver ore in their own mines send the samples to the authorities for classification, after which the tax is determined accordingly.) After the ore is mined, it is sold to the smelters in pecks; the high grade fetch-

Figure 14–1. Mining silver ore.

Figure 14–2. Smelting silver ore.

Figure 14–3. Separating lead from silver.

Figure 14–4. Refining silver.

ing some six or seven taels per peck, the medium three or four taels, and the low one or two taels. (The more the ore glitters, the less the silver content, since all the goodness has been dissipated at the surface.)

The ore is put into a furnace after having been concentrated by hand-picking and washing. The furnace is built of earth in the shape of a large platform and is about five feet high, with the bottom covered with porcelain fragments and charcoal ash. Each furnace receives two *tan* of ore, which is surrounded by charcoal made from chestnut wood and packed against the inner surface of the furnace. A brick wall, over one *chang* high and equally wide, is built [against the back of the furnace], and the blast bellows are placed behind the wall and operated by two or three men, who are shielded from the heat by the wall. When the charcoal is burnt off, more can be added with a long iron fork [Figure 14.2].

With the fire fanned by the bellows, the ore is melted into a mass at the proper temperature, but the silver is still alloyed with lead. Approximately one hundred catties of this alloy can be produced from two *tan* of ore. After cooling, the alloy is refined in a separation furnace, also known as a "frog" furnace, which is packed with charcoal of pine wood. It has an opening so that the temperature of the furnace can be estimated by observing the color of the flame. The air for the furnace is provided by bellows or fans [Figure 14.3]. With sufficient heat, the lead will sink to the bottom (the litharge formed on the bottom of the furnace can be reconverted into metallic lead by treating it in another furnace). With lighted willow twigs, [the operators] can look into the furnace from time to time through the opening, and silver is obtained when all the lead has been eliminated. The newly concentrated silver, called "crude silver," does not show a thread-like pattern on the surface. Even after another concentration [i.e. cupellation] only a single star will appear [on the surface of the silver], which the Yunnanese call *ch'a-ching*. After the addition of a bit of copper, the concentrated silver is [refined] by melting it with lead. It is then cast in a mold. The thread-like pattern is achieved (it becomes visible only after the molten silver is poured into the mold, because the latter's walls retain the precious properties of the metal).

At Ch'u-hsiung, a different method is used. Because of the low lead content of the local ore, lead is purchased from other places for the purpose of smelting. To each 100 catties of silver ore, 200 catties of lead are first added in the furnace. Both are then melted into a solid mass and placed in the frog furnace for the separation of silver from lead, which is the same method as described above.

Silver is produced thus and by no other means. The baseless conjectures and annotations in the magicians' and naturalists' books are therefore very objectionable indeed. The essence of the earth is generally so distributed that gold and silver are not found within 300 *li* of each other, which is an indication of the principles of Nature. As to the "panned particles" obtained by the lowly laborers who wash the sweepings of dust and rubbish with water and then refine them with fire, thus getting between 0.03 and 0.06 ounce [of silver] after a day spent at it, they are actually the bits sheared off scissors and axes, stuck to the bottoms of shoes and thence mixed with dust and are found in street or household sweeping or in refuse along river banks. Silver is not produced in the surface layer of the earth.

For practical use, silver can be adulterated only with copper or lead. When ingots are to be made from small pieces of silver, or when impure silver is to be purified, the refining is done in a crucible placed over a tall furnace. The silver is placed in the crucible together with a little niter. [When melted,] the copper and lead which adhere to the bottom of the crucible are called "silver rust," while that which is beaten out onto the ash pit is called "furnace bottom." Then the "rust" and the "bottom" are put together into a refining stove consisting of an earthen pot in which a fire has been made. Lead will melt first and flow out; copper and silver [when they flow out] can be separated with iron rods. All this goes to show something of the possibilities of natural creation and human effort. The refining stove is illustrated as follows [Figure 14.4].

SUPPLEMENT: CINNABAR SILVER

Of all the tricks of magicians, that of making cinnabar silver fools the ignorant most easily. The method consists of putting equal parts of lead, red mercuric sulphide, and silver into a jar, which is then sealed and stored in a warm place for twenty-one days. During this time the essence of silver is absorbed by the mercuric sulphide, and smelted "silver" is obtained. This silver, however, is stuff without spirit, a mass of dead substance. If refined with lead it will decrease in weight with each smelting and after several firings will altogether disappear, thus [causing the maker] to lose the money spent for the mercuric sulphide and charcoal. It is set down here lest the foolish, in their hope to get rich, are still ignorant of these practices.

COPPER

The commonly used red copper is either obtained from the mountain or prepared by smelting the copper ore in a furnace. Mix copper with smithsonite [$ZnCO_3$] or zinc, brass is formed; with arsenic and other drugs, white copper; [11] with alum and niter, blue [or aluminum] bronze; with tin from South China, bronze; and with zinc, casting brass. All these alloys are derived from copper, which is obtainable wherever copper mines are found.

According to *Shan-hai ching* [Geographic Classic] there were 437 copper producing mountains in China. This is an estimate probably based on fact. Among the present sources of supply in China, Szechuan and Kweichow are foremost in the west while in the southeast there are imports from overseas. There are, in addition, many copper mines at Wuchang in Hukuang and Kuang-hsin in Kiangsi. Ores of a very low grade, known as Meng-shan copper, are produced in such places as Heng-chou and Jui-chou [in modern

Figure 14–5. Digging a shaft to obtain copper and lead ores.

Figure 14–6. Smelting an ore containing both copper and lead; molten lead and copper flowing out from two separate holes.

Hunan and Kiangsi]. This copper can be used for casting, but cannot be refined into goods of high quality.

Where there is copper in a mountain, the earth [on the surface] is mixed with stones. The ore is mined through a shaft several *chang* deep [Figure 14.5]. The run-of-mine copper ore is encased in boulders having the shape of ginger-root. Specks of copper are also present in these boulders, which, unlike those found around silver ore, can also be smelted for copper. The copper ores within the boulders are of various sizes and shapes—some shiny, some dull, some resemble brass, others are like ginger-shaped iron.

Before smelting, copper ore is washed to eliminate earth particles. When heated in the furnace, that which melts and flows out is called "natural copper," or "stone marrow lead."

There are several kinds of copper ore. In one, copper alone is present without lead or silver and is extracted with a single process of smelting in a furnace. Another kind contains a combination of copper and lead. Smelting is carried out by placing the ore in a furnace on the side of which are two holes one above the other. Lead melts first and flows out from the upper hole; copper melts later and flows out from the lower one [Figure 14.6]. The Eastern Barbarians [Japanese] have an ore containing both copper and silver. When it is smelted the silver coagulates on the surface [of the molten mass] while the

Figure 14–7. Smelting zinc ore in earthen jars.

Figure 14–8. Ploughing up pieces of iron ore.

copper sinks to the bottom. The latter is imported to China in the shape of rectangular bars and is called Japanese copper. After being bought by the inhabitants of Chang-chou [in Fukien], this copper is refined to separate the silver from the copper, and the latter is then beaten into thin cakes and sold in the same way as Szechuan copper.

In converting copper into brass, the fuel used is the "self-bellowing coal" (this coal is found in powder form which, when made into cake-shaped briquettes with a clay binder, will burn all day without the use of bellows. It is produced in Yuan prefecture and Hsin-yü district in Kiangsi). Place 100 catties of these coal briquettes in the furnace, then into an earthen crock put ten catties of copper and six catties of smithsonite, and place the crock inside the furnace so that the metals will melt. This method was later modified by the use of zinc to replace the smithsonite, which is apt to be lost in the process of melting. In this modified method, six catties of copper are melted in a crock with four catties of zinc; brass is obtained when the alloy cools and can be hammered or molded for purposes of manufacture.

To make bronze musical instruments, [copper is] mixed with lead-free tin from South China. Such articles as gongs and bells are made of an alloy consisting of eight parts copper and two parts tin. For the making of small bells and cymbals the bronze is of an even finer quality.

For casting, the lowest grade of metal consists of copper and zinc in equal parts, or even six parts of zinc to four parts copper. The higher grade is known as "third-fire yellow brass" or "fourth-fire refined brass," which contains seven parts copper to three parts zinc.

Only pure copper can be used for the adulteration of silver, yet the two will not combine as soon as they are exposed to zinc, arsenic, or alum. When mixed with copper the color of silver will change at once from white to red; the two metals can be completely separated, however, by a refining process.

SUPPLEMENT: ZINC

Zinc [*wo-ch'ien,* lit. Japanese lead], a term of recent origin, does not appear in ancient books.[12] It is extracted from smithsonite, and is produced primarily in the T'ai-hang Mountains of Shansi, followed by Ching-chou [in Hupei] and Heng-chou [in Hunan]. Fill each earthen jar [retort] with ten catties of smithsonite, then seal tightly with mud, and let it dry slowly so as to prevent cracking when heated. Then pile a number of these jars in alternate layers with coal and charcoal briquettes, with kindling on the bottom layer for starting the fire [Figure 14.7]. When the jars become red-hot, the smithsonite will melt into a mass. When cooled, the jars are broken open and the substance thus obtained is zinc, with a twenty per cent loss in volume. This metal is easily burnt off by fire if not mixed with copper. Because it is similar to lead, yet more fierce in nature, it is called "Japanese lead."

IRON

This metal is found in iron mines. The ore rests lightly on the earth's surface instead of in deep caverns and is found mostly in low rolling hills instead of high mountains.

There are several kinds of iron ore, such as the lump and granular forms. The black "lump" ore [magnetite, Fe_3O_4] is found on the surface of the soil and has a shape resembling the weight of a steelyard. The lumps look like iron, but will disintegrate at the touch like a clod of earth. Those on the surface of the ground can be picked up for purposes of smelting and refining. [To get the ore beneath the earth surface] the earth is ploughed over after rain and the ore pieces lying several inches deep in the soil are picked up [Figure 14.8]. These pieces will increase with each ploughing and are inexhaustible [*sic*]. Large quantities of the lump ore are found in Kansu in the northwest and Ch'üan-chou in the southeast. On the other hand, granular ore [hematite, Fe_2O_3 and/or limonite, $2Fe_2O_3.3H_2O$] is found in large amounts in Tsun-hua near Peking and P'ing-yang in Shansi. Granular ore becomes observable when its earth coating is partly removed. When the ore is washed and smelted in a furnace, the resulting iron is the same as that processed from the lump ore. [Figure 14.9 shows the concentration of granular ore by washing.]

There are two kinds of iron. One is pig iron, and the other, wrought iron. The former is the direct product of a [blast] furnace, while the latter is made from further treatment of pig iron. Steel is produced by refining a mixture of both pig and wrought iron.

The furnace for smelting iron is built of a salt and mud mixture and is usually constructed against a hillside, or surrounded by a large wooden frame. Great care and much time must be spent on fashioning the salt mud [furnace wall], as the slightest crack in it will bring the whole work to failure. Each furnace has the capacity for some 2,000 catties of ore together with hard wood, coal, or charcoal, depending on the local condition in south or north China. Usually four or six men are required to work the bellows. When the ore melts and becomes iron, it flows out from a hole [i.e. iron notch] on the side of the furnace which has been sealed with mud until casting time. A complete day and night is divided into six equal periods, and one casting takes place in each period. The discharge hole in the furnace is sealed up again after each casting, and the smelting is resumed.

If the pig iron is intended for casting purposes, the molten iron is discharged directly into pig beds molded in the form of long bars and round pieces. If it is intended for the making of wrought iron, the molten iron is run into a square ditch built a few inches below the furnace and a few feet away from it, surrounded by a low wall, where a few persons holding willow sticks and a certain amount of dry and finely screened earth stand ready. When the molten iron flows into the square ditch, one man quickly throws the earthen material into the molten mass,[13] while the others vigorously stir it with the willow sticks,[14] and soon wrought iron is obtained [Figure 14.10]. Two or three inches of each willow stick will be burnt off at the end of an operation, so that new ones will

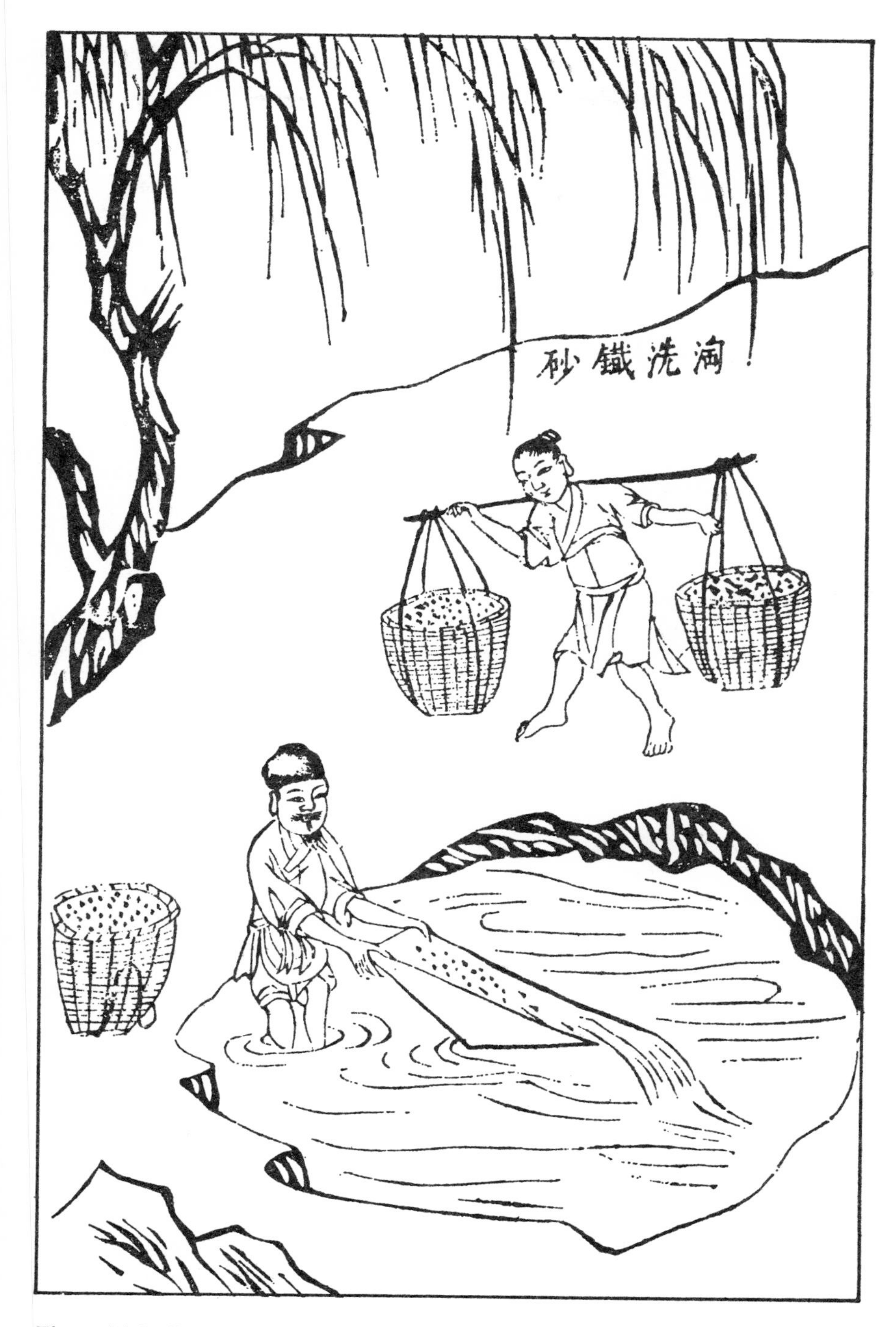

Figure 14–9. Concentrating iron ore by washing.

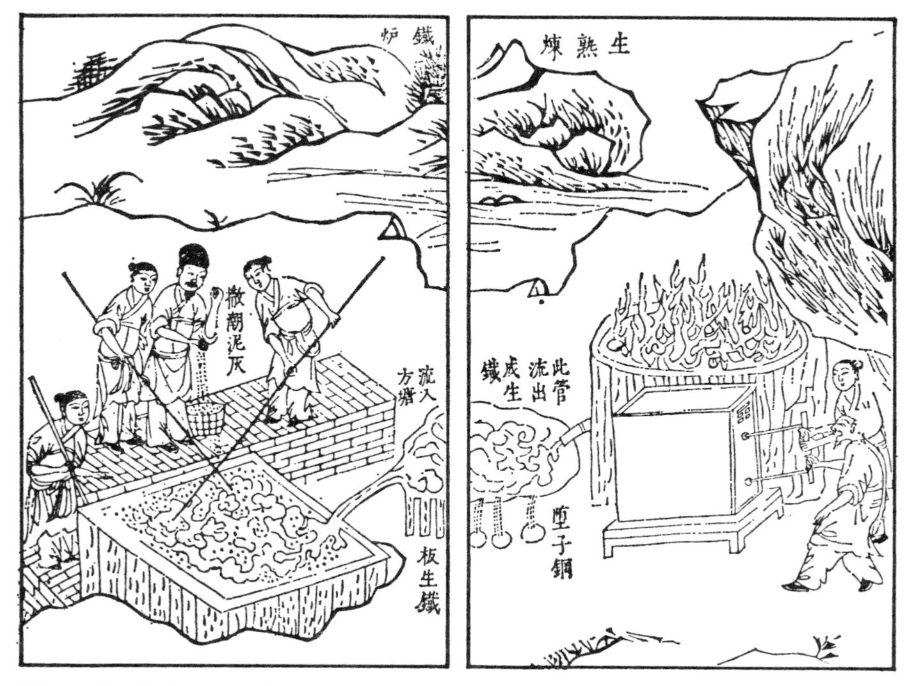

Figure 14–10. Smelting iron ore to make pig iron and wrought iron. Blast furnace at right and puddling pit at left.

have to be used at the next turn. When the wrought iron is somewhat cooled [and becomes pasty] it is either cut into square blocks in the ditch, or lifted and hammered into round pieces before it is sent to market. This method, however, is not used in the refineries of Liu-yang [in Hunan].

For making steel, wrought iron is hammered into thin strips the width of a finger and about one and a half inches long. These strips are wound tightly into bundles, over which pig iron is placed (it is interesting to note that in south China pig iron is termed "pendulum-shaped raw steel"), and over the whole pile are placed worn-out straw sandals (with plenty of mud on them to prevent burning). Mud is also applied to the bottom of the pile after which the mixture is fired with bellows in a furnace. When the temperature is high enough, the pig iron melts first and gradually diffuses into the wrought iron. The uniformly mixed material is taken out for hammering. This process of melting and hammering is repeated many times [before steel is obtained]. It is customarily called "mixing the steel" or "dripping the steel." [15] In Japan some knives and swords are made of a fine steel that has been processed some one hundred times, so that if a sword is hung by the eaves in the sun [the reflection from] it will light up the entire room inside. But this steel is not made of the mixture of pig and wrought iron and is sometimes considered to be low grade. Some of the barbarians [Japanese?] use "ground juice" for quenching the swords ("ground juice" is a kind of naphtha not produced in China) and claim that the steel so treated can cut jade. I have, however, never yet seen such a performance.

If hard spots, known as "iron cores," show up in the iron so that it can not be hammered properly, smear some sesame oil over them [while the iron is hot] and they will disappear. When iron occurs in the shady side [*yin*] of a mountain, lodestone can be found on the sunny side [*yang*], although there are several places where this is not so.

TIN

Tin is produced in many places in southwestern China, but in very few in the northeastern parts of the country. Tin is called *ho* in ancient books, because it was produced most abundantly in Lin-ho Commandery [in modern Kwangsi]. Eight-tenths of today's tin supply comes from Nan-tan and Ho-ch'ih in Kwangsi, followed by Heng-chou and Yung-chou [both in Hunan]; large quantities are also produced in Ta-li and Ch'u-hsiung [in Yunnan], but these places are too remote and not easily accessible.

There are two kinds of tin ore: mountain tin [tin stone deposits on hillsides] and stream tin [alluvial deposits]. Mountain tin includes two varieties, the "melon tin" and the "granular tin." Melon tin is about the size of a small gourd, while granular tin is the size of beans. Both kinds are close to the earth surface and are therefore obtainable without deep underground mining. Sometimes a tin ore vein in the ground becomes full, causing the hillside to collapse,[16] so that people can freely pick up the exposed tin ore. Stream tin is produced in the rivers of Heng-chou, Yung-chou, and in the river at Nan-tan, Kwangsi. Stream tin is black and powdery, having the consistency of double-bolted flour. The inhabitants of Nan-tan recover the tin ore from the river by first working from the south northward, and then from the north southward alternately at ten-day intervals [Figure 14.11]. Panned in this way, the ore supply will continue without exhaustion. One day's sluicing and smelting, however, will not yield more than one catty [of tin], which is not much considering the cost of the fuel.

Mountain tin produced at Nan-tan is found in the shady side of the mountain there. As there is no water for concentration of the ore, some hundred lengths of bamboo are connected to form an aqueduct. Water is conducted here from the sunny side of the mountain and the gangue materials of the ore are washed away [Figure 14.12]. The concentrated ore is then smelted in a furnace.

For smelting tin, a blast furnace is fed with several hundred catties each of tin ore and of charcoal in alternate layers. When the right temperature is reached and the ore does not melt immediately, a small amount of lead added to the mixture will induce it to flow out freely. Sometimes [instead of lead] the residual waste of tin refining is used for this purpose. The bottom of the furnace is a horizontal basin covered with charcoal ash and porcelain fragments in which an iron pipe is placed that leads the molten tin to flow into a low receiving pit outside the furnace [Figure 14.13]. When it first comes out of the furnace, the metallic tin, being pure white in color, is very brittle and breaks into pieces when hammered. Some lead must be added to soften it, after which the tin [pewter] can be used for making utensils. If tin objects bought in the shops contain too much lead and purification is desired, these objects should be immersed in vinegar and let boil eight or nine times, thereby eliminating the lead.

The process described above is the only way of producing tin. The magicians' books mention obtaining "grass tin" from the grass *Portulaca oleracea*. This is nonsense. The notion that arsenic is the outcropping of tin is also erroneous.

LEAD

There are more lead-producing mines than there are copper or tin. Three kinds of lead can be distinguished: The first is derived from the ore of silver mines, with the lead enclosing the silver. A preliminary smelting will produce a lead-silver mass; with a further smelting the lead is isolated from the silver and sinks to the bottom [of the furnace]. This is called "silver-mine lead" and is found most abundantly in Yunnan. The second kind is derived from the ore of copper mines. When smelted in a furnace, the lead flows out first, followed by copper. This is called "copper-mine lead" and is produced mostly in Kweichow.

The third kind is produced from the ore of lead mines. The miners dig tunnels into the hills and with the aid of oil lamps search for the ore veins. The path is usually twisted, as in silver mines. When the ore is washed and smelted, the lead so obtained is called "grass-joint lead," and is produced in greatest quantity in such places as Chia-ting and Li-chou in Szechuan.

Aside from these three, there is produced at Ya-chou [in western Szechuan] a pod-shaped, or tadpole-shaped ore, called "sinker lead," which is found in mountain streams. Furthermore, many other kinds of lead ore are produced elsewhere, such as the "copper-mixed lead" of Shang-jao in Kuang-hsin prefecture and of Lo-p'ing in Jao prefecture [both in Kiangsi], and the *yin-p'ing* lead of Chien-chou [in Szechuan].

The "silver-mine lead" is obtained by smelting the so-called "bottom" [or the residual litharge of silver refining] in a furnace. The "grass-joint lead" is obtained by smelting the lead ore in a furnace, from the side of which is a pipe that lets the molten lead flow into a long and narrow earthen ditch. The product is commonly known as "carrying-pole lead" or "mined lead," so as to differentiate it from the lead recovered from silver refining.

The price of lead is low, yet it is an amazingly versatile metal. Out of lead is made white-lead powder and litharge. Silver is purified, and tin softened, with lead.

Figure 14–11. Recovering tin ore from the river at Nan-tan, Kwangsi.

Figure 14–12. Concentrating mountain tin ore by washing, Ho-ch'ih, Kwangsi.

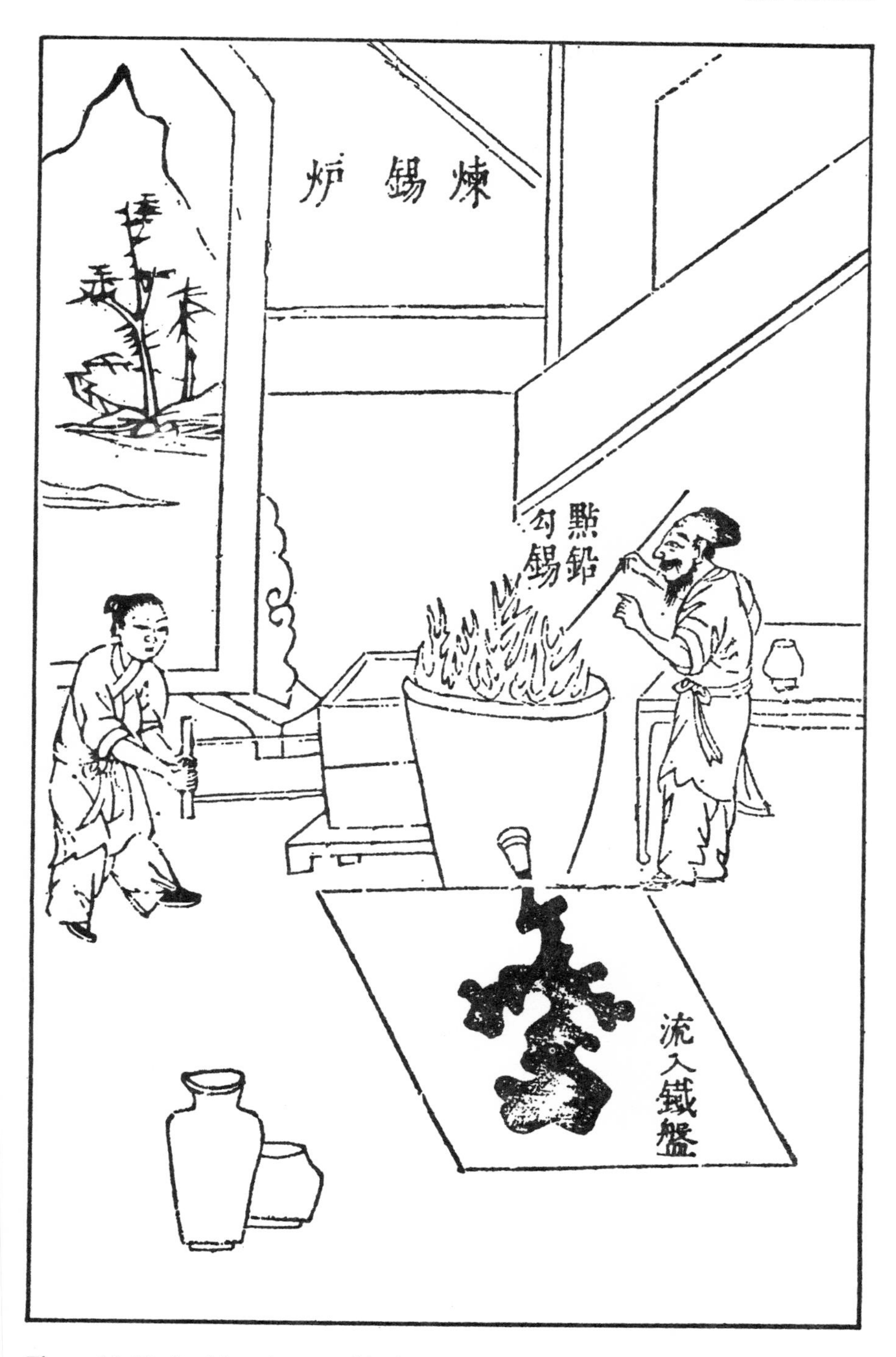

Figure 14–13. Smelting tin ore with the addition of lead.

SUPPLEMENT 1: CERUSE

To make white lead [$2PbCO_3.Pb(OH)_2$], melt one hundred catties of metallic lead and, after cooling, cut it into thin sheets which are rolled into small cylinders and put into a wooden vat. One bottle of vinegar is fixed in the middle of the vat, and another bottle is placed beneath the [perforated part of the] vat.[17] The outside seams of the vat are sealed securely with salt-mud and paper. The temperature inside the vat is kept moderately warm for seven days by means of a small fire [lit. "a four-ounce fire"]. When the vat is opened at the end of the period, the lead sheets are covered with a frost-like white powder, which is brushed into a large water jar. The cylinders that have not turned into frost are returned to the wooden vat to undergo a further reaction and are scraped again after another seven days, [the process being repeated] until all the lead is gone. Whatever residue there is can be used as raw material in making litharge.

[To make ceruse,] to every catty of white lead powder add two ounces of bean flour and four ounces of ground powder of oyster shells, and stir them in a jar of water. After decantation of the mixture, several layers of paper are placed over fine ashes that have been molded so that ditches run through them, and the powder sediments are spread on the paper. When the powder is nearly dry, the paste is cut into either tile-like slices or thick pieces. When these are completely dry they can be stored and marketed. This powder is also called Shao-powder, since in ancient times it was produced only in Ch'en-chou and Shao-chou [in Hunan and Kwangtung] (the commonly used term "ch'ao-powder" is not correct). Nowadays Shao-powder is made in all provinces.

When ceruse is mixed with colors for painting, the white color always shows through; when used as a cosmetic powder by women, it will turn the natural complexion greenish-sallow. If put in a charcoal fire, ceruse will melt and be reconverted into lead—this is known as "all colors lead to blackness" [since the color of pure lead is bluish-gray].

SUPPLEMENT 2: LITHARGE

The ingredients for making lead litharge are: ten ounces of native sulphur, one ounce of niter, and one catty of lead. Melt the lead first. While it is in the molten state, add some drops of vinegar, then add a piece of sulphur while the molten mass is steaming. Shortly afterward, a small bit of niter is added. When the steaming subsides, more vinegar is added and the process is repeated with the niter and sulphur being added little by little. Litharge is obtained when the mass turns into powder. If the residue left from making white lead is used as the raw material, the process then consists of roasting it with niter and alum without vinegar.

To convert litharge back into lead, heat the former over a slow fire with a greenish-white liquid. The liquid is poured off when it turns yellow, and the rest is lead.[18]

NOTES

1. The five metals are gold, silver, copper, iron, and tin; the term "five metals" is often used to denote metals in general.

2. The earlist specimens of both gold and copper have been found in the excavations of the An-Yang ruins of Shang dynasty (ca. 1783–1122 B.C.). Much later, in the *Tso Chronicles,* a work written around the sixth century B.C., gold is rated as the highest of all metals. During the subsequent period of the Warring States (403–221 B.C.) there was a considerable growth of trade along with the emergence of a money economy; as gold began to assume importance in the market place, it was often mentioned as a measure of a man's wealth. At this time a standard unit, the *i,* was employed to determine the weight of gold, one *i* consisting of 20 (sometimes 24) ounces.

During the Former and Later Han dynasties (206 B.C.–A.D. 220) gold, measured in catties, was widely used as currency. Large quantities were granted by the emperor to members of the aristocracy on special occasions. It was recorded that Wang Mang accumluated 5,000,000 ounces of gold during his reign (A.D. 9–23). As pointed out by Needham, some of the Chinese gold of this period probably came from Europe and Siberia. Writing in the first part of the seventeenth century, Ku Yen-wu, a contemporary of Master Sung, contrasts the abundance of gold during Han times with the lack of it in later ages, and blames the situation on the Buddhist custom of decorating temples and images with large amounts of gold. The value of gold relative to silver steadily increased during the Ming dynasty. In the fourteenth century, the exchange rate was 4:1, it went up to 7.5:1 in the early fifteenth century, and toward the beginning of the seventeenth century, about the time when *T'ien-kung k'ai-wu* was written it stood at 8:1.

3. "Dog-head gold" was apparently large nuggets.

4. The difference in color may be caused by the fact that crude gold bullion is far more contaminated by admixture with silver and white base metals.

5. "Southern barbarians" probably refers to the inhabitants of the aboriginal regions in Kwangsi.

6. *Ling-piao lu* is an abbreviated version of the T'ang dynasty work *Ling-piao i-lu* (Account of Unusual Matters in Ling-nan) by Liu Hsün.

7. Touchstone was described by Theophrastus in about 300 B.C. A touchstone must be completely black and free of sulphur. In testing gold that contains silver, or vice versa, the touchstone is first rubbed with the metal to be tested, and then with a few selected touch needles of known composition. By comparing the color and appearance of the streaks and finding the needle whose streak comes closest to that of the metal, the composition of the latter can be approximated.

8. The process described herein is essentially a method of cementation, which was first described by Agatharcides in 113 B.C. and was probably in use as early as 700 B.C.

9. The earliest written reference to silver is found in the Tso *Chronicles.* Silver was first used for coins during the reign of the Emperor Wu (140–87 B.C.) of the former Han dynasty, and there is evidence that the government issued silver currency during the Later Han period. It was not until the T'ang (618–907) and the Northern Sung (960–1127) dynasties, however, it was widely used as money.

The production of silver in the Ming period was strictly controlled by the government. At the beginning of the dynasty, the use of silver and gold as currency was prohibited, but by the time the present work was written, the restrictions were no longer enforced owing to the depreciation of paper currency. Thus, from the middle of the sixteenth century, silver was legal tender throughout the country. Mining and smelting, however, were supervised by officials, toward the last part of the dynasty generally eunuchs sent from the Court. The government usually received one-thirtieth to one-twentieth of the product as tax; mines were opened or closed at the direction of the government, often depending on whether or not the latter urgently needed the funds realized through the silver taxes. Laws against private mining and smelting were severe. In 1438, for example, the penalty for illegal smelting of silver was death, and the families of the guilty were sent into exile.

The locations of new silver mines were often arbitrarily chosen by the officials (eunuchs) in charge, whose primary concern was their own gain rather than the replenishment of the government treasury. If the mine proved to be unproductive, the required amount would then be made up by the wealthy families in the locality. These and similar abuses are recorded in many works and compilations, among them the *History of Ming* (*Ming shih*), chüan 81

10. Chen-yuan was known in the T'ang dynasty as Yin-sheng fu, "Silver-producing Prefecture."

11. According to Chang Hung-chao, white copper was first recorded under the name *wo* in the *Book of Poetry,* a word dating back to the early part of the Chou dynasty (ca. 1122–771 B.C.). The term "white copper," however, was loosely used by the early Chinese to denote more than one kind of white metal, possibly including arsenic-copper, copper-nickel, and copper-zinc-nickel alloys. Arsenic-copper alloy was probably known to the Chinese in the Sung Dynasty (A.D. 960–1279), as indicated by the Ho Sui's procedure for changing copper into cheap silver. Copper and arsenic in certain proportions yield an alloy of almost silver-white appearance, capable of a very high polish but only slightly ductile, so that it was probably used only in the cast condition. It is used very little in China nowadays.

12. According to the Chinese scientist Wang Chin, the use of zinc in China can be roughly classified into four periods. First, from the Chou through the Sui dynasties (1122 B.C.–A.D. 617), zinc was unintentionally introduced into cast coins as an impurity of lead and was not identified as a separate metal. Second, in the T'ang dynasty (A.D. 618–907), smithsonite was recognized as a zinc mineral and was purposely combined with copper to yield brass. A considerable amount of brass was produced at that time for making ornaments. Third, from Sung to early Ming (tenth to late fourteenth centuries), smithsonite and/or brass were intentionally used for minting coins. Fourth, from the middle period of Ming through Ch'ing (ca. 1460–1911), smithsonite was converted into metallic zinc, which was in turn used for casting coins and making brass.

13. From Joseph Hall's "wet puddling" process for the manufacture of wrought iron, it can be inferred that the earthen materials, consisting largely of iron silicate rich in iron oxides, were employed not only to hasten oxidation but also to increase the yield of wrought iron. Silica is essential in all puddling as it combines with iron oxide to form a fusible oxide slag which permits the wrought iron to adhere in masses.

14. The procedure was essentially a puddling process, with the square ditch serving as refining hearth. Through vigorous stirring with willow sticks, the different portions of the molten pig iron are brought to the surface and exposed to the air, thereby resulting in the oxidation of silicon, phosphorous, and carbon. The contact between air and molten iron is further facilitated by the fact that the willow sticks undergo destructive distillation, the resultant steam and gases producing a considerable bubbling or boiling action in the square ditch.

15. The characteristic Chinese process of steelmaking was decarburization direct from cast iron, not the addition of carbon to pure iron. This method, known for many centuries as the "hundred refinings," depended upon the discreet use of an oxidizing blast of cold air and developed side by side with the more drastic process of fining cast iron to wrought iron. The method seems to have been in full use from the second century B.C. on.

From the fifth century A.D., however, large quantities were made by the method described here. A reconstruction of the medieval Chinese process has yielded good eutectoid steel.

16. "Collapse of the hillside" seems to be a description of natural erosion exposing outcroppings of harder tin ore (probably cassiterite).

17. *Tseng,* the vat constructed on the principle of the double boiler, was originally used for steaming rice. The *Tseng* has seven holes on the bottom of its upper half; if a bottle of vinegar is placed beneath the perforated part, the interior is filled with acid vapors when the entire vat is heated.

18. The greenish-white liquid is probably a reducing agent.

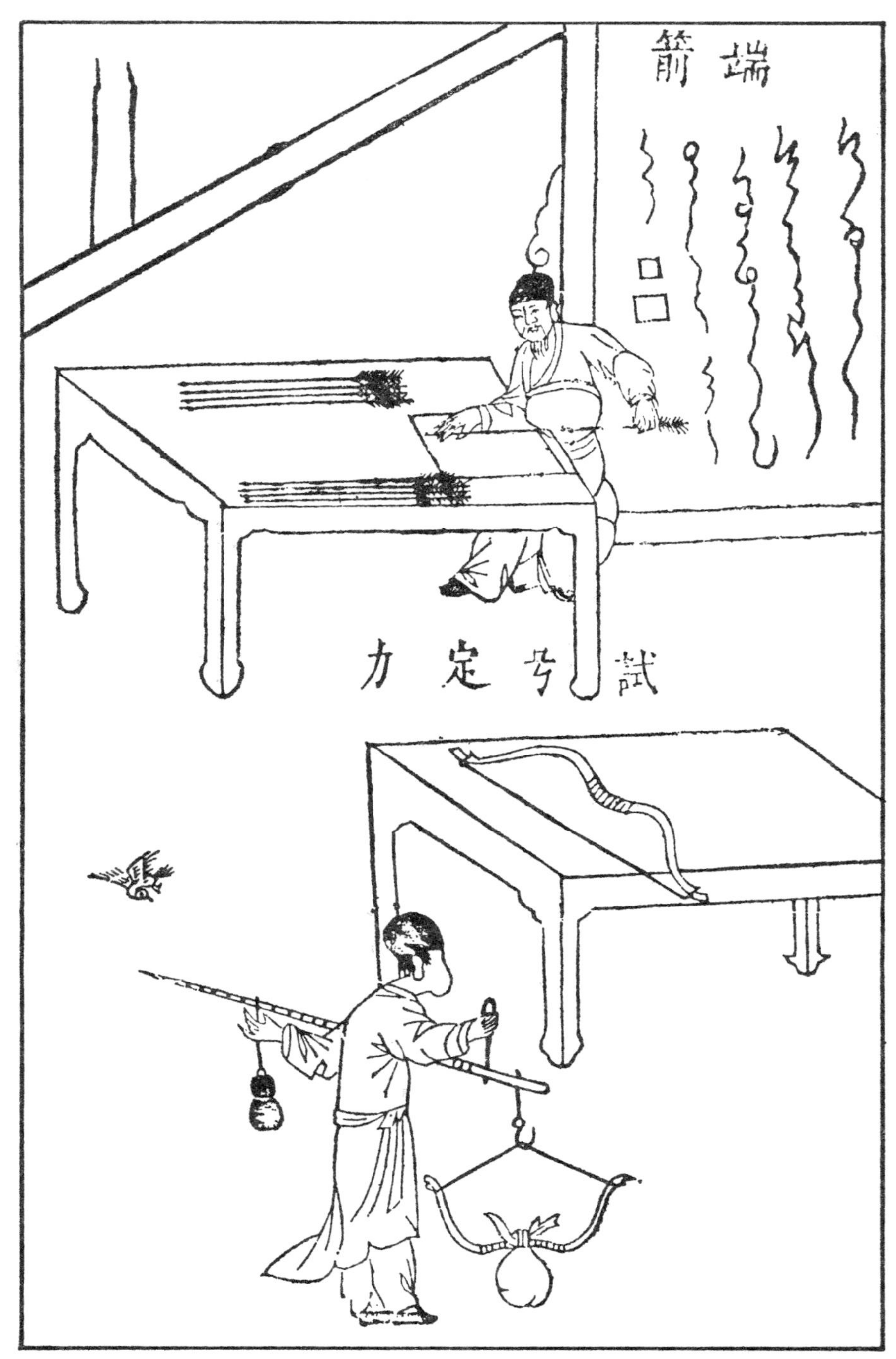

Figure 15–1. Straightening the shaft of an arrow (top) and determining the weight or pull of a bow (bottom).

15

WEAPONS

Master Sung observes that the ancient sages would never use armed force unless it was absolutely necessary. The *Miao* people were still rebellious at the end of Emperor Shun's fifty-year reign, [and had to be pacified by Emperor Yü of the succeeding Hsia dynasty]. How would it be possible for any intelligent king and sacred emperor to exist without military power? Bows and arrows were traditionally used for controlling the country. The Taoists, having the ancient Emperor Ko-t'ien's philosophy [of governing without force], state that life-taking weapons are implements of evil design. This was a word of caution. The technique of making firearms originated with the Western and Southern Barbarians and was later introduced into China proper. Since that time, many varieties of new firearms have been designed and manufactured, and today Chinese military leaders consider [the development of firearms] as the most important thing. Is this concept correct? This ingenious invention, however, would not have reached such an advanced state of development were it carried out by human endeavor alone.

BOW AND ARROW

The central body of a bow is made of bamboo and oxhorn. (The northeastern barbarians, having no bamboo, use pliable wood as a substitute.) The two ends of a bow are made of mulberry-twig wood. When unstrung, a bow curves in such a manner that the bamboo lies at the inside and the horn at the outside surface. When the bow is drawn, the positions of bamboo and horn are reversed. The bamboo part is a whole piece, while the horn part is made of two pieces joined together. The end of the mulberry-twig wood is carved [into a nock] for receiving the string, while the body of the wood is mortised securely into the bifurcated bamboo. One side of the bamboo is made into a smooth surface for fastening the horn.

The first thing to be done when making a bow is to prepare a flat strip of bamboo. (The bamboo should be felled in the autumn or winter, because it tends to decay and be spoiled by insects in the spring and summer.) The prepared bamboo strip, having a slightly narrow center and two relatively wide ends, is about two *ch'ih* long. One side is first coated with glue on which the horn is fastened, while the other side is covered with a layer of glue, and the whole thing reinforced by winding a sinew around it. The horn used here is made by joining the toothed ends of two pieces of oxhorn together. (This type of long oxhorn is not available to the northern barbarians, who have to use four pieces of sheephorn joined together to form a horn plate. In Kwangtung province, the horns of both water buffalo and yellow cattle are used by the bow maker.) The horn plate of a bow is strengthened by covering it with ox sinew and glue, and the coating of glue is in turn covered by birch bark, which is soft and resilient, and gives the name "warm grasp" to the bow.

The birch [*Betula japonica*] tree, known as *hua* in China, is found chiefly in Liao-yang in Manchuria, but is also found in abundance in the Tsun-hua district [of Hopei province] in north China, as well as in the Lin-t'ao commandery [of Kansu province] in west China. The birch tree also grows in Fukien, Kwangtung, and Chekiang provinces. Birch bark, being soft as cotton, is used to cover the handgrips of bows. Birch bark is also used for covering the hilts of knives and the poles of spears. The extremely thin variety of bark is employed as the inside lining of knife scabbards and sword sheaths.

The sinew taken from the spine of an ox is rectangular in shape and weighs about thirty ounces. After an ox is killed, this sinew is first dried in the sun, then soaked and softened in water, and finally separated into flax-like fibres. The resulting sinew fibres are twisted into bow strings by the northern barbarians, because of their lack of silk fibres. In China proper, however, the sinew fibres are used for protecting and reinforcing the body of bows. The fibre is also made into cords for bowing cotton.

Glue is made from the bladders and intestines of fish. The boiling [of bladders and intestines in water for the manufacture of glue] is largely carried out at Ning-kuo commandery [in modern Anhui province]. In Chekiang, where the *Seiaena schlegeli* fish is obtained from the East sea, the bladder that remains from the making of dried salt fish is converted into glue, which is stronger than any kind of metal or iron. The northern barbarians also manufacture a glue by boiling sea-fish bladders [in water]. This glue is as strong as the Chinese products, though it differs from the latter in [chemical and physical] properties. It is not a mere coincidence that these raw materials are created by Heaven, since the lack of one of them will prevent the making of a good bow.

The newly made crude-bow is placed on the rafters or a shelf in a room and is dried slowly from underneath by a constant fire on the floor. The drying period varies from a minimum of ten days to a maximum of two months. The completely dried bow, being taken out of the room, is first polished, and then reinforced with ox sinew, glue, and Chinese wood lacquer to result in the best product. If the bow manufacturer cannot wait for such a long period of drying, [the residual moisture] will cause the finished product to deteriorate at a later date.

Silk filaments from worms fed on the leaves of thorny trees [*che* trees] are used for the making of bow strings, because they are stronger and tougher than [mulberry silk]. To make a string, a silk thread is tightly wound around a core of more than twenty silk threads in three sections of more than seven *ts'un* each in length, leaving two gaps each about 0.1 to 0.2 *ts'un* long. Such a cord can be folded into thirds when the bow is not strung. In the past, the northern barbarians used ox sinews for making all their bow strings, which could be easily spoiled by rain and fog. Therefore, they generally avoided war with China during the summer season. Nowadays, however, silk bow strings are also widely used by the northern barbarians. Bow strings may be coated with yellow wax, although it is not absolutely necessary.

The nock at each end of a bow for receiving the string is covered with a piece of extremely thick ox leather or soft wood. This covering material, called cushion, serves the same purpose as the pegs of a lute. When the string is drawn and snapped back to its original position, a tremendous force is developed toward the inside surface of the bow. The bow is protected from such a destructive force [by covering the nock at its two ends with a shock-absorbing cushion], otherwise it will be damaged.

Bows of various weights or pulls are made to suit the strength of individual archers. A bow of 120-catty pull is for the strong bowman, while the still heavier bows are for the very few people who have the "tiger's strength." The bow for the average bowman is 10 to 20 per cent less in pull than the 120-catty one, and that for the weak bowman is 50 per cent less. In spite of differences in pull, all these bows, when fully drawn, can propel arrows to hit a target. On the battle front, however, the strong archers are needed for the piercing of human chests and thin wooden shields. The weak archers are esteemed for the good marksmanship of hitting a bull's eye or a willow leaf, thus conquering through skill instead of main force. To determine the pull of a bow, the maker steps on the bowstring and presses it down toward the ground. The centural part of the bow is hung on the hook of a steelyard, and the force applied for a maximum bending of the bowstring, known as weight or pull, is measured by balancing the weight suspended on the marked beam of the steelyard. [See the lower diagram of Figure 15.1.] In its crude form, a bow for the strong archer consists of roughly 7 ounces of horn and bamboo as well as 0.8 ounces of sinew, glue, and silk thread. The weight of these materials is 10 to 20 per cent less in a bow for the average archer, and 20 to 30 per cent less for the weak archer.

To preserve a finished bow, mildew and moisture must be avoided by all means. (The mildew season starts first in the south and then moves north. It begins roughly at "grain-rains" time [ca. April 20] south of the Mei Range, at "grain-filling" time [ca. May 21] south of the Yangtze River, in June north of the Yangtze River, and in July in Hopei and Shantung provinces. No place, however, can surpass Huai-an and Yangchow [both in Kiangsu province] as the most mildew-inducing spots in the land.) To protect bows from mildew, drying ovens or boxes, each heated underneath with a charcoal fire, are usually set up in the homes of military officers. (This drying device is used to cope with not only the early summer mildew, but also the spring dew and the autumn rain.) A drying oven is too expensive for the soldier, who places his bow above the cooking stove. Any negligence in this drying procedure will cause the bow to disintegrate. (In recent years, the southern provinces have been ordered not only to manufacture bows, but also to send them to the north. On the arrival of the bows in north China, a great number of them have been rejected. This is caused by the rapid deterioration of bows in the absence of fire, but no official has offered to explain this to the central government in Peking.)

The raw material used for making arrow shafts varies with geographical location. Bamboo is employed in south China, willow in north China, and birch in the land of the Northern Barbarians. The arrow shaft is about two feet in length, while the arrowhead is roughly one inch long. A bamboo arrow is manufactured by gluing three or four bamboo strips together, which are subsequently trimmed and polished with the aid of a knife into a perfectly smooth shaft. After its two ends are wound with silk threads and painted with lacquer, the shaft is known as *san-pu-ch'i* or "three unevennesses." The so-called "arrow bamboo," grown in Chekiang and Kwangtung provinces, can be trimmed directly into a shaft without the trouble of preparing and gluing bamboo strips. A minimum amount of peeling and cutting is needed for making shafts from the straight twigs of willow and birch trees. A bamboo shaft, being naturally straight, does not need to be straightened. In contrast, a wooden shaft, tending to curve when dry, must be straightened during the manufacturing process by drawing the shaft through the straight groove. [This groove is] carved in a section of wood several inches long, which is known as an "arrow straightener." [See the upper diagram of Figure 15.1.] During this treatment, both ends of the shaft are shaped to their proper size. The butt of a shaft is grooved to fit securely on the bow string, while an arrowhead is mounted on the upper end.

Arrowheads are made of iron. (The flint arrowheads, mentioned in an ancient book, *Tributes of Yü,* are local products of no practical value.) The Northern Barbarians make their arrowheads in the shape of a peach-leaf spearpoint; the Li tribesmen of Kwangtung make theirs like flat blades; and in China Proper the arrowhead resembles a three-edged awl. A whistling arrow is equipped with a hollow-centered wooden whistle approximately one inch long. When the arrow is in propulsion, air is forced into the whistle and produces a shrill sound. Such an arrow, referred to by the famous Taoist philosopher Chuang-tzu [ca. 300 B.C.] as *Hao-shih* or "whirring dart," [was used by ancient Chinese armies and bandits as a signal to begin the attack.]

The trajectory and speed of a traveling arrow are controlled by its tail fin of three feathers, which are glued around the butt end of the arrow shaft. (Mildew and moisture also cause this glue to deteriorate. Diligent officers and soldiers periodically dry their arrows with fire.) The best arrow-feathers are taken from the wings of an eagle. (Having a long tail and two short wings, an eagle is similar to, but larger than, a falcon.) The feathers from a horned falcon or hawk are next best, while feathers from owls and sparrow hawks are still lower in grade. It is difficult for the arrow makers of south China to obtain falcons and hawks, much less eagles; they therefore use the feathers of wild geese and even those of wild swans as tail fins for the urgently needed arrows. An eagle-feather finned arrow, having the ability to resist a gust of wind, not only travels faster than a falcon-feather finned arrow, but also reaches its true position after only ten paces of traveling. Most of the Northern Barbarian's arrows are finned with eagle feathers. In comparison, even the best made falcon-feather arrow may still waver slightly in the wind. Arrows with the feathers of wild geese and wild swans are difficult to control in their discharge from the bow and tend to travel astray in the wind. For these reasons, arrows made in the south are inferior to their counterparts in the north.

HAND CROSSBOW

The hand crossbow is a weapon for defending army camps, but not for fighting at the battle fronts. The crosswise bow and the lengthwise stock of a hand crossbow are called wing and body, respectively. The gear or lock attached to the stock for drawing and releasing the bow string is known as the trigger mechanism. Wood is used for making the stock, [which is] approximately two feet long. The upper end of the stock, having a cavity exactly 0.1 inch below its lengthwise front surface, is fastened transversely to [the central body of] a bow. (A slightly thicker partition between the cavity and the front surface of the stock would hinder the movement of the bow string.) On the other hand, there is no dimensional specification for the partition between the cavity and the back surface of the stock. A straight, shallow groove is carved into the front surface of the stock for supporting the arrow.

When the body of the crosswise bow is made from a highly resilient wood, it is known as *pien-tan* or "a coolie's carrying-pole" on account of its great strength. Another way of making the bow body is to reinforce the lower surface of an ordinary wood strip with several layers of laminated bamboo. (These plied bamboo plates decrease gradually in length with their increasing distance from the wood.) The resultant hand crossbows are designated as three-ply, five-ply, and the maximum, seven-ply [see Figure 15.2].

[While the arrow is being fitted and aim taken], the bow string must be held under tension by a tooth-like crossbow lock, which is fastened securely to a carved slot in the lower part of the stock. The lock is free to turn when its attached trigger is raised, thus releasing the bow string. The necessary force for drawing a bow string depends entirely on the strength of the bow. A strong archer is able to pull the string of a stout hand crossbow by holding the bow body to the ground with one foot. This [action] is termed in the *History of the Former Han* as *chüeh-chang*. The arrow propelled from such a powerful bow travels at the highest speed.

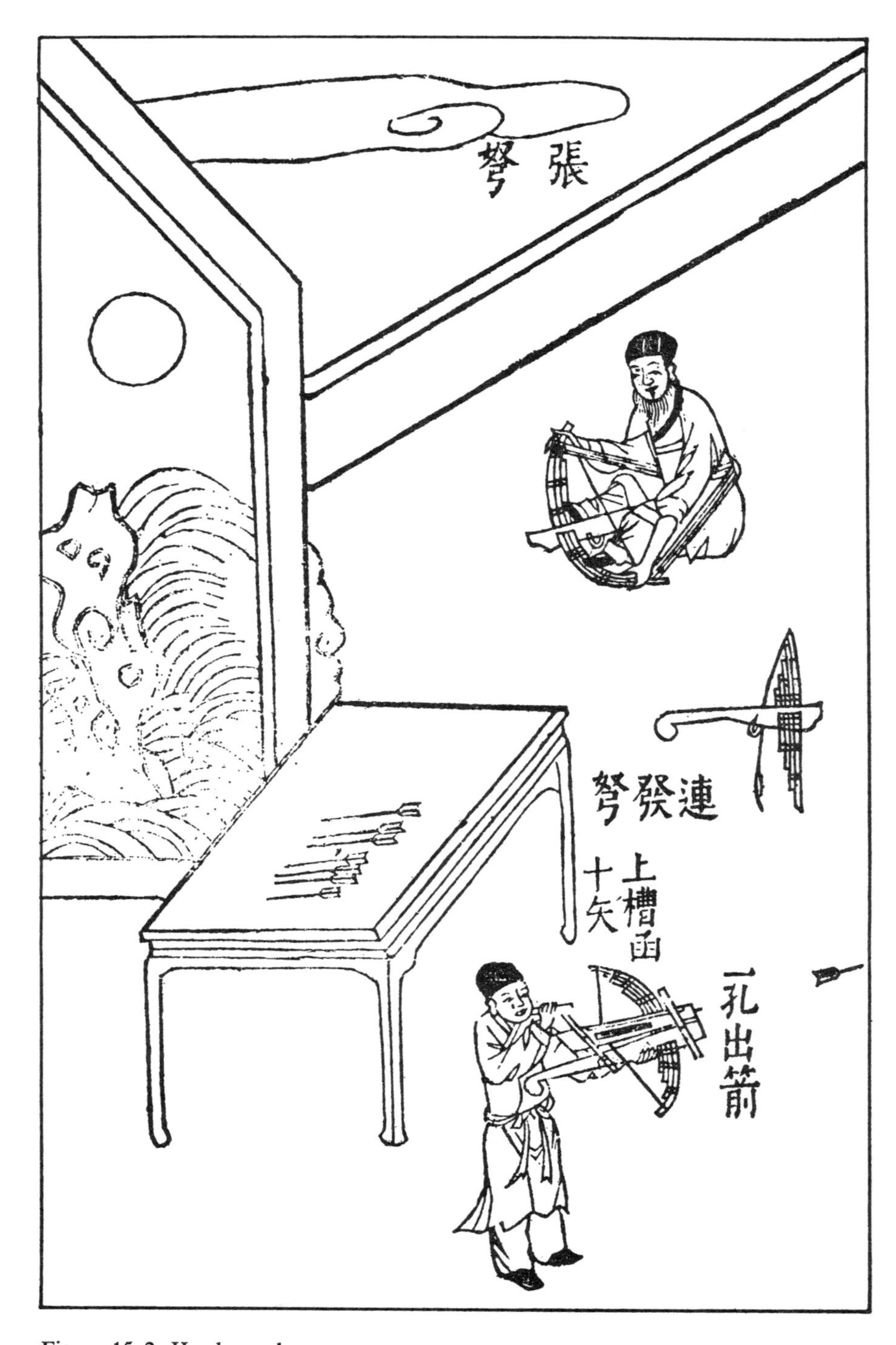

Figure 15–2. Hand crossbow.

The bow string is made with a ramie cord as the core, which is then wound with goose feathers and covered with yellow wax. Such a string is taut only when it is attached to a bow body and becomes slack again after being detached from the bow. [It can be inferred, therefore, that the interstices between the loosely connected threads of] the ramie cord are capable of receiving both the head and tail of the inserted goose feathers. [This passage is rather unclear. The bow string may have been merely sheathed in quills to keep it dry—*trans.*]

Arrow tail fins are prepared by inserting bamboo leaves into a split arrow stem which is subsequently wound with [silk threads]. In making poison arrows for shooting wild beasts, the tubers of wild *aconitum* are boiled in water. The resulting liquid, being highly viscous and poisonous, is smeared on the sharp edges of arrowheads. These treated arrowheads are effective in the quick killing of both human beings and animals, even though the victim may shed only a trace of blood.

Compared with the more than 200-pace traveling range of the arrow from a strong bow, the arrow from the strongest hand crossbow is powerful only within fifty paces and will fail to penetrate a piece of pongee at one foot beyond this limit. On the other hand, the crossbow arrow travels ten times as fast and penetrates twice as deep as the bow arrow. Weapons of the present [Ming] dynasty, such as the mighty-arm crossbow [*shen-pi nu*] and the vanquisher crossbow (*k'o-ti nu*), are able to discharge simultaneously two or three arrows. Another weapon, called *Chu-ko* crossbow, distinguishes itself by having a straight trough on the front end of the stock for holding ten arrows. [See the lower diagram of Figure 15.2.] The crosswise bow body is made of a highly elastic wood, while the stock is equipped with a wooden mechanical contrivance. [Like the goat's-foot lever used in Europe,] this contrivance is raised by hand to pull the bow string onto the crossbow lock. [The drawn string is subsequently released] to send forth an arrow, which leaves an empty spot on the stock that is occupied immediately by another arrow which has descended from the trough. To discharge this newly placed arrow, the bow string is again pulled by raising the mechanical contrivance, and then released. In spite of its ingenious design and highly skilled construction, the *Chu-ko* crossbow has a very weak projectile force, as demonstrated by its short range of slightly more than twenty paces. For this reason, it is used only to ward off burglars, not as a military weapon.

The concealed crossbow (*wo nu*) is used by the mountain people for hunting and is placed in key areas frequented by wild beasts. With its trigger mechanism attached to a pulling cord, the crossbow is hidden beside a track, and the arrow is discharged as animals appear within shooting range. The result of such an operation, however, is limited to the killing of only one animal.

SHIELDS

The shield and the spear were the oldest known weapons, which the ancient Chinese paired together in their references to them. The use of the shield was greatly increased at a later time, since each infantryman and cavalryman held a short sword in the right hand, and in the left hand a shield to ward off arrows. In contrast, in ancient chariot warfare, a warrior, with both hands holding a spear, halberd, or lance, had no time to handle a shield, and was protected from arrows by the special shieldman on his chariot. The shield is not more than three feet long, with its upper inside surface attached to a woven willow loop one foot in diameter. The shield's triangularly shaped top has a five-inch-high apex in the center, while its bottom is fastened to a bamboo pole that serves as a handle. A medium-sized shield, known as *tun,* or buckler, is used by the foot soldier to ward off arrows and halberds. It is commonly called *p'ang-p'ai.*

MATERIALS FOR MAKING GUNPOWDER AND INCENDIARY WEAPONS

At present, gunpowder and firearms are widely discussed by every person who aspires to become an official, and they are the subjects of books written for presentation to the throne. These books [1] are not necessarily based on experimental results. Nonetheless, I include a few pages of rough description. Gunpowder is manufactured by using saltpeter and sulphur as the principal, and the ash [charcoal] of grass and wood as the auxiliary, components.[2] It is believed that saltpeter and sulphur are respectively negative and positive in character. A combination of the positive (*yang*) and the negative (*yin*) forms gunpowder. [The fire and shock waves resulting from] the explosion of gunpowder out of a compact space will blast nearby persons and things into total destruction. Saltpeter is an upward projecting agent. The gunpowder used for straight shooting, therefore, is composed of 90 per cent saltpeter and 10 per cent sulphur. In comparison, sulphur is a lateral blasting agent. The gunpowder employed for making mines and bombs consists of 70 per cent saltpeter and 30 per cent sulphur. Ash, being the auxiliary component of gunpowder, is produced by burning the wood of willow, pine, or birch root, bamboo leaves, hollyhocks, bamboo roots, or egg-plant stalks. Of these plants, bamboo leaf is the most fiery.

To wage pyrochemical warfare, various types of incendiaries are used to produce "poisonous fire," "divine fire," "magical fire," "scorching fire," and "spraying fire." The incendiary for "poisonous fire" consists of white arsenic and sal ammoniac [NH_4Cl] as the principal components, and the subsidiary ingredients of "gold juice" (*chin-chih*), "silver rust" (*yin-hsiu*), and human manure. The incendiary for the "divine fire" is composed chiefly of cinnabar [HgS], orpiment [As_2S_3], and realgar [As_2S_2]. The incendiary for "scorching fire" is prepared from a mixture of borax, porcelain powder, *Gleditschia japonica* (*ya-tsao*), and *Xanthoxylum piperitum* (*ch'in-chiao*). The incendiary for "flying fire" is manufactured from a mixture of cinnabar, orpiment, calomel (*ch'ing-fen*), *Aconitum* (*ts'ao-wu*), and *Croton tiglium* (*pa-tou*). For attacking a fort with fire, the incendiary is a mixture of *t'ung* oil and rosin. The above is a brief summary of incendiary weapons. Some people claim that the smoke of burning wolf dung, being black in daylight and red at night, rises straight into the air, and that the ash of a river whale can be inflamed by wind. These two materials, however, must be tested before a detailed description of their pyrochemical properties can be made.

SALTPETER

Saltpeter, known also as niter, is found not only in China but also in the barbarian countries. The Chinese saltpeter industry is located in the northwestern part of the country. If a trader from southeastern China fails to pay for the official certificate that permits him to deal in saltpeter, he is judged guilty of illegal trading. Saltpeter and salt are derived from the same source. The formation of saltpeter crusts on soil is attributed to the efflorescence of the underground nitric salts in dry weather [through the capillary action of the soil]. When the surface soil is thin and adjacent to water, the effloresced product is salt. In contrast, saltpeter is formed on the thick surface soil adjoining mountains. Saltpeter derives its name *hs'iao* [to dissolve] from being highly soluble in water.

After the eighth full moon of the lunar calendar, people north of the Yangtze River and the Haui River could sweep their earthen floor every other day and obtain small amounts of saltpeter earth, which may be used in the subsequent extraction. Most of the Chinese saltpeter is produced in three provinces. The crude saltpeter from Szechuan is known as *ch'uan̊ hsiao,* from Shansi as *yen-hsiao* or saltpeter salt, and from Shantung as *t'u-hsiao* or saltpeter earth. When the crude saltpeter is swept up from the earth (and occasionally from the wall), it is put into a jar to be soaked in water for one night. After the floating impurities are skimmed off, the mixture is poured into a large pot. More water is then added and boiled until the niter is completely dissolved. As soon as the solution is sufficiently concentrated, it is transferred into another vessel in which saltpeter crystallizes out overnight. The crystals floating on the top are called *mang-hsiao* or "wheat-beard" niter; and the longer crystals, *ma-ya-hsiao* or "horse-teeth" niter. (These are the extracted essence of the locally produced crude saltpeter.) The impure crystals in the lower part of the crystallizing vessel are known as *p'u-hsiao.* If purification is desired, the mixture is again dissolved in water and boiled with the addition of a few turnips. This is then poured into a basin and let stand overnight for recrystallization. The resulting crystals are as white as snow and are called *p'en-hsiao* or "basin" niter.

Both "horse-teeth" niter and "basin" niter may be used for making gunpowder. Before use, however, the niter should be roasted either on a piece of newly made tile or in an earthen pot, depending upon the amount to be roasted. The resulting dry niter is ground in a stone mortar, but not with an iron roller, because any spark produced accidently by the latter could start a catastrophic explosion. The niter powder is mixed with a predetermined

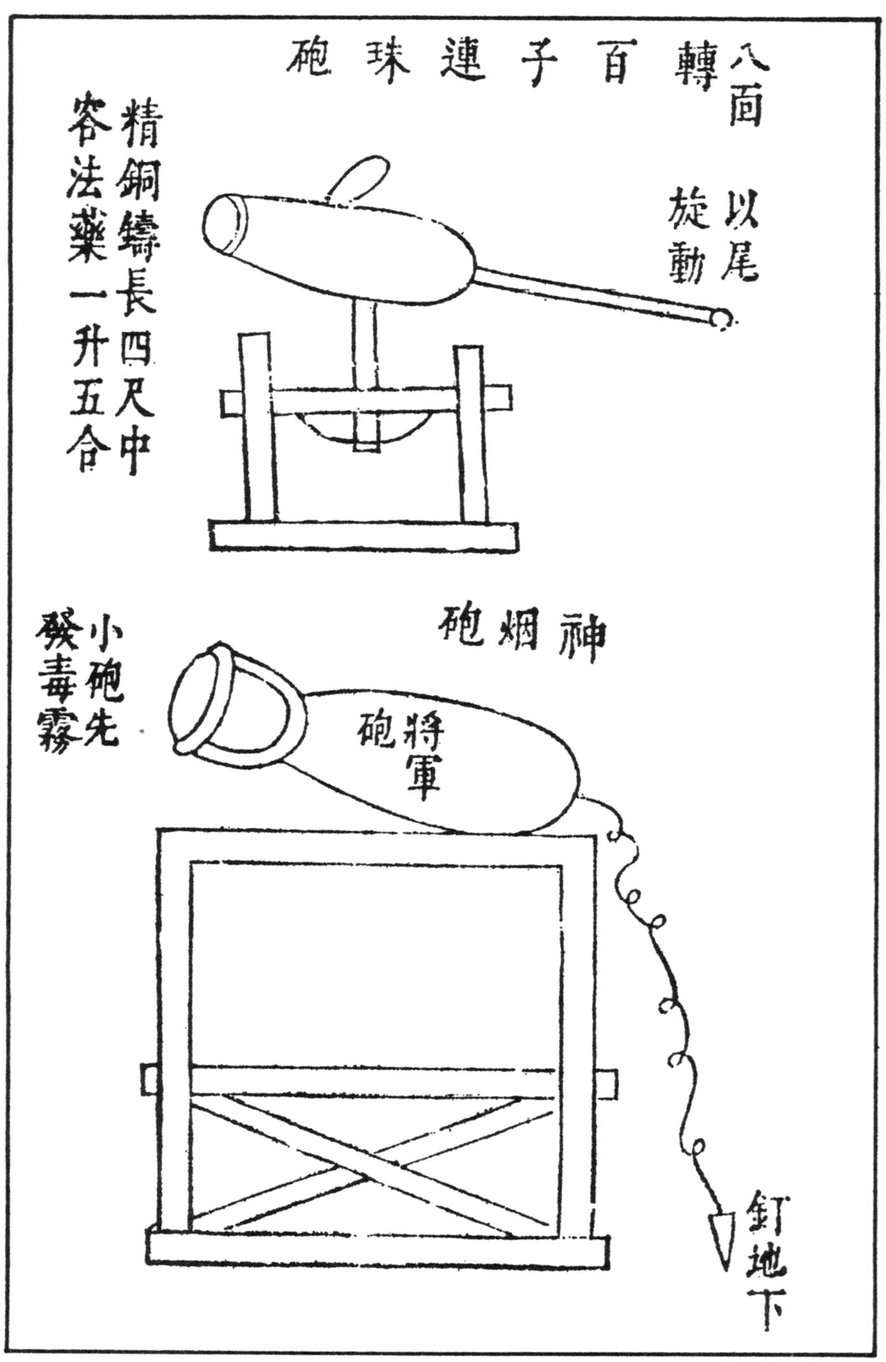

Figure 15–3. The eight-directions-rotating string-of-100-bullets cannon (top) and the poison-mist-divine-smoke cannon (bottom).

amount of sulphur, in accordance with the kind of gunpowder to be prepared, and ground again. Ash is added to the mixture at the last.

On account of the hygroscopic property of the roasted niter, a new shipment of freshly prepared gunpowder should be used in large cannons.

SULPHUR (*See Chapter 11, Calcination of Stones, for details.*)

The explosive property of gunpowder is the result of the mixing of sulphur with saltpeter. The exportation of sulphur from China to the Northern Barbarians' country is strictly prohibited, since the latter is rich in saltpeter but poor in sulphur. In the firing of a cannon, the fuse is made by mixing saltpeter with charcoal but not with sulphur. The addition of sulphur prevents the fuse from burning through the ignition-nipple of a cannon. It is difficult to crush sulphur with a roller. A mixture of one ounce of sulphur and 0.1 ounce of saltpeter, however, can be easily rolled into fine powder.

FIREARMS

The occidental cannon [3] (*hsi-yang p'ao*), being spherical in shape like a copper drum, is cast of wrought copper. When it is fired, men and horses within half a *li* are shocked to death. (The cannon is equipped with a mechanical contrivance [probably some sort of carriage] that enables it to travel on level ground, stopping when a ditch is encountered. Immediately after the fuse is lighted, the operator must run back and jump into a deep hole, because his life can be saved only when the cannon is fired on a high spot.)

The red-barbarian cannon (*hung-yi p'ao*), being cast of iron and about ten feet long, is used for the defense of cities. It is packed with several bushels of gunpowder and an iron ball. Upon firing, the iron ball is projected over a distance of two *li,* so as to pulverize any substance in its path. When its gunpowder is ignited by a fuse, the cannon is drawn back by a 1,000-catty recoiling force. This is remedied by anchoring the cannon against a wall, which is apt to be crumbled.

[Some of the commonly used cannons and guns are as follows:] "Great commander" or *ta chiang-chün;* "second commander" or *erh chiang-chün* (actually a smaller sized red-barbarian cannon; however, it is considered in China as large artillery); Portuguese calivers or Feringi or *fo-lang-chi,*[4] (being used on board warships); "three-barrel pistol" or *san-yen-ch'ung;* and "string-of-100-bullets cannon," *po-tzu lien-chu-p'ao.* [See Figures 15.3 and 15.4.]

The land mine [Figure 15.5] is buried in the ground. After its fuse is lighted by an inflamed tinder kept in a bamboo tube, the mine explodes to result in an upward blast. The land mine used for lateral blasting is made by increasing the proportion of sulphur in the gunpowder. (The mine's fuse is made of *fan-yu* [possibly bitumen or asphalt], while the cannon's muzzle is covered with a basin.)

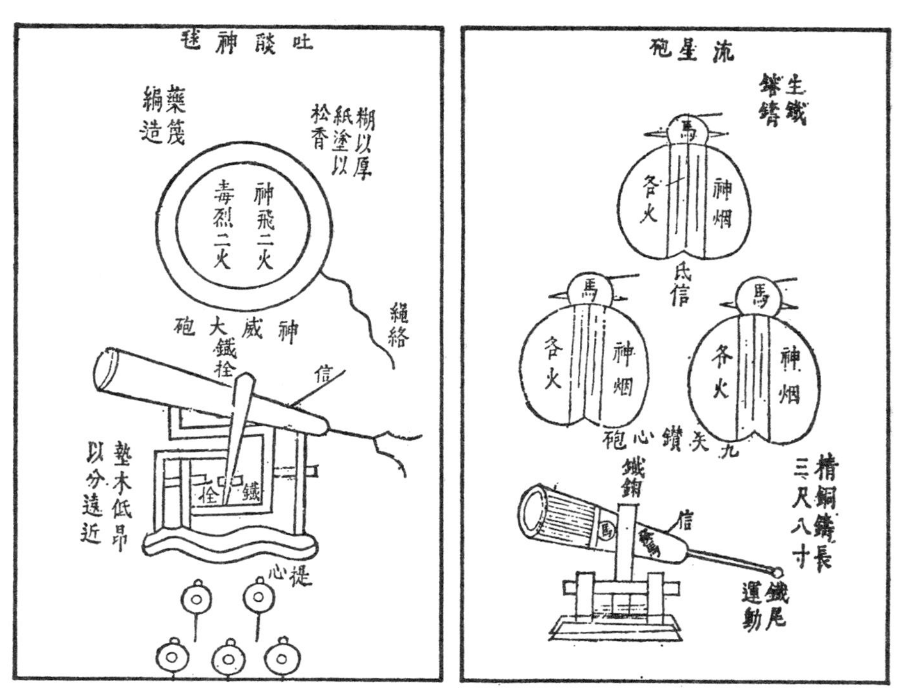

Figure 15–4. The divine-fire-ball (right-hand top), the divine-frightening cannon (right-hand bottom), the meteorite or shooting-stars cannon (left-hand top), and the nine-arrow-heart-piercing cannon (left-hand bottom).

The [stationary] submarine mine, known as *hun-chiang-lung,* is made by covering a cannon [which has been affixed to a plank of wood] with a lacquer-painted leather bag. It is sunk to the water bottom [by attaching two stones to the plank, as shown in Figure 15.6]. The trigger of a suspended flint-and-sickle inflaming device [probably a sort of flintlock] in the leather bag is connected with a rope leading to the shore. When the rope is pulled, the trigger is automatically released and thus is a passing enemy ship destroyed. In spite of its effectiveness, however, it is still a clumsy firearm.

The bird pistol,[5] called *niao ch'ung* is about three feet long [as shown in Figure 15.7]. An iron barrel containing gunpowder is inserted into a wooden stock, which can be conveniently held by hand. The barrel is made by hammering separately three pieces of red-hot iron, each wrapped around a chopstick size cold iron bar. The three resulting iron tubes are combined into a crude barrel through the heating and pressing together of their adjoining ends. The muzzle of the barrel is polished with a four-edged steel reamer, of the size of an ox gristle, so as to obtain an extremely smooth finish. This is necessary for the free discharging of pellets. The stock end of the barrel, containing gunpowder and pellets, is larger in diameter than the discharging end [i.e. choke]. Each pistol is loaded with 0.12 ounce of gunpowder and 0.2 ounce of lead and iron pellets. A lighted hemp wick instead of fuse is used to fire the pistol, (even

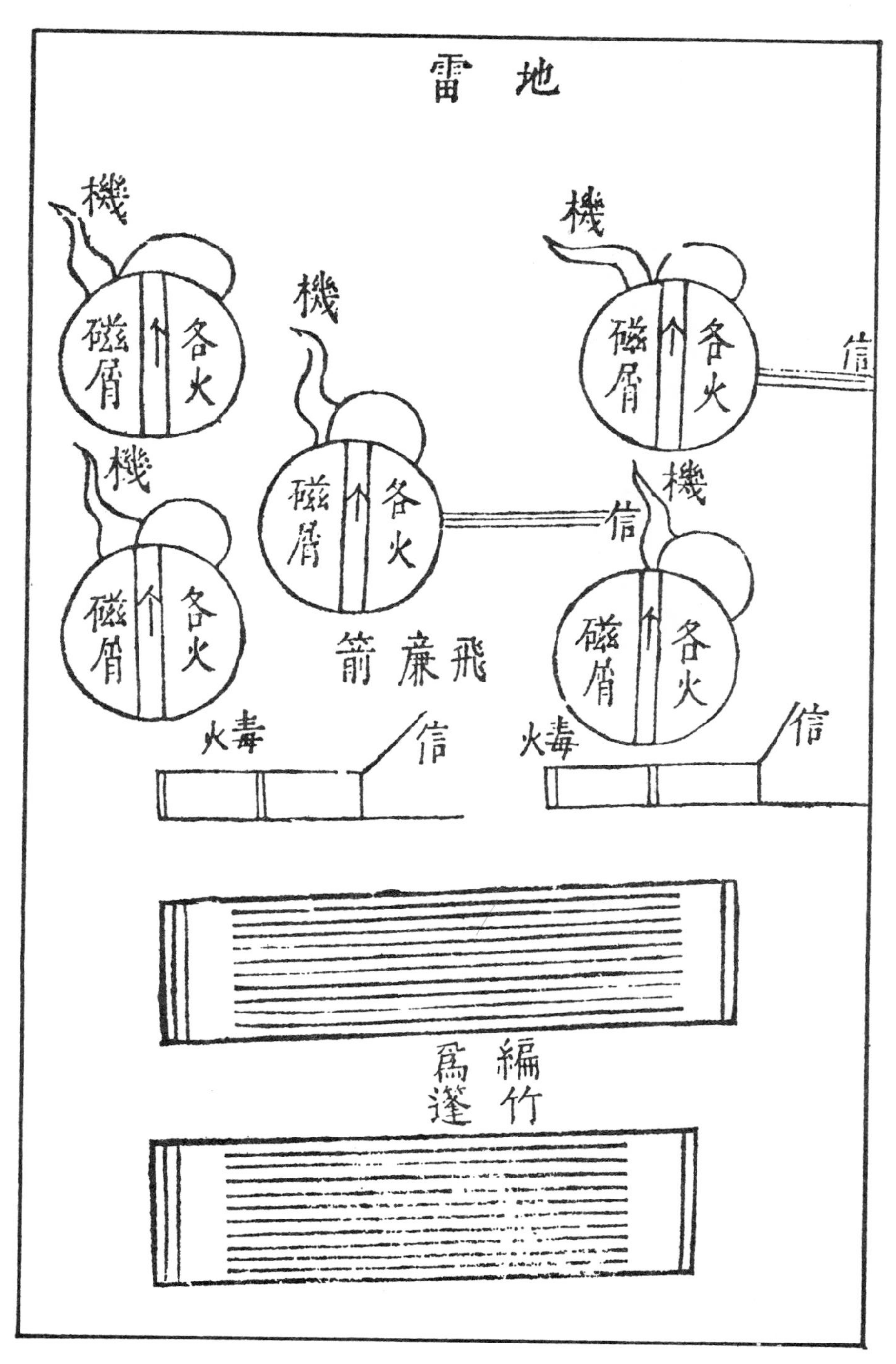

Figure 15–5. A land mine.

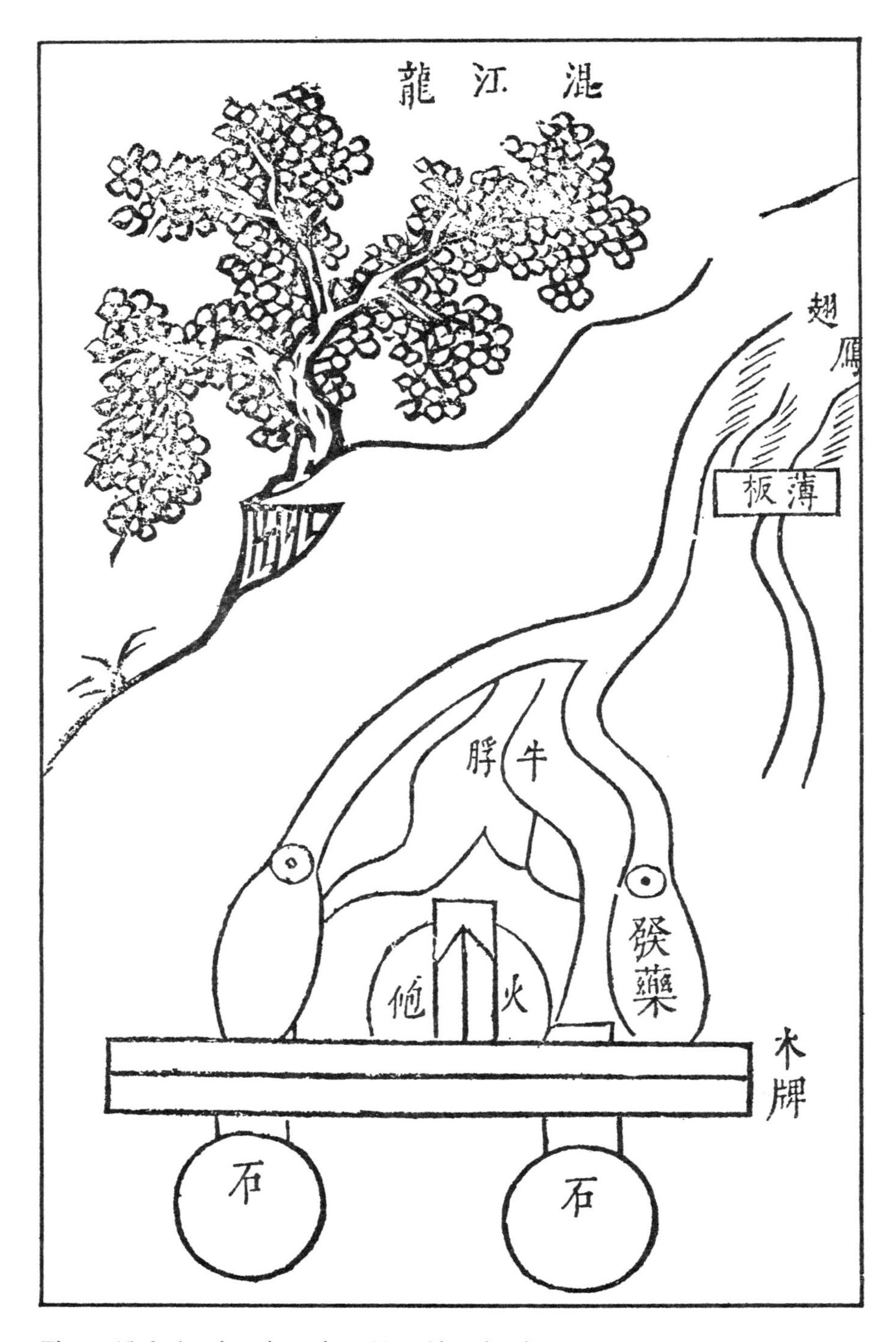

Figure 15–6. A submarine mine (Hun-chiang-lung).

Figure 15–7. The bird pistol (niao ch'ung).

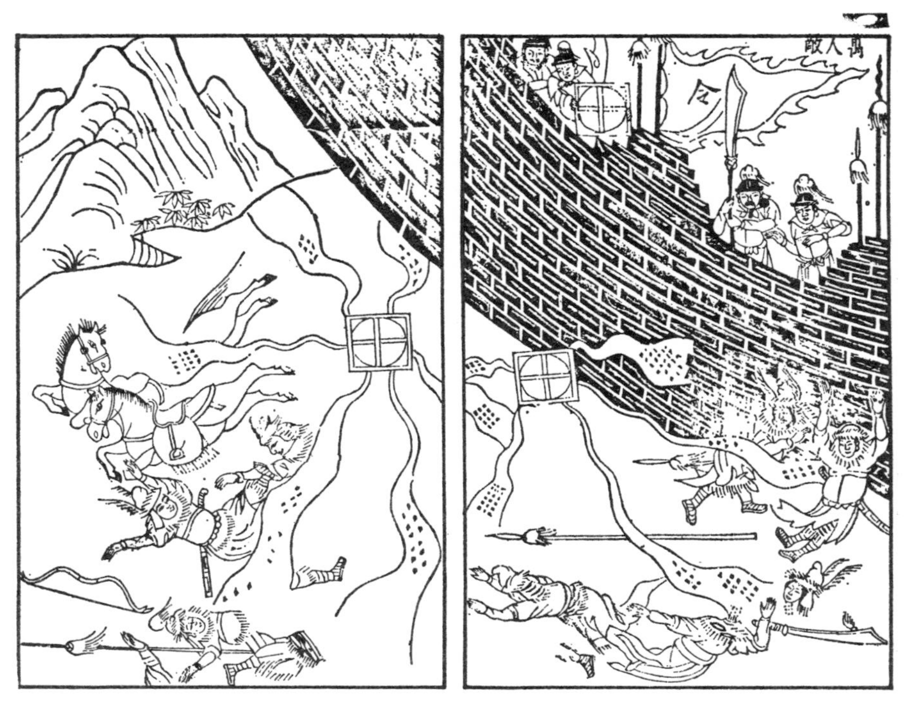

Figure 15–8. The killer-of-myriads (Wan-jen ti).

though a fuse is still employed in south China). To fire at an enemy, the shooter holds a pistol in his left hand, and uses his right hand to pull [the trigger of] the gun-lock, which brings the lighted hemp to the top end of the nipple orifice filled with gunpowder. The pellets thus projected will shatter birds into pieces within a distance of 30 paces, kill them without tearing their bodies at more than 50 paces, and are extremely weak at 100 paces. In comparison, the pellets discharged from a bird gun, called *niao ch'iang,* can travel more than 200 paces. The bird rifle or gun is longer and contains more gunpowder than the bird pistol, even through they are similar in shape and construction.

The "killer-of-myriads" or *wan-jen-ti* is a toxic incendiary bomb for fighting enemies from the top of a garrison wall. It is conveniently used [in the present dynasty] to defend the remotely located small cities, in which the cannons are either weak in firing power or too heavy and clumsy to be effective weapons. The sulphur and saltpeter in a killer-of-myriads are ignited to project incendiary flame and toxic smoke in all directions, thus killing many men and horses instantly. The killer-of-myriads is made of a dried hollow sphere of clay [Figure 15.8], which is filled with a mixture of sulphur-saltpeter gunpowder, poisonous firepowder, and divine firepowder through a small hole in the sphere. The relative proportion of these three powders can be varied by the manufacturer. After the sphere is affixed with a fuse, it [the sphere] is enclosed in a wooden frame to prevent any breakage in the fall from wall to ground. Another method of making the killer-of-myriads is to coat the inside surface of a wooden barrel with clay paste [and to fill the inner space with the above powders].

When a city is being attacked by the enemy, the defender on the city wall lights the fuse of a killer-of-myriads, and throws it downward to the ground. Incendiary flames and toxic smoke are projected from the small hole of the ignited bomb, which rotates in eight directions. When the bomb rotates toward the city, its flame and smoke are screened by the city wall from hurting the defenders. When it rotates outward, however, the enemy and their horses are both destroyed. This is the best weapon for city defense. The time required for a manufacturer to learn the nature of gunpowder and the method of making weapons is usually less than ten years, depending upon his intelligence. Those who are in charge of defense work must consider this point carefully.

NOTES

1. Four should be mentioned: *Wu-pei chih* (Treatise on Weapons and Military Equipment) by Mao Yuan-i; *Teng-t'an pi-chiu* (Necessary Knowledge for a Military Commander) by Wang Ming-ho; *Chün-ch'i t'u-shuo* (An Illustrated Guide to Weapons) by Pi Mou-kang; and *Wu-pien* (Manual of Weapons) by T'ang Ching-ch'uan. The first two, dealing chiefly with the technical aspects of military pyrotecnics, are particularly interesting.

2. If the date A.D. 1044 attributed to the original edition of the *Wu-ching tsung-yao* is correct, mixtures containing saltpeter, sulphur and charcoal, usually with other ingredients such as oils, vegetable matter, arsenic compounds etc., were known at that time. These were generally deflagrating mixtures, and some were explosive, though probably not as powerful as modern gunpowder. True gunpowder, of more or less modern composition, was known in the later part of the Mongol Yuan dynasty (A.D. 1260–1368), at which time it was also known in Europe. It is uncertain whether it was developed by the Chinese or Mongols.

3. Wei Sheng (died 1164), a gallant leader of the Southern Sung armies, launched "fire stones" to a distance of 200 paces. In the manufacture of his "fire drug" or gunpowder, niter, sulphur, and willow charcoal were employed. When the Mongols laid siege to the city of Kaifeng in 1232, the defender terrified them by means of a "heaven-quaking thunderer" (*chen-t'ien-lei*). This instrument is described in *Wu-pei chih* as an iron tube or vessel filled with a drug. When fired, the device gave forth a sound like that of thunder and could be heard beyond the distance of 100 *li* (about 33 miles). No

armor could withstand its shock. This was probably a metal bomb, and it was not until some years later that primitive cannon appeared in China.

4. "Portuguese calivers" were on board Portuguese ships that visited Canton in A.D. 1517. Through some confusion of terms, the guns received the name of *fo-lang-chi,* meaning Franks or Frankish, which was probably the name given to the Portuguese themselves by their Arab or Malay interpreters.

5. The Chinese bird pistol or bird-beak gun was probably introduced to the Chinese by the Japanese, who in turn derived their knowledge from the Portuguese seafarers, according to Davis and Ware. Lang Ying, a writer of the sixteenth century, says in his *Ch'i-hsiu lei-kao* that: "As regards the bird mouth (bird-beak) wood guns, the Japanese invaded Chekiang (in 1522) during the reign of Kia Tsing [i.e. Chia-ching], and on some of their number being taken prisoner, their weapons were obtained and the prisoners made to give instruction in the method of manufacturing them." Mayers suggests that the term bird-mouth, *niao-tsui,* probably refers to the bell-shaped muzzle of the early blunderbuss, but may perhaps be the origin of the term *niao-ch'iang* or bird gun, which now applies to muskets of all sorts. Davis and Ware believe that the term "bird-beak" refers to the shape of the lock, possibly a flintlock. It seems, however, that a matchlock instead of a flint type is described in the present book. And a shotgun would automatically be called a "bird gun."

16

VERMILION AND INK

Is there anything in the world, queries Master Sung, to excel learning and writing? The red-hot fire is the source of the deepest black [i.e. lampblack for making ink] and the whiteness of mercury can be metamorphosed into the reddest vermilion: such are the unfathomable wonders of Creation! Then, in the proper functioning of the government, the [emperor's] vermilion rescripts are applied to documents of state, [written] in ink, hence law and order are proclaimed across the land; while in myriads of ink-printed volumes, vermilion is used [to punctuate and comment], and so the truth of the universe is clarified.[1] These are the treasures that belong to a study where there is no room for such things as pearls and gems.

As to the painters, who re-create Nature in their work, they use every kind of color—some are used in their natural state, others are mixtures of different elements. In all, these represent the changes that occur through invoking the water and fire essences in Nature. How can this be possible without the aid of forces divine?

VERMILION

Cinnabar, mercury, and vermilion are essentially the same material and differ from each other only in grade, grain size, [and method of preparation]. The best cinnabar ore, which is produced in Ch'en-chou [in Hunan], Chin-chou (now called Ma-yang [Hunan]), and western Szechuan, contains mercury yet is not used for making the latter. This high-quality ore, which is used in polishing arrowheads, mirrors and the like, fetches a price three times higher than that of mercury; its sublimation into mercury will only result in reduction of the amount to be gained. It is therefore sold as pure cinnabar, while mercury is distilled only from cinnabar ore of a lower grade. Mercury, in turn, can be converted into vermilion.

Top-grade cinnabar ore is reached after digging some 100 feet below the earth surface. The first appearance of its seam has the look of white stones, which are called cinnabar bed. Some ore pieces close to the bed are the size of hen's eggs. Low-grade ore that is not used medicinally, but only in powdered form as a painter's color or for the production of mercury, is not necessarily surrounded by a bed of white stones. It is obtainable only a few dozen feet beneath the earth surface, the bed occasionally being mixed with dark or yellow stones, or sand. When the seam is filled with cinnabar, there will be cracks in the surrounding stone or sand formations. This kind of ore is produced mostly in Ssu-yin and T'ung-jen in Kweichou province, but is also found in Shan-chou and Ch'in-chou [in Shensi and Kansu, respectively].

When the mined ore is low-grade and whitish in color [because of its gangue minerals], it is used exclusively for making mercury, but is not pulverized into vermilion. If, however, the ore shows a glittering red substance though the color is principally white, it is placed in a large iron trough, where it is pulverized into fine powder by a roller. The ground ore is transferred into a large jar and mixed with water for hydraulic classification. [After settling] for three days and nights, the substance that floats on the top of the water is poured off into another jar and called "second-grade vermilion," while the part that sinks to the bottom is dried and called "first-grade vermilion" [Figure 16.1].

For the sublimation and condensation of mercury, either the white low-grade cinnabar ore or the "second-grade vermilion" obtained from the top of the water jar is used. The [ore is] moistened with water and made into large rolls which are put into pots at thirty catties per pot. Thirty catties of charcoal are used under each pot as fuel. When the ore is ready for distillation, the pot is covered with another, [which is inverted], a hole having been opened in the middle. Around the sides, salted mud is used to seal up the seams. A curved iron distilling tube, securely wound with hemp strings and plastered with salted mud, is attached to this upper pot. When the fire is lighted, one end of the tube is inserted into the pot through the opening (the place of insertion is also tightly sealed), while the other end is placed into a medium-sized [earthenware] jar filled with two bottles of water. The vapors and gases rising from the pot are conducted through the tube to the water, where they are stopped [Figure 16.2]. After ten hours of heating, the ore is converted to mercury and spread over the pot. It is swept off after cooling for one day. This is a most mysterious process of Heaven [2] (there is an unreasonable commentary in the *Materia Medica* [in connection with the distillation of mercury], that a hole is to be dug in the ground into which should be placed a bowl filled with water).

Vermilion is obtained through the conversion of mercury [into mercuric sulfide, HgS, with the aid of sulphur], hence it is termed "mercuric red." Either

Figure 16–1. Grinding and hydraulic classification of mercury ore.

Figure 16–2. Sublimation and condensation of mercury ore.

Figure 16–3. Making vermilion through sublimation of mercury with sulphur.

Figure 16–4. Preparing and collecting oil lampblack.

widemouthed earthen jars, or double pots, may be used for this process. First, each catty of mercury is mixed with two catties of red sulphur stones (out of which sulphur is made) and ground together until a fine, lumpless powder is produced. The mixture is then roasted and put into a jar which is covered securely with an iron bowl. An iron rod is next placed over the bowl [which serves as leverage] and the whole thing is tied up with wire. The seams are sealed with salted mud. Three [large iron] nails are planted in the ground under the jar, so that the latter rests on a tripod. A fire is lighted [under the jar] and lasts for as long as it takes to burn three incense sticks. Meanwhile, water is repeatedly applied to [the top of] the bowl with an old writing brush [Figure 16.3]. This results in the formation of vermilion powder, which adheres to the [upper part of the inner surface of the] jar, with the part that sticks to the rim of the jar having the most brilliant shade of red. The vermilion is scraped out of the jar after being thoroughly cooled and is ready for use. The red sulphur stone will sink to the bottom of the jar and can be used again. Each catty of mercury will yield 14 ounces of top-grade vermilion [3] plus 3.5 ounces of the second-grade. The yield is determined by the amount of sulphur added.

Vermilion produced by pulverizing natural cinnabar and that [obtained] by the conversion of mercury are similar in function. Only the natural vermilion, not the artificially made variety, is used in painting for imperial and aristocratic families. For use in the study, vermilion is molded into sticks after mixing with glue. The color will show clearly if the stick is rubbed on a stone ink slab. If [the stick is] rubbed on one made of tin, the color will turn black.[4] Lacquer workers must mix vermilion with *t'ung* oil in order to make the color bright; if mixed in varnish it will appear dark and lustreless.

There are no other sources for mercury and vermilion. Tales that mercury is obtainable from a certain grass, or from the sea, are pure nonsense and believed only by the gullible. Vermilion cannot be reconverted into mercury. This is because at this point the wonders of Divine Nature are exhausted.

INK

Ink is made of lampblack, of which one-tenth is made from burning *t'ung* and vegetable oils or lard, and nine-tenths from burning pine wood. The most valuable ink sticks of our dynasty [5] are produced in Hui-chou [of Anhui province]. There are some manufacturers who, wishing to avoid the problems of transporting quantities of oil, send their agents to stay in the prefectures of Ching, Hsiang [in Hupei], Ch'en and Yuan [in Hunan], where they can buy *t'ung* oil at a low price and return with lampblack made on the spot. If, when used for writing on paper, the black ink shows a tinge of red when placed in slanting sunlight, the lampwick used in making the lampblack was first soaked in the juice of *Lithospermum officinale* before it was burned. One catty of oil, after burning, will yield more than one ounce of fine quality lampblack [Figure 16.4]. The deft artisan is able to tend two hundred such lamps, but if the hands are slow and [the gathering of the lampblack is] delayed, the soot may ignite under the heat of the lamps, resulting in the loss of both the raw material and the product.

Ordinary ink is made from pine wood after all the resin has been eliminated. The least amount of resin left in the wood will result in a non-free-flowing quality in the ink produced. To get rid of the resin, a small hole is cut near the root of the tree, into which a lamp is placed and allowed to burn slowly. The resin in the entire tree will gather at the warm spot and flow out [Figure 16.5, right-hand diagram]. The pine tree is then felled and sawed into pieces in preparation for burning. [For making pine wood lampblack], a rounded chamber of bamboo is built. Resembling in appearance the curved rain-shield on small boats and built of sections, it has a total length of more than 100 *ch'ih*. The external and internal surfaces of this chamber and the connecting joints are all securely pasted with paper and matting, but small holes are made at certain intervals for the emission of smoke. The floor of the chamber is constructed of brick and mud with channels for smoke built in. After the pine wood has burned for several days, the chamber is allowed to cool and [workers] will now enter and scrape out the lampblack [Figure 16.5 left-side diagram]. That which is obtained from the last one or two sections of

Figure 16–5. Removing resin from pine tree (right) and making pine wood lampblack (left).

the chamber is of "pure" quality and is used as raw material for the manufacture of the best ink. The lampblack obtained from the middle sections is of "mixed" quality and is used in ordinary ink. That from the first one or two sections, however, is scraped and sold only as low-grade lampblack; it is further pounded and ground by printers and used [in printing books]. In addition, lacquer workers and plasterers also use the coarse grade as black paint.

[Prior to] the manufacture of ink, the pine wood lampblack is soaked in water for a long time to result in a floating fraction of fine size particles and a sunken fraction of coarse ones. After [the sized lampblack] is mixed with glue, the amount of hammering administered will determine whether [the finished product] will be brittle or sturdy.[6] Precious ingredients, such as gold dust or musk essence, can be added at will to either oil or pine wood lampblack. I have described here only the bare outlines of the materials and general processes of inkmaking. Learned readers are certainly acquainted with books such as *Mo-ching* [Ink Classic] and *Mo p'u* [Categories of Ink] wherein detailed information is available.[7]

SUPPLEMENT

White lead powder or ceruse: pure white color. See Chapter 14, "Metals."

Litharge: reddish-yellow color. See Chapter 14, "Metals."

Indigo: pure blue. See Chapter 3, "Dyes."

Purple powder: bright red; the more valuable variety is obtained by mixing equal parts of white lead powder and vermilion, while that of coarser quality is made with the safflower juice used by dyers.

Deep blue [smalt]: a dark blue color; see Chapter 18, "Pearls and Gems" [should be Chapter 7, "Ceramics"].

Copper green [copper acetate]: a deep green color. It is derived from yellow copper sheets which are painted with vinegar and covered with chaff. The whole thing is slightly heated and the copper green [formed on the metal surface] is scraped off every day.

Stone green: see Chapter 18, "Pearls and Gems" [Verdigris made from Malachite].

Ochre: Orange-red color. It is found in most mountains, but the product of Tai [Northern Shansi] is best in quality.

Orpiment: also called "yellow stone-kernel." It is found inside rocks that have purple exteriors, but are yellow in the middle.[8]

NOTES

1. There is evidence that either cinnabar or vermilion was used as a pigment as early as the Shang dynasty (1176–1122 B.C.), and Bagchi believes that Chinese vermilion was introduced into India during the first century B.C. Chinese records, including *Pao-p'u tzu, Chou-i ts'an-tung ch'i,* and *Chin-tan ta-yao,* show clearly that vermilion was prepared and used by Chinese alchemists for prolonging life and transmuting metals. The oral traditions of Taoism attribute alchemy to the Yellow Emperor and also to Lao-tzu (ca. 604–500 B.C.), but the late Chou period (ca 400–255 B.C.) would seem to be historically sounder.

2. A number of early Chinese recipes for extracting mercury are given in various alchemical books, including *Pao-p'u tzu* (written by Ko Hung in the Tsin dynasty, A.D. 281–361) and *Tao tsang* (first printed in the Sung dynasty A.D. 1186–91). From the modern point of view, the method of the present book may be explained by the fact that mercury is not easily oxidized, hence roasting the sulfide mineral produces metal instead of oxide.

3. The process of forming vermilion by sublimation of mercury together with sulphur, as related in this chapter, is still described in English and German technical books as the "Chinese method."

4. Dark-brown stannous sulfide, SnS, would be formed by rubbing cinnabar or vermilion on a tin slab. According to Carter, cinnabar was used in China as a red ink from the time of the Han dynasty, and probably earlier. This substance is still used for taking impressions from seals, a practice that may have begun previous to the invention of black ink.

5. Lampblack ink, known in English as "India ink," was probably first made by Wei Tan, known also as Wei Chung-chiang, in the Wei dynasty (A.D. 220–65).

6. The manufacture of Chinese ink sticks involved mixing, cooking, pounding, rounding, shaping, molding and drying. In the process of mixing, the lampblack was incorporated with the solution of glue and medicinal decoction to form a thick paste. The mixture was then kneaded into small pieces which were wrapped in cloth and steamed in a cooking vessel. After steaming, the pieces were taken out and pounded with mortar and pestle. This alternate steaming and pounding process was repeated until a homogenous and flexible mass had been obtained. The pounded mass was rolled, still hot, into long strips which, in turn, were cut into small pieces, each weighing about one tael and four mace. After being moistened in a porcelain jar, these pieces were hammered individually for about four hundred strokes to harden the mass and make it smooth. The hammered mass was rolled, with the addition of camphor and musk, at an even temperature. The rolled mass was first rounded or shaped and then molded into an ink stick. After drying, the ink stick was rubbed with a piece of coarse cloth and polished with wax. Finally, it was wrapped in paper and stored.

7. *Mo-ching,* by Chao Kuan-chih of the Sung dynasty, and *Mo-p'u* by Li Hsiao-mei of the same period. Another work, *Mo-fa chi-yao* [Essentials of Ink Making], by the Ming author Shen Chi-sun, gives special attention to the manufacture of ink from oil lampblack.

8. Orpiment (As_2S_3) and realgar (As_2S_2) were known in the Chou dynasty. In modern times, large quantities of orpiment have been exported from Tali, Yunnan.

17

YEASTS

Master Sung considers that the bad effect of alcoholic drink on society has resulted in an increased number of crimes and litigations. Yet originally, wine was entirely blameless. It was when [people were] sacrificing to Heaven or commemorating their ancestors, or while performing music and singing at court or in the country, that wine and sweet-wine, made from yeast and malt respectively, were offered. Thus, [wine] was first made by sages, as was plainly written down in records.[1]

[In making yeasts,] the essence of the "five grains" is changed. After the [ground] grains are mixed with water [and kneaded into] solid forms, they are transformed into yeast or malt by the breeze. The kind used by physicians is known as medicinal yeast, while that which preserves food and dishes is red in color. Since ancient times, numerous blends of principal and supplementary materials have been used [for making yeasts]. These help to promote health and long life and get rid of illnesses: it is not possible to mention in detail all the benefits that they bring. Had not the Fiery Emperor and the Yellow Emperor invented the use of yeasts, how could the lesser minds of later ages achieve these skills?

WINE YEAST

Wine yeast, or leaven, is needed for the fermentation of wine, for without it no wine can be made, even with the best of grains. In ancient times, yeast was used for making wine and malt for making sweet-wine. In later times, however, the manufacture of sweet-wine was discontinued because its taste was thought to be too weak, hence the art of malt-making also became lost.

[Wine] yeasts can be made from wheat, rice, or [wheat] flour, according to local conditions, due to differences between north and south China. The principles, however, are the same. Of these, "wheat yeast" can be made from either wheat or barley. In hot summer weather, the wheat grains in the hulls are washed with well-water and dried in the sun, ground into a meal, and shaped into cakes after being mixed with the water in which the grains have been washed. After the cakes are wrapped in paper-mulberry leaves, they are either hung in the air or covered with rice stalks to induce the growth of the yellow mould. The yeast is ready for use after forty-nine days of such culturing.

"Flour leaven" is made by cooking five catties of white [wheat] flour and five pints of soy beans in the juice of smartweed until pasty, which is then mixed with five ounces of powdered smartweed [2] and ten ounces of apricot seed paste. The mixture is kneaded into cakes, then either hung up or covered with rice stalks for fermentation, as previously described. Another variety is made by culturing of yellow moulds in cakes made of glutinous rice flour and smartweed juice; neither the method nor the length of fermentation differ from those described before.

There are innumerable varieties of the chief and supplementary materials as well as flavoring herbs that go into wine-making. They range from a few to some hundred ingredients [for an individual wine], which differ in different parts of the country. In recent times, the kernel of barley is used as the chief material to mix with yeast for making the Job's-tears wine in Peking, while lentil is the principal material for preparing the green lentil wine in Ning-po and Shao-hsing of Chekiang province. These are the two leading wines of the country. (They are described in the *Chiu ching* or *Wine Classic.*)

If, at the wine-yeast maker's home, the culturing of moulds is not properly done, or not regularly tended to, or if the implements are not properly cleaned, then a few pellets of the badly prepared yeast can spoil whole bushels of the [wine maker's] grain. For reliable yeast, therefore, the wine maker must procure it from those who have good reputations and are well-known as leaven manufacturers.

The yeast for making the yellow wine [3] in Hopei and Shantung comes from Anhui, and is shipped to North China by boat or cart. The red wine produced in South China is made with the same kind of yeast as that from Anhui. It is called "fire yeast" in general, except that the kind sold in Anhui is shaped into square blocks, while that of South China is made into round cakes or balls.

Smartweed is the spirit of yeast; and grains are the body. [In yeast's manufacture,] however, some old wine-mash must be added so as to bring the spirit and body together. It is not known when the old wine mash first originated—this is similar to the use of old vitriol wastes in the calcination of vitriol stone [i.e. "copper coal," as cited in Chapter 11 of the present book].

MEDICINAL YEAST

Medicinal yeast is for medical use by physicians and is different from wine yeast. First originating in the T'ang dynasty, this yeast has never been used in

Figure 17–1. Washing fermented rice in a mountain stream.

the making of wine. It is made of white [wheaten] flour only. To every 100 catties of flour [certain amounts of] the natural juices of artemisia [*Artemisia apiacea*], smartweed [*Polygonum posumbu*], and burweed [*Xanthium strumdrium*] are added, and the resultant mixture is shaped into cakes. The cakes are wrapped in hemp leaves or paper-mulberry leaves, and covered to induce fermentation, as in making soybean sauce. After [the cakes] are covered with yellow mould, they are dried in the sun and stored. As for the addition of other drugs, this is done in accordance with the wish of the individual physician as there is no fixed prescription for it.

RED YEAST

The making of red yeast is an innovation of more recent times. Its principle lies in deriving wondrous powers out of the odorous and the rotten, and the method [of making it] involves the transformation of the essences [of grains]. When thinly spread on fish or meat—the things in the world most easily decomposed—yeast will help them to retain their fresh qualities even at the height of summer and for as long as ten days no fly will come near them, and their color and taste will remain as fresh as before.[4] This is indeed a marvelous drug!

Grains of the common [i.e. non-glutinous] rice, whether of the early or late variety, are pounded and hulled to the most excellent whiteness and then soaked in water for seven days. When the odor has become unbearable, the grains are taken to a river and rinsed clean with the free flowing water (only the running water of mountain streams should be used; water from large rivers will not do) [Figure 17.1].

After washing, the odor still remains, but when the material is steamed in a pot it will change and give off a most fragrant aroma. When half cooked, the rice is taken out of the pot and quickly immersed in cold water. When the rice has cooled off, it is steamed again; this time being allowed to cook thoroughly.

The cooked rice is placed together, several *tan* to a heap, for the addition of leaven. For making red yeast, the leaven must be manufactured from the best red colored wine mash at a proportion of one peck of mash to three pints of the natural juice of smartweed mixed in alum water. Two catties of this leaven are added to every *tan* of steamed rice while the latter is still hot, then mixed quickly by several pairs of hands until it has cooled. The mixture is allowed to stand for a considerable length of time under constant observation, so that the rice can be definitely fermented by the leaven. The occurrence of fermentation in the mixture is indicated by a slight rise in its temperature. [In actual preparation], the steamed rice, mixing with leaven, is put in large bamboo baskets and washed once with an aqueous alum solution. The mixture is then divided into separate woven bamboo trays and placed on shelves in order to catch the breeze [Figure 17.2]. From now on, the air will be the determining factor for culturing the yeast, fire and water exerting practically no

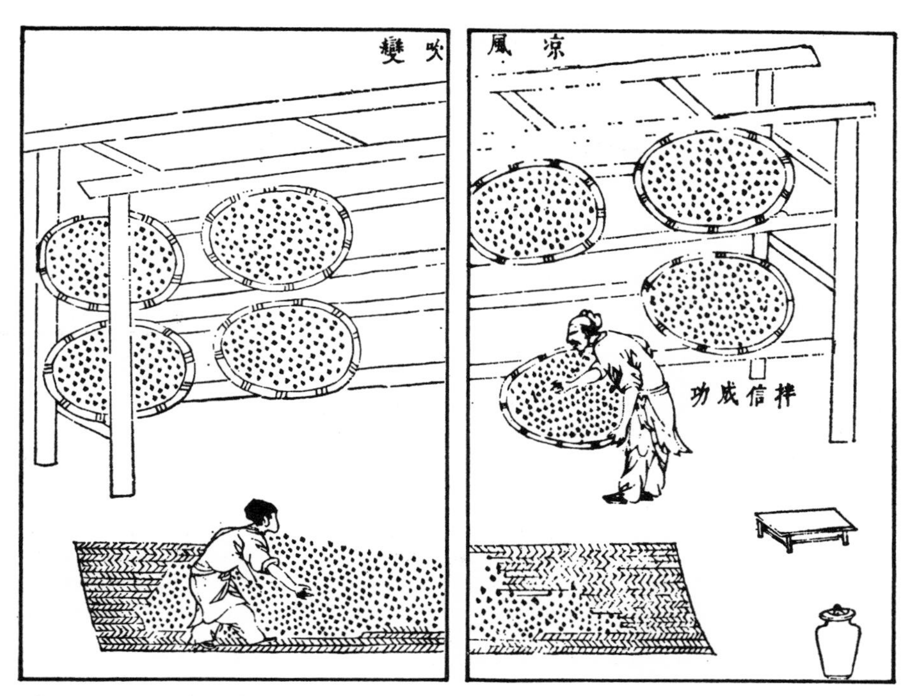

Figure 17–2. Using air to ferment steamed rice in bamboo trays.

influence. Each bamboo tray contains about five pints of the steamed rice. The room [in which the trays are shelved] should be large and high-ceilinged, so as to keep the pressure of heat from the roof, and the room should face south so as to escape the [strong] afternoon sun. The material is stirred about three times in every two-hour period. For seven days, the people tending the yeast will stay constantly near the trays, never daring to sleep soundly and rising several times during the night.

At first the rice is snowy white, but after one or two days the color turns pitch black. From black it turns to brown, from brown to rust, from rust to red, and at its brightest the red color again changes into a light yellow. With the help of air currents, the substance will go through all these stages of change right before one's eye, and this process is called "cultivation of yellow yeast." The yeast produced through such a process is twice as valuable and potent as ordinary yeast. The rice is washed once with water between the black and brown stages, and once more between the brown and red stages. After it has turned red, however, it is not washed again.

In making this yeast, it is necessary that the workers' hands, and the trays and mats used, be absolutely clean. The slightest bit of dirt will bring the entire operation to ruin.

NOTES

---

1. The present book and *Materia Medica* both relate that winemaking was practiced in the time of the Yellow Emperor (2697–2598 B.C.). *Chan-kuo ts'e* (Policies of the Warring States) relates that rice wine was first made by I-ti, the daughter of a mythical ruler, for Emperor Yü (2205–2198 B.C.). *Shuo-wen,* however, attributes the invention of rice wine to Tu K'ang or Shao K'ang who lived about the same time as I-ti. The invention of wine by the ancient Chinese in the time of Yü or earlier is partly supported by the discovery of large numbers of wine vessels at the An-yang excavation.

The technique of fermenting grape wine was introduced into China Proper from Chinese Turkestan during the Han dynasty. Ssu-ma Ch'ien (163–85 B.C.) relates in his *Historical Records* that the rich stored ten thousand piculs of grape wine for several decades without its spoiling. *Materia Medica,* on the other hand, attributes the origin of grape wine to the T'ang (618–906) dynasty. It is said that a distilled spirit, called *shao chiu* or *paikan chiu,* was introduced into China from Central Asia in the Yüan dynasty (1279–1368). The foreign origin of the distillation process is evidenced by the name of one distilled spirit, *A-la-ku,* which would seem to be a transliteration of the Near Eastern term "arrack."

2. Powdered smartweed has long been used in China for making wine yeasts. It is prepared in the summer by drying and pulverizing the leaves of the wild *liao* (*Polygonum posumbu*) or smartweed before it has blossomed. This powder contains enzymes as diastase capable of converting starch into sugar. The natural juice and the prepared powder of smartweed were also used in China for the treatment of diseases.

3. *Huang chiu* or "yellow wine" is the general name of a rice or millet wine widely used in China. It has many local varieties, the most famous being the "Shao-hsing wine" prepared from glutinous rice at Shao-hsing in Chekiang province.

4. "Red yeast" may have contained *Penicillium* mold, which thrives in fermented soybean mash, moldy bread, and moldy cantaloupe.

18

PEARLS AND GEMS

Is it actually true, queries Master Sung, that the splendor of the mountains is contained in jade and the glamor of water concentrated in pearls? Or are these but random notions of the imagination? In the world of Nature's creations, the bright and clear things are usually the opposites of the dull and turbid, and the rich and fertile have as opponents the things that are withered and harsh. The valuable and the valueless qualities are supplementary to each other in the same locality.

A distance of twenty thousand *li* separates Ho-p'u and Yü-t'ien,[1] yet in one place pearls reign supreme, and in the other jade predominates. Once these gems have become known to man, they gain great favor among the people, who use them to shine and sparkle in their temples and mansions. Of the countless treasures of China, gems have succeeded in occupying the place of honor. Is it possible that the glory of China's mountains and the glamor of her waters have all become concentrated in human bodies and that the essence of excellence is limited in the world?

PEARLS

Made by mussels [or oysters], pearls are conceived inside the mollusks through the shining of moonlight; the most valuable ones being the oldest. It is incorrect to hold, as some people do, that pearls are also contained in snake intestines, dragon jaws, and shark skins.

The pearls produced in China come from the two [pearl] beds of Lei-chou and Lien-chou.[2] Before the Three Dynasties, the Huai River region and Yang-chou were also considered as southern territories, which were closer to the pearl-producing localities. It is recorded in the *Tributes of Yü* that "the Huai Barbarians presented pearls." These pearls might have been obtained [by the Huai people] through trade, as the tributes were not necessarily local products. Although it has been reported that under the Chin dynasty [1127–1234] pearls were obtained from the Pu-yü March,[3] and in Yuan dynasty [1271–1368] from Yang-ts'un and Chih-ku-k'ou,[4] that is but faulty information transmitted erroneously through the years. No pearl was ever found in those places! As for the record that [during Yuan] pearls were produced at the Hu-lü-ku River, the fact is that it is not in China, but lies in barbarian [Mongol?] territory.

When a mussel [or an oyster] conceives a pearl, it does so by creating substance out of immaterial matter. While the common small aquatic species are often eaten by other fishes and therefore are short-lived, the mussel [and oyster] are protected by their hard shells. Even if they were swallowed whole [the eater] would not be able to digest them. Hence only the mussel [and oyster] can live to be centuries old and create the priceless treasure [of pearls]. When a pearl has been conceived, the mussel [or oyster], even if it lives in water a thousand fathoms deep, will open its shell when the full moon shines, letting the moonbeams fall on the pearl, of which the form is made out of the essence of the moon. The Harvest Moon especially delights aged mussels—on a clear night they will float with opened shells all night long, following the course of the moon and turning in every direction to absorb the moonlight. The reason why other shore areas [besides Lei-chou and Lien-chou] do not produce pearls is because they are pounded and shaken by surfs, making it impossible for mussels to rest quietly.

The pearl bed of Lien-chou stretches from the Tu-lan Sands of Wu-ni to Ch'ing-ying, covering a distance of 180 *li.* That of Lei-chou covers the area between Tui-lo Island and the outskirts of Shih-ch'eng, a distance of some 150 *li.* The pearl mussels [and pearl oysters] are gathered by the Tan people, who are most devout in offering sacrificial animals to the "Sea God" in the third month of every year.[5] These people eat raw sea food and are able to distinguish the various colors of the sea water after diving into it. Spots occupied by the flood-dragon [as indicated by the color] are left strictly alone.

Pearling boats are constructed somewhat wider and rounder than other vessels. Large numbers of straw mats are taken along on an expedition. When a boat comes across an eddy, mats are tossed into the water, thereby gaining a safe passage. The diver is tied at the waist with a long rope, [which is attached to a winch and attended by people] on the boat, and dives into the water carrying a basket. The open end of a curved tin pipe is attached to his nose and mouth enabling him to draw breath, and his ears and neck are covered with a neckwear of cured leather. The farthest depths that pearl divers reach are about four or five hundred *ch'ih,* where they pick up the mussels [and oysters] and put them in the baskets. When a diver feels short of breath, he will pull on the waist rope and will be quickly hoisted out of the water [Figure 18.1]. Some, whose luck happens to be adverse, are drowned. A diver who has just emerged from the water must be wrapped immediately in a hot boiled woolen blanket,

Figure 18–1. Divers collecting mussels and oysters.

Figure 18–2. Collecting mussels and oysters with a dip-net drag.

otherwise he will die of cold shivers. In the Sung dynasty, an official by the name of Li designed and constructed a contrivance, [here termed "dip-net drag," to collect pearl mussels and oysters from the bottom of river and sea without the use of divers]. The frame of the contrivance is made of iron and consists of a toothed drag [attached under a hoop]. Stones are hung from the two corners of the [hoop, causing the contrivance to dip into the bottom more or less deeply]. A wooden pole is fixed vertically on top of the [hoop, and is handled by the fisherman in the stern to change the course of dragging and depth of dipping]. A sack, made of hemp rope, is fastened to the [hoop and trails behind it]. The contrivance is tied to the sides of the boat. The boat sails with the wind, [while the toothed drag digs up the mussels and oysters which, due to the motion of the boat,] roll into the sack [Figure 18.2]. But even with this method wrecks and drowning occur. Both methods [i.e. divers and dip-net] are employed nowadays by the Tan people.

Like jade that is still encased in its rock crust, the value of a pearl inside a mussel [or oyster] is unknown, and becomes manifest only after it has been taken out and examined. The large pearls range from 0.5 to 1.5 inch across. There is a variety known as "pendant pearl," which is slightly oval in shape, somewhat resembling an inverted cooking pot, with one side highly lustrous suggesting gold plating. One of these is worth as much as a thousand taels of silver. This pearl, since ancient days, has been labeled "bright moon" or "light at night." Actually, these beautiful names have been accorded to the pearls because they glimmer with a thread of light if held against the sun on a fair day, not because there are pearls that really shine in the dark of night.

Next in value are the "running pearls" which, if placed in a plate with a level bottom, will roll around unceasingly. They fetch about the same price as the "pendant pearls" (one "running pearl" placed inside a corpse will prevent decomposition, hence the royal families are willing to buy them at a high price). Next are the "shining pearls," which have good luster but are not very round in form. Next are the "snail shell pearls," followed in order by the varieties known as "official rain," "tax," and "green tally." There is a "young pearl" that looks like the grain of the large millet, an "ordinary pearl" that looks like a pea, and the odd-shaped and fragmentary pearls are called *chi*. [All these grades of pearls], from "light at night" to the *chi,* call to mind the human ranks, from prince to ordinary subjects and servants.

Pearls in Nature are limited in number, and become exhausted if they are gathered too frequently. If left undisturbed for a few decades, the mussels [or oysters] will have a chance to exist in peace and increase their progeny, and thus be able to create the precious gems in large quantities. This is the reason for the saying, "pearls return where pearls [mussels] are." This is a fixed law of the universe and [any increase in pearl output] is not due to the moral suasion of virtuous officials (in the present dynasty, 28,000 ounces were gathered in one expedition in the Hung-chih reign [1488–1505], but only 3,000 ounces were obtained at one time in the reign of Wan-li [1573–1619], which was not worth the cost).

GEMS

Gem stones are produced in [underground] pits. Most of the gem stones occur in the various Western Barbarian regions, while in China they are produced only at Chin-ch'ih-wei and Li-chiang in Yunnan province.

All large and small gem stones are encased in rock beds, being similar to jade as enclosed by its gangue crust. Gold and silver are formed underneath a cover of accumulated earth, but gems are different. They are exposed from the bottom of their pits directly to the air, and are formed by absorbing the essence of the sun and moon, which give the gems their brilliance. This is the same principle that causes jade to form in rushing streams, and pearls in watery depths.

The gem-producing pits are waterless even at great depths, this being one of the clever arrangements of Nature. They are filled, however, with a foggy "gem vapor" which remains in the pits, and this gas is fatal to man after prolonged exposure.[6] The gem miners, therefore, often work in groups of ten or more, with half that number descending into the pits and the other half remaining above ground. The man who goes down into the pit is tied at the waist with a long rope; two bags also hang from a belt at his waist [Figure 18.3]. As he nears the gem stones, he quickly picks them up and drops them into the bags (gem pits are free of snakes and insects). A large bell is also tied to his waist. When he begins to feel suffocated by the gem vapor he immediately rings the bell, and the people above ground pull him up by the rope. The miner, even if not ill, will have become unconscious [Figure 18.4]. This is remedied by a drink of boiling-hot water and abstinence from food for three days. He will then slowly recover his normal health.

There are large, medium, and small stones in the miner's bags, ranging from the size of a bowl, to that of a man's fist, to that of a bean. Their respective qualities cannot be known until they have been filed and opened by the gem polisher.

Those stones belonging to the category of red-yellow gems are: *mao-ching* or "cat's-eye" or "tiger's-eye" [chrysoberyl], *mo-ho ya* [carnelian], *hsing-han-sha* [probably aventurine, sometimes called goldstone], *hu-p'o* [amber], *mu-nan* [probably beryl], *chiu-huang* [topaz], and *la-tzu* [ruby or rubellite]. Tiger's-eye is yellow with a reddish tinge. The highest priced amber is called *i* (worth five times the cost of gold in price) and is red with a dark shade. Seen in daylight it is black, but in lamplight it is bright red. Beryl is pure yellow in color, and ruby pure red. Some foolish writer in days of old marked down amber together with the China-root fungus in association with pine trees. What nonsense![7]

Within the category of blue-green gems there are *pi-pi chu* [sapphire], *tsu-mu lü* [emerald], *ya-hu shih* [possibly oriental bluish topaz], *k'ung ch'ing* [malachite], and the like (when the internal substance of malachite has been taken, its crust is roasted and crushed to become *tseng ch'ing* [possibly malachite powder]). As for the *mei kuei* or "round" gems [probably garnet or mica] of the sizes of beans or green lentils, they are of all colors—red, green, blue, and yellow. The *mei kuei* gems occupy the same rank among gem stones as that of *chi* among pearls. In addition to the previously cited *hsing-han-sha* [aventurine], there is a better variety called *chu-hai chin-tan* or "boiling-sea elixir." [This kind of reddish gem stone,] however, is produced only in the Western Regions and cannot be found in the mines of Yunnan province.

Amber lends itself most easily to simulation. The better grade is made of melted sulphur, the lower grade of ox or sheep's horn boiled in dark red juice, the result showing a red shade when held before the light. But it is now very easy to distinguish a piece of false amber (genuine amber produces a paste when rubbed against a hard surface). As for the notion that [amber] is capable of attracting lamp-wick grass [*Juncus effurus* var. *decipiens*], it is baseless talk that serves only to confuse people.[8] With the help of the human spirit, all materials can attract the small, light particles. The mistaken notions of Naturalists are hereby stricken out of the record so as to prevent damage to books [i.e. learning].

JADE

All the best-quality jade imported into China is produced at Ts'ung-ling in Yü-t'ien (the latter is the name of a western state during the Han dynasty; in later times it was known for a time as Beshibalig, and for some time was under the rule of the Ch'ih-chin mongols. Its exact name is not known.) This place is known as Lan-t'ien, which is a special term for the jade-producing area in Ts'ung-ling. It has been mistakenly identified with Lan-tien district in Hsi-an prefecture. A river springs from A-ju Mountain and divides into two at Ts'ung-ling, one called the White-Jade River [Figure 18.5] and the other, Green-Jade River [9] [Figure 18.6]. The book *Travels in Western Regions,* by the Tsin dynasty [265–419] author Chang K'uang-yeh, mentions a Black-Jade River, which [mention] is erroneous.

Crude jades [or jade-containing rocks] are not buried deeply under the earth, but instead are formed in places where precipitious mountain streams rush by. Because of the swiftness of the current, however, miners do not obtain jade from its original deposit. Rather, they wait until the streams are swollen in the summer, when the crude jades will be carried away by the current and can be gathered from the river, perhaps 100 or even 200 or 300 *li* from their original deposits. Since jade is formed by the shining essence of moonlight, those who gather it along the river banks often keep watch there under the autumn moon. When an especially brilliant patch of light is spied, there the crude jades are sure to be found. After traveling together with the current [for a considerable distance], the jade-containing rocks are still mixed with oridnary rocks in the shallow water and must be picked up, examined, and identified.

The White-Jade River flows southeastward, the Green-Jade River northwestward. There is a place called Wang-yeh in I-li-ba-li where numerous crude-jade rocks are found in the river. According to local custom, unclothed

Figure 18–3. Miner in a gem pit.

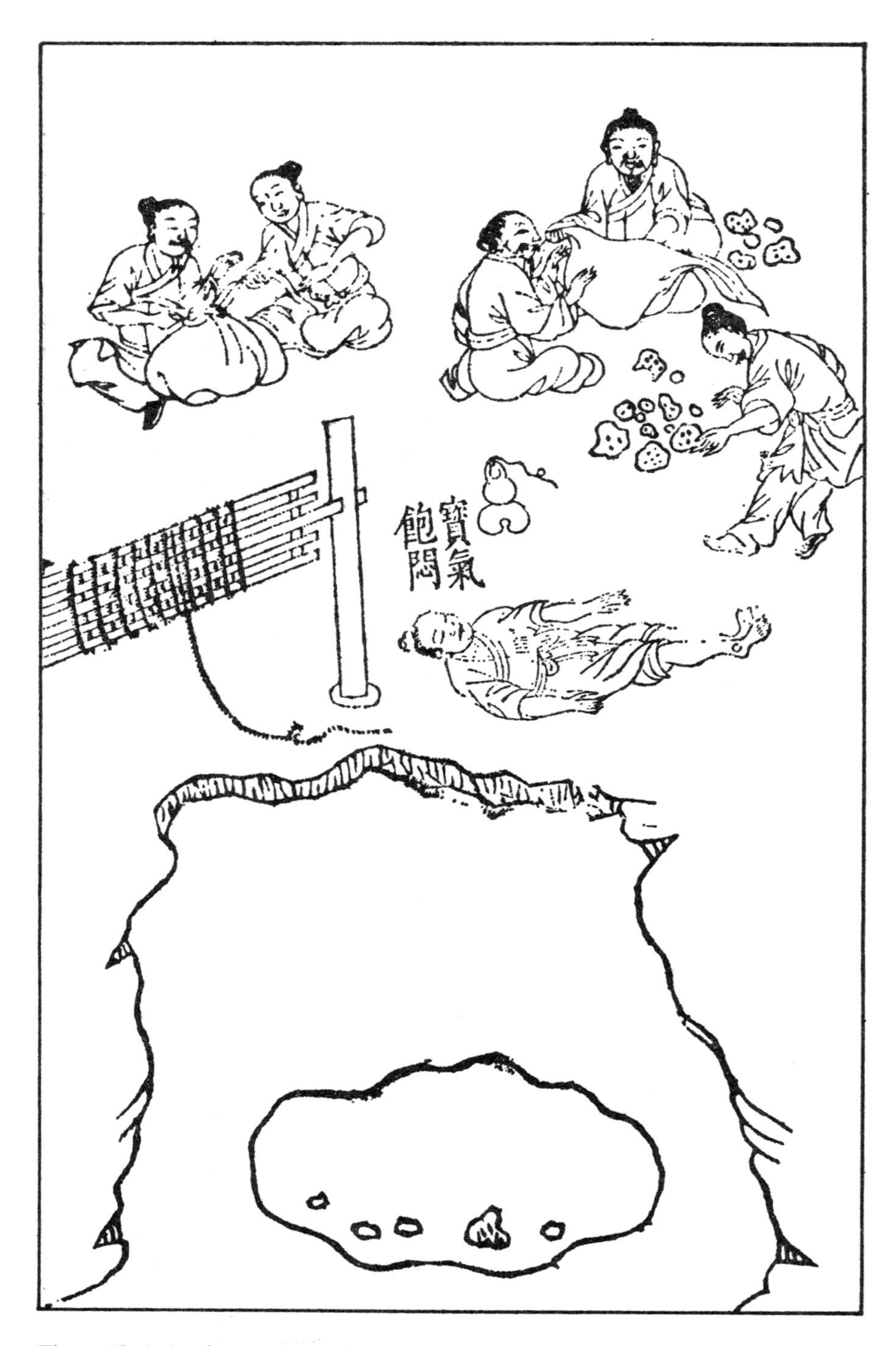

Figure 18–4. A miner overcome by gem vapor.

women are sent to gather these. The theory is that, [since both women and jade are of negative nature or *yin*,] they will attract each other. As a result, the jades will not disappear and can be easily picked up by the women [Figure 18.6]. This is probably an indication of the barbarians' ignorance. (Jade is little valued in the barbarian country and it is allowed to flow down the river for hundreds of *li*. If not sold because of the inaccessibility [of buyers], it is readily abandoned as useless.)

There are only two colors in jade, white and green; the latter known as "vegetable jade" in China. As for the so-called red jade or yellow jade, they are varieties of unusual stones, spinel and the like, which are not jade even though they cost no less than the latter.

The veins of jade are embedded in mountain boulders. Before these boulders are pushed out from their original locations [and broken into fragments or pebbles] by water, the jade within them is soft like cotton wool. Once in the stream, it begins to harden and becomes even harder after it is carried into the river mud and fanned by the wind. It would not be correct, however, to say that there is literally a kind of "soft jade" that can be polished for use.[10] The outer layer of jade within the rock is called "jade skin," and is used for making such things as inkslabs, etc., which do not cost much. Sometimes the piece of jade in the rock measures more than one *ch'ih* in diameter and is without the slightest blemish. In ancient times this was made into sceptres used by kings and emperors. This sort of jade is a rarity, known as "the jade worth the value of cities." Pieces measuring some five or six inches across are used to make cups and bowls and are considered great treasures. Besides the above, the only unusual jade is produced in So-li in the Western Ocean. Under ordinary light, this jade appears white in color, but under the sun red color is reflected from it, and on rainy days it turns blue. We may call this "uncanny jade." It is part of the Imperial Palace treasures.

In the T'ai-wei Mountain of Northwestern Korea, there is another variety known as "Thousand-year jade-containing rocks" in which is encased "sheep's fat jade," which is as good in quality as the fine Ts'ung-ling jade. In addition, descriptions of other [kinds of jade] have been recorded in books, but I have not been able to corroborate these through seeing them or hearing about them.

The [hand picked] crude jades are brought, with the aid of either boat or camel caravans, to Kan-chou and Su-chou [11] by the local Turbaned Moslems, having passed through Chuang-lang and Chia-yü Passes. (The tradition of these local Moslems requires that a new layer of turban be added to a man's head with each additional year, so in old age a man's headdress is extremely clumsy. These people are therefore called Turbaned Moslems. Even their king is careful not to show his hair. When asked the reason for this practice, they answer that the sight of hair brings famine in that year. This is utterly ridiculous.) Here [i.e. in Kan-chou and Su-chou] the merchants from China will buy the crude jades [from the Moslems] and take them eastward into China Proper, thence to Peking. The crude jades are then examined by jade artisans, their prices fixed, and cutting and polishing operations commenced.

(Although the valuable jade pieces are concentrated in the capital, the finest craftsmanship is to be found in Soochow.)

For cutting jade, a round iron disk is made [and mounted on a frame connected with pedals underneath] and a basin of sand is placed beside it. The disk is turned by pedals; at the same time sand is sprinkled on it so as to cut through the jade [Figure 18.7], which will then break open cleanly. In China, such jade-cutting sand is produced in two districts; Yü-t'ien in Shun-t'ien Prefecture and Hsing-t'ai in Chen-ting Prefecture. This type of sand is not found in the river, but is located near springs. This sand, like flour, has extremely fine particle sizes. When it is used in the cutting or polishing of jade, [the jade surfaces] will not be damaged.[12] Once the jade is cut, fine workmanship follows; the best tool for the latter being knives made of *pin-t'ieh* [probably meteoric nickel-iron]. (*Pin-t'ieh* is found in the meteorites around the Hami Garrison [in Sinkiang province]. It is obtained by breaking open the meteorites.) The odd pieces of jade resulting from cutting, carving, and polishing can be used to make inlay ware. Extremely fragmented pieces are ground, screened, and mixed with ash to fill cracks in lutes. This is why the sound of the lute has a jadelike quality. Where the design is too delicate for the use of knife or chisel, the pattern is first traced on the jade with the liquid from the warts of a toad, and then carved. The way in which different things in Nature are able to control each other is indeed mysterious.

The difference between genuine jade and jade simulated from inferior agates is similar to that between silver and tin. It is very easy to detect. Recently, however, artificial jade has been made from high-grade white porcelain, which is first ground into a fine powder, then mixed with the juice of the *Ampelopsis serjamiaefolia* bush, and next made into articles. When dried, these objects will gleam like jade. This is the most skillful method of making imitation jade.

The origin of pearls and jade is different from that of gold and silver. While gold and silver take form deep under the earth's surface after having received the essence of the sun, pearls, gems, and jade absorb the moon essence and do not need the earth's covering. The gems in their pits exposed to the blue skies above; pearls, denizens of deep waters; and jade, existing beside precipitous streams; all need only the covering of atmosphere and water. For [the protection of] the pearls, there is a "shell fortress" under water, presided over by the shell goddess and guarded by immortal dragons, which no man dares violate. When destiny points to certain pearls that have to do service in the human world, they are pushed out by the shell goddess and so are taken by pearl divers. It is likewise impossible to get the jade that has just begun to form, for it is obtainable only after the jade god has pushed it out into the river, which is a thing just as strange and mysterious as the fortress for pearls.

SUPPLEMENT: AGATE, QUARTZ CRYSTAL, AND LAPIS LAZULI

Agate is neither gem nor jade. It is produced in many places in China in over a dozen varieties. Most is made into hairpins, buttons, and the like. Some

Figure 18–5. The White-Jade River.

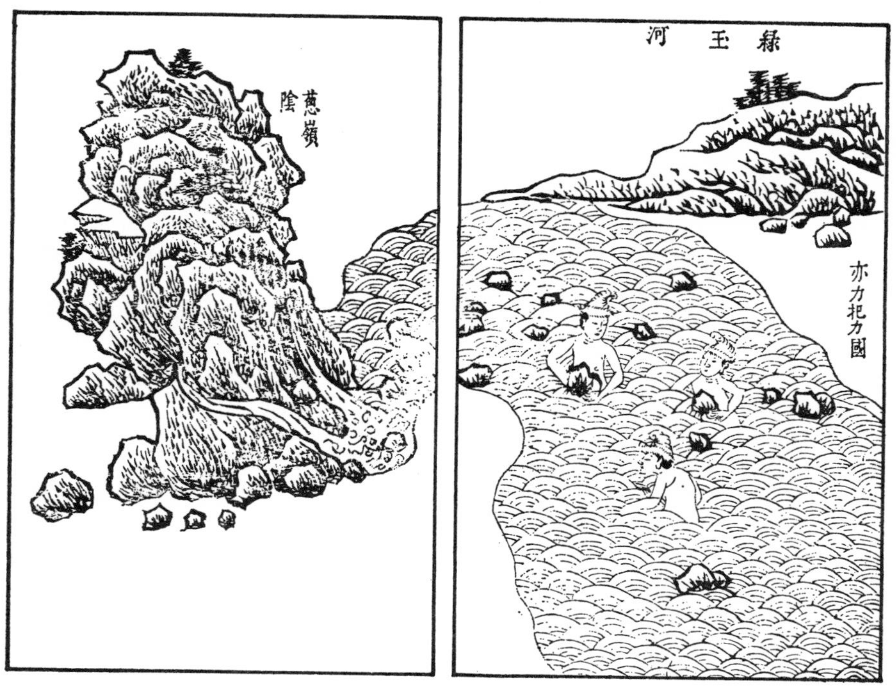

Figure 18–6. The Green-Jade River.

Figure 18–7. Cutting jade.

agates are used to make *go* pieces, while the largest pieces are fashioned into room screens and table tops. The best quality agate comes from the deserts of the Ch'iang territory west of Ningsia, but as it is abundantly produced in China itself, the merchants do not trouble themselves with purchasing it from distant places. Most of the agate on the Peking market is the product of Chiu-k'ung [lit. "nine hollow"] Mountain, at Jui-chou in Ta-t'ung, and of Ssu-chiao [lit. "four-cornered"] Mountain at Hsüan-fu.[13] There are different varieties of agates, including "two-colored" agate, "landscape-designed" agate, and "red-flower" agate. Further, "watery" or "milky" agate and "banded" agate are produced in Shen-mu and Fu-ku, and are sold widely in other places. The above is a brief summary [of agates]. To test an agate, rub it against a piece of wood. If it remains cold, then it is genuine. Although it is easy to make artificial agate, few people attempt it, since the price of the genuine article is not very high.

Less quartz crystal than agate is produced in China. The quartz crystals used in south China nowadays come from Chang-p'u in Fukien (from a mountain called Copper Mountain). Those used in north China come from Huang-chien [lit. "yellow top"] Mountain in Hsüan-fu. Those used in central China, come from Hsing-yang-chou in Honan (the best being black in color). They also come from Hsin-kuo-chou in Hukuang Province (from P'an-chia Mountain). Black crystal is produced only in north China, none of it in the south. There are deposits in other mountains, many of which are not yet mined, or were previously mined but later closed by the strict order of the government (such as the case in Kuang-hsin [Kiangsi province], where the mining of quartz crystals was abandoned for fear of the palace eunuchs).

Quartz crystals are formed in remote mountain caverns. When a mountain stream flows down rock crevices, it washes over the crystals unceasingly and comes out of the cave in a never ending current; half a *li* away from the latter the surface of the water still presents a boiling appearance, as though there were oil droplets in it. Before leaving the caverns the quartz crystals are soft as cotton but are hardened after exposure to the wind. When an artisan has found a suitable piece of crude quartz crystal, he can shape it roughly right on the mountain, and then bring it back for finer carving and polishing. By this method the process is made ten times easier.

Lapis lazuli is similar to the Chinese quartz crystal and Cambodian quartz-crystal prism (*huo-ch'i chu*) [14] in that all are lustrous and transparent. Lapis lazuli is produced not in China, but in the Western Regions.[15] It comes in many different colors. The Chinese, enjoying its attractive appearance, have striven to produce its artificial counterpart. Thus, *liu-li wa* or "lapis tiles" are made by first baking the tiles and then [coating them over] with glaze. This results in a green or yellow color. "Lapis bowls" are made by dissolving sheep's horns with heat. They are used as oil containers and candle shades. "Lapis lamps" are made by smelting [and casting a mixture of] nitre, lead, and pearl fragments [into thin plates] and piecing these together with copper wires. This same [molten] mixture can also be fashioned into "lazurite vases" or sacks. (The nitre used here has been refined to have horse-tooth-shaped crystals on its surfaces.) All sorts of dyes and colors can be freely used to tint [the lazurite ware]. The makers of lazurite lamps and beads are the people of Shantung province, north of the Huai River. This is because nitre is produced there.

Nitre is a non-substantial material and is capable of transforming itself into nothing when heated with fire. In contrast, black lead is a heavy-bodied material [and consequently is difficult to destroy with fire]. When these two materials are heated together, nitre will strive to induce lead to become nothing, while lead, on the other hand, will strive to keep nitre in the present world. The result of their melting in the same pot is the creation of a translucent and lustrous appearance. This is the manifestation of the wonders of the universe through common earthly occurrences. With this, I close my volume on the Divine Creations.

NOTES

1. Ho-p'u, in the Lien-chou prefecture in southern Kwangtung and on the shore of the South China Sea, is one of the renowned pearl-producing centers of China. Yü-t'ien, in the present Sinkiang province, which has been famous for its jade since ancient times.

2. Lei-chou and Lien-chou, prefectures in southern Kwangtung, have long been known for their pearl production. Fishing for pearl-producing mussels and oysters in Lien-chou was started at the latest during the Han dynasty (206 B.C.–A.D. 219), and became famous by the era of the Three Kingdoms (A.D. 220–264). In the Ming dynasty, Lei-chou and Lien-chou were the most important pearl-producing areas in China, as indicated in the "Book of Money and Commodities" in the *History of Ming*. Significantly, Hai-nan Island, located near these two prefectures, was named Chu-yai, or "pearl cliff," in the year 110 B.C., because of its wealth of pearl-producing mussels and oysters.

3. "P'u-li March" in the Chinese text is a misprint of "P'u-yü-March," as pointed out by Yabuuchi Kiyoshi.

4. Yang-ts'un and Chih-ku-k'ou are in modern Hopei province, the latter being the port of Ta-ku outside Tientsin.

5. Tan people are often called "boat people" in south China, since they live mostly in boats along the rivers and sea coasts of the provinces of Kwangtung, Kwangsi, and Fukien. Records indicate that for centuries pearl diving was a special occupation of the Tan.

6. The poisonous gas in an unventilated pit is most likely to be carbon monoxide or methane.

7. Chang Hua states correctly in his *Po-wu chih* (A.D. 290, Tsin dynasty) that amber is the fossilized resin of pine trees.

8. The fact that amber can attract light articles (static electricity) was noted in *Lun heng* (by Wang Ch'ung A.D. 82 or 83) and in *Hua-yang Kuo chih* (by Ch'ang chü, A.D. 347).

9. Yü-t'ien is a state located in an area known as Ho-t'ien since the eighteenth century (not to be confused with the Yü-t'ien District set up in late nineteenth century some 150 kilometers to the east). Ho-t'ien is situated in southwestern Sinkiang just below the northern foothills of the Karakorem Range and near the upper reaches of a water course, the Ho-t'ien or South Ts'ung-ling River. To the west is the Kashgar or North Ts'ung-ling River. The two join the Yerkand River and form the Tarim River system in south-central Sinkiang. These rivers have their headwaters in the Ts'ung-ling range, which is the general term for the eastern part of the Pamir Heights. The White-Jade River and the Green-Jade River, as mentioned in the present book, are probably the Ho-t'ien and the Kashgar.

Hsi-an (Sian), also known as Ch'ang-an in various periods of Chinese history, is the capital of Shensi province. Lan-t'ien mountain, located east of Hsi-an, has long been known for its production of jade.

10. Soft jade usually refers to nephrite. In the present case, however, it may represent talc, particularly the grayish variety, stealite, which is often encountered in early Chinese carvings.

11. Kan-chou and Su-chou are old transcontinental trading centers and garrison points, to which the modern province Kansu owes its name.

12. Jade-cutting sand, produced in the Yü-tien district near Peking, is identified with fine-size quartz by Chang Hung-chao. This sand was used extensively in the Ming dynasty but is seldom used by the modern Chinese lapidary because of its relative softness.

13. Ta-t'ung is a prefecture in northern Shansi. Hsüan-fu, located just beyond the Great Wall north of Peking, was the frontier market town of Kalgan.

14. *Huo-ch'i chu* is identified by Chang Hung-Chao (*Lapidarium Sinicum*, pp. 55–58) with the kind of quartz-crystal beads or prisms that are capable of focusing sun light to start a fire. Chinese records show that since the Han dynasty *huo-ch'i chu* was sent from Cambodia as a tribute and thus bears that country's name.

15. Western Regions, in this case, refer to modern Sinkiang province, India, Iran, and possibly Siberia.

BIBLIOGRAPHY A

Chinese Sources

Chang Ch'i-yün 張其昀 *The Essence of Chinese Culture* 中國文化要義 The China News Press, Taipei, 1957.

Chang Hua 張華 *Po-wu chih* 博物志 (ca. A.D. 290). Ssu-pu pei-yao edition, Chung-hua Book Co., Shanghai.

Chang Hung-chao 章鴻釗 *Metals and Stones as Treated in Laufer's Sino-Iranica* (in Chinese), Memoirs of the Geological Survey of China 地質專報 Series B, no. 3. The Geological Survey of China, Peking, 1925.

——— ———, *Shih-ya* 石雅 (Lapidarium Sinicum: A Study of the Rocks, Fossils, and Metals Known in Chinese Literature), Memoirs of the Geological Survey of China 地質專報 Series B, no. 2. The Geological Survey of China, Peking, 1927.

Chao Hsi-huo 趙希鵠 *Tung-t'ien ch'ing-lu* 洞天清錄 Sung dynasty, ca. 1240.

Ch'en Hsü-ching 陳序經 *Tan-min ti yen-chiu* 蛋民的研究 (A Study of the Tan People). Commercial Press, Shanghai, 1946.

Ch'i Ssu-ho 齊思和 "Chou-tai hsi-ming li k'ao" 周代錫命禮考 (Investiture Ceremonies of Chou), *Yenching Journal of Chinese Studies* 燕京學報 no. 32 (June, 1947), Yenching University, Peiping.

Chia Ssu-hsieh 賈思勰 *Ch'i-min yao-shu* 齊民要術 (5th century A.D.). Ssu-pu pei-yao edition, Chung-hua Book Co., Shanghai. This work has been rendered into modern Chinese in Shih Sheng-han's 石聲漢 book (see below).

Ch'in Han-chang 秦含章 *Jang-tsao chiang-yu chih li-lun yü chi-shu* 釀造醬油之理論與技術 (The Manufacture of Soy Bean Sauce). Commercial Press, Shanghai, 1947.

Chinese Society of Engineers 中國工程師學會 ed. *San-shih-nien-lai chih Chung-kuo kung-ch'eng* 三十年來之中國工程 (Chinese Engineering in the Past Thirty Years). 2nd ed. Chinese Society of Engineers, Nanking, 1948.

Chou T'ing-ch'ung, Ho Chia-ken, Chang Ch'ang-shao 周廷沖, 何嘉根, 張昌紹 "T'ung-yu chih tu-hsing shih-yen" 桐油之毒性試驗 (Toxicological Studies on T'ung Oil), *Science* 科學 vol. 29, no. 4 (April, 1947), The Science Society of China 中國科學社 Shanghai.

Chou Wei 周緯 *Chung-kuo pin-ch'i shih-kao* 中國兵器史稿 (A Draft History of Chinese Weapons). San-lien Book Co., Peking, 1957.

Chu Co-ching 竺可楨 "Climatic Pulsations During Historical Times in China," *Geological Review* (in Chinese), vol. 16 (1926), pp. 274–286.

Chu Yen 朱琰 *T'ao-shuo* 陶說 1774. 6 vols. Trans. in S. W. Bushell, *Description of Chinese Pottery and Porcelain,* Oxford, 1910.

Ch'ü Shou-yueh 曲守約 "Chung-kuo ku-tai ti tao-lu" 中國古代的道路 (Roads in Ancient China), *Tsing Hua Journal of Chinese Studies* 清華學報 (New Series) vol. 2, no. 1 (May, 1960), pp. 143–152.

Ch'ü T'ung-tsu 瞿同祖 *Chung-kuo feng-chien she-hui* 中國封建社會 (Feudalism in China). Commercial Press, Shanghai, 1936.

Ch'üan Han-Sheng 全漢昇 "Nan-Sung tao-mi chih sheng-ch'an yü yün-hsiao" 南宋稻米之生產與運銷 (The Production and Distribution of Rice in the Southern Sung Dynasty), *Bulletin of the Institute of History and Philology, Academia Sinica* 中央研究院歷史語言研究所集刊 vol. 10 (Shanghai, 1938), pp. 403–432.

Fan Wen-lan 范文瀾 *Chung-kuo t'ung-shih chien-pien* 中國通史簡編 (A General History of China). Hsin-hua Book Co., Shanghai, 1950.

Fang Pin-kuan 方賓觀 and others, comp., *Chung-kuo jen-ming ta-tz'u-tien* 中國人名大辭典 (Cyclopedia of Chinese Biographical Names). Commercial Press, Shanghai, 1921.

Fu Ch'in-chia 傅勤家 *Chung-kuo tao-chiao shih* 中國道教史 (A History of Taoism in China). Commercial Press, Shanghai, 1937.

Fu Chu-fu 傅築夫 "Chung-kuo li-shih shang yin ti wen-t'i" 中國歷史上銀的問題 (The Problem of Silver and Silver Currency in Chinese History), *Social Science* 社會科學 vol. 2, no. 1 (May, 1935), Central University, Nanking.

Hamada Kosaku 濱田耕作 *Ku-yü kai-shuo* 古玉概說 (A General Treatise on Ancient Jades), transl. from the Japanese by Hu Chao-ch'un 胡肇椿 Chung-hua Book Co., Shanghai, 1940.

Ho Shih-chin 何士晉 *Kung-pu ch'ang-k'u hsü-chih* 工部廠庫須知 (Manual for the Workshops and Storehouses of the Board of Works). Ming dynasty, 1615.

Ho Yü-tu 何宇度 *I-pu t'an-tzu* 益部談資 part 1 in *Hsueh hai lei-pien* 學海類編 vol. 117, ca. 1595.

Hsi Shu 席書 *Ts'ao-ch'uan chih* 漕船志. (An Account of Rice Tribute Boats), dated 1501 and 1544. Hsüan-lan-t'ang ts'ung-shu ed., vols. 51–56.

Hsia Yü-keng 夏與賡 *Shan-ts'an t'u-shuo* 山蠶圖説 (Illustrated Description of Wild Silkworms), 1906 ed.

Hsiang Yuan-pien 項元汴 *Li-tai ming-tz'u t'u-p'u* 歷代名瓷圖譜 (Notable Porcelain of Past Dynasties), revised and annotated by Kuo Pao-ch'ang 郭葆昌 and J. G. Ferguson 福開森. Chih-chai Publishing Co., Peiping, 1931.

Hsü Chung-shu 徐中舒 "Ko-i yü nu chih suo-yuan chi kuan-yü ts'u-lei ming-wu chih k'ao-shih" 弋射與弩之溯原及關於此類名物之考釋 (The Origin of Spear and Crossbow), *Bulletin of the Institute of History and Philology*, Academia Sinica 歷史語言研究所集刊 no. 4, pt. 4, 1934.

Hsueh-yin 雪垠 "Cheng-chou ti-hsia ti tsu-kuo ku-wen-hua" 鄭州地下的祖國古文化 (The Ancient Chinese Civilization Excavated at Chengchou), *China News* 中國新聞 (March 16, 1955).

Hu Hou-hsüan 胡厚宣 *Yin-hsü fa-chueh* 殷墟發掘 (The Excavations at the Yin Sites). Hsüeh-hsi sheng-huo Publishing Co., Shanghai, 1955.

——— ———, "Yin-tai nung-tso shih-fei shuo" 殷代農作施肥説 (Agricultural Fertilization in Yin Dynasty), *Historical Research*, Academy of Sciences 中國科學院, 歷史研究, vol. 1, no. 1 (Peking, 1955), pp. 97–106.

Huang Ch'ang-ku 黃昌穀 "Kang-t'ieh chin-t'u-hsueh t'i-yao" 鋼鐵金圖學提要 (An Outline of Metallography of Iron and Steel), *Science* 科學, vol. 6, no. 1 (January, 1921).

Ku Chi-kuang 谷霽光 "Sung Yuan shih-tai tsao-ch'uan shih-yeh chih chin-chan" 宋元時代造船事業之進展 (The Development of Shipbuilding during the Sung and Yuan Periods), *Wen-shih tsa-chih* 文史雜誌, vol. 4, nos. 4–5 (September, 1944), Chungking.

Ku Yen-wu 顧炎武 *Jih-chih-lu chi-shih* 日知錄集釋 (The Jih-chih-lu with Commentaries). 1872 ed., Hupei, China.

Kuo Mo-jo 郭沫若 *Chung-kuo ku-tai she-hui yen-chiu* 中國古代社會研究 (A Study of Ancient Chinese Society), Hsin-wen-i Publishing Co., Shanghai, 1951.

Kuo Pao-chun, Hsia Nai and others 郭寶鈞, 夏鼐, 等 *Hui-hsien fa-chueh pao-kao* 輝縣發掘報告 (Report of Excavations at Hui-hsien, no. 1). The Institute of Archeology, Academy of Sciences 中國科學院考古研究所 Peking, 1956.

Lan P'u 藍浦, *Ching-te-chen t'ao-lu* 景德鎮陶錄 (The Ceramic Wares of Ching-te-chen), revised by Cheng T'ing-kuei 鄭廷桂. 1815. 10 vols. Translated by G. R. Sayer, *The Potteries of China*, London, 1951.

Lao Kan 勞榦 "Chien-tu-chung so-chien ti pu-po" 簡牘中所見的布帛 (Textile Fabrics as Seen in the [Han] Wooden-slat Letters), *Hsueh-shu chi-k'an* 學術季刊 vol. 1, no. 1 (Taipei, October, 1953), p. 14. Summarized in *K'o-hsueh hui-pao* 科學會報 (Science Bulletin), Chinese Association for the Advancement of Science, vol. 1, no. 4 (Taipei, October, 1953), p. 14.

——— ———, "Lun Chung-kuo tsao-chih-shu chih yüan-shih" 論中國造紙術之原始 (On the Origin of the Art of Making Paper in China), *Bulletin of the Institute of History and Philology,* Academia Sinica 中央研究院歷史語言研究所集刊 no. 19 (1948), pp. 489–498.

Li Chao-hsiang 李昭祥 *Lung-chiang ch'uan-ch'ang chih* 龍江船廠志 (An Account of the Lung-chian Shipyard). Preface dated 1553. In Hsü Hsüan-lan-t'ang ts'ung-shu, vols. 117–119.

Li Chi 李濟 editor-in-chief, *An-yang fa-chueh pao-kao* 安陽發掘報告 (Preliminary Reports of Excavation at Anyang), 4 parts, The National Research Institute of History and Philology, Academia Sinica, Peiping, 1929–1933.

Li Shih-chen 李時珍 *Pen-ts'ao kang-mu* 本草綱目 (Materia Medica). Ming dynasty, 1596. The Complete Library 萬有文庫 Series I, Commercial Press, Ltd., Shanghai, 1930.

Li Shu-hua 李書華 *Chih-nan-chen ti ch'i-yuan* 指南針的起源 (The Origin of the Compass). Ta-lu Magazine Society, Taipei, 1954.

Lo Chen-yü 羅振玉 *Heng-nung chung-mu i-wen* 恒農冢墓遺文 (Record of the Heng-nung Tomb), 1918.

Lo Jung-pang 羅榮邦 "Chung-kuo chih ch'e-lun-ch'uan: ya-p'ien chan-cheng chung ti chi-chieh-hua ch'uan-chih chi-ch'i li-shih pei-ching" 中國之車輪船：鴉片戰爭中的機械化船隻及其歷史背景 (China's Paddle-Wheel Boats: Mechanized Craft Used in the Opium War and Their Historical Background), *Tsing Hua Journal of Chinese Studies* 清華學報 New Series, 2, no. 1 (Taipei, May, 1960), pp. 189–215.

Mao Tso-pen 茅左本 *Wo-men tsu-hsien ti ch'uang-tsao fa-ming* 我們祖先的創造發明 (Our Ancestors' Creations and Inventions). Hsin-hua Book Co., Shanghai, 1951.

Pai Shou-i 白壽彝 *Chung-kuo chiao-t'ung shih* 中國交通史 (A History of Communications in China). Commercial Press, Shanghai, 1937.

Peng Chia-yuan 彭家元 "T'u-jang hsing-chih yü nung-shou kuan-hsi" 土壤性質與農收關係 (Relation Between Soils and Crops), *Science* 科學 vol. 8, no. 7 (August, 1923), pp. 725–736.

P'eng Tse-i 彭澤益 *Chung kuo chin-tai shou-kung-yeh-shih tzu-liao 1840–1949* 中國近代手工業史資料 (Sources on the History of Modern Chinese Handicrafts 1840–1949), 4 vols. San-lien Book Co., Peking, 1957.

Shen Kua 沈括 *Meng-ch'i pi-t'an* 夢溪筆談 1089–1093.

Shih Sheng-han 石聲漢 *Ts'ung Ch'i-min yao-shu k'an Chung-kuo ku-tai ti nung-yeh k'o-hsüeh chih-shih* 從齊民要術看中國古代的農業科學知識 (Ancient Chinese Agricultural Science as Seen in *Ch'i-min yao-shu*). Science Publishing Co., Peking, 1957.

Shu Hsin-ch'eng 舒新城 and others, *Tz'u-hai* 辭海 (A Dictionary of Chinese Words and Idioms). Chung-hua Book Co., Shanghai, 1947.

Ssu-ma Ch'ien 司馬遷 and Ssu-ma T'an 司馬談 "Book on Money and Commerce" 貨殖傳, *Shih chi* 史記 vol. 129, ca. 90 B.C.

Sung Hsi-shang宋希庠 *Chung-kuo li-tai ch'üan-nung k'ao* 中國歷代勸農考 (On the History of Advising Farmers in China). 2nd ed., Cheng-chung Book Co., Shanghai, 1946.

T'an Tan-chiung譚旦冏 *Chung-hua min-chien kung-i t'u-shuo* 中華民間工藝圖說 (Illustrated Accounts of Native Chinese Industries). Chung-hua Ts'ung-shu Commission, Taipei, 1956.

Teng Chih-ch'eng鄧之誠 *Chung-hua erh-ch'ien-nien shih* 中華二千年史(Two Thousand Years of Chinese History), vols. I and II. Commercial Press, Shanghai, 1934.

Tseng Yang-feng曾養豐 *Chung-kuo yen-cheng shih* 中國鹽政史 (A History of the Salt Administration in China). Commercial Press, Shanghai, 1936.

Wan Kuo-ting萬國鼎 *Chung-kuo t'ien-chih shih* 中國田制史 (History of the Land Systems in China), vol. 1. Cheng-chung Book Co., Nanking, 1934.

Wang Chen-to王振鐸 "Chih-nan-ch'e chi-li-ku ch'e chih k'ao-cheng chi mo-chih" 指南車記里鼓車之考証及模製 (Research and Reconstruction of the South-pointing Carriage and the Odometry-drum Carriage), *Bulletin of History Study*, Peiping Research Institute, 國立北平研究所,史學集刊 no. 3, 1937.

———, *Han-tai k'uang-chuan chi-lu* 漢代壙磚集錄 Peiping, 1935.

———, "Ssu-nan-chen yü lo-ching-p'an" 司南針與羅經盤 (The Compass and the Diviner's Board), *Chinese Journal of Archaeology*, Academia Sinica, Institute of History and Philology 中央研究院,歷史語言研究所,中國考古學報 no. 3 (May, 1948), no. 4 (December, 1949).

Wang Chia-yin 王嘉蔭 *Pen-ts'ao kang-mu ti k'uang-wu shih-liao* 本草綱目的鑛物史料 (Historical Notes on Minerals in the Chinese Materia Medica). Science Publishing Co., Shanghai, 1956.

Wang Chin 王璡 "Chung-kuo ku-tai t'ao-yeh chih k'o-hsueh kuan" 中國古代陶業之科學觀 (The Scientific Aspects of Ancient Chinese Ceramic Industry), *Science* 科學 vol. 6, no. 9 (Shanghai, September, 1921), pp. 869–882.

——————, "Wu-shu-ch'ien hua-hsueh ch'eng-fen chi ku-tai ying-yung ch'ien hsi hsin la k'ao" 五銖錢化學成份及古代應用鉛錫鋅鑞考 (The Chemical Composition of Wu-shu Coins and the Use of Lead, Tin, Zinc, and Lead-tin Alloy in Ancient Times), *ibid.*, vol. 8, no. 8 (Shanghai, August, 1933), pp. 839–854.

Wang Hsiao-t'ung 王孝通 *Chung-kuo shang-yeh shih* 中國商業史 (History of Commerce in China). Commercial Press, Shanghai, 1936.

Wang I-chüeh 王義珏 "Chih ko yuan-liao chiao-pei chih hua-hsueh fen-hsi" 製革原料角棓之化學分析 (The Chemical Analyses of a Chinese Gall Used in Leather-Making), *Science* 科學 vol. 7, no. 6 (Shanghai, 1922), pp. 597–601.

Wang Kung-mu and others 王恭睦等 *K'uang-wu-hsueh ming-tzu* 鑛物學名詞 (The Terminology of Minerals). Ed. by the National Bureau of Compilation and Translation 國立編譯館 Commercial Press, Shanghai, 1934.

Wang Yin-chang 王寅章 "Wei-tseng jang-tsao" 味噌釀造 (The Manufacture of Wei-tseng), *Ta Kung Pao* 大公報 January 16, 1948.

Wu Ch'eng-lo 吳承洛 *Chung-kuo tu liang heng shih* 中國度量衡史 (History of Chinese Weights and Measures). Commercial Press, Shanghai, 1937.

Wu Jen-ching 吳仁敬 and Hsing An-ch'ao 辛安潮 *Chung-kuo t'ao-tz'u shih* 中國陶瓷史 (A History of Chinese Pottery and Porcelain). Commercial Press, Hong Kong, re-issue of 1954 (1st ed. 1935).

Wu T'ing-k'ang 吳廷康 *Mu-t'ao-hsüan ku-chuan t'u-lu* 慕陶軒古磚圖錄 (Illustrated Account of Ancient Bricks from Mu-t'ao Pavilion). 1815.

Yabunchi Kiyoshi 藪内清 *Tenkō kaibutsu no kenkyū* 天工開物の研究 (Studies of T'ien-kung k'ai-wu). Kōsei Sha, Tokyo, Japan, 1953.

Yang Ta-chin 楊大金 *Hsien-tai Chung-kuo shih-yeh chih* 現代中國實業誌 (The Modern Chinese Industries), 2 vols. Commercial Press, Shanghai, 1940.

Yeh Shao-chün 葉紹鈞 ed., *Tuan-chü shih-san-ching ching-wen* 斷句十三經經文 (Punctuated Text of the Thirteen Classics). Kaiming Book Co., Shanghai, 1934.

BIBLIOGRAPHY B

Books and Articles in Other Languages

CHAPTER 1. THE GROWING OF GRAINS

Ames, J. W. and Gaither, "Barnyard Manure: Production, Composition, Conservation, Reinforcement, and Value," *Ohio Agriculture Experiment Station Bulletin 246* (1919).

Bear, F. E., *Soils and Fertilizers.* 3rd ed. John Wiley & Sons, Inc., New York, 1942.

Bishop, C. W., "The Rise of Civilization in China with Reference to Geographical Aspects," *The Geographical Review,* vol. 22, No. 4 (October, 1932).

Buck, J. L., *Land Utilization in China.* University of Chicago Press, 1937.

Castellani, F., "Kansas Limestone Aids Farmlands," *The Explosive Engineering* (July–August, 1945).

Chamberlain, J. S. and C. A. Browne, *Chemistry in Agriculture.* The Chemical Foundation, Inc., New York, 1926.

Chang, C. C., *An Estimate of China's Land and Crops.* University of Nanking, Nanking, China, 1936.

Chi Ch'ao-ting, *Key Economic Areas in Chinese History, as Revealed in the Development of Public Works for Water-Control.* Allen & Unwin, Ltd., London, 1936.

Collings, G. H., *Commercial Fertilizers.* 3rd. ed. The Blakiston Co., Philadelphia, 1941.

Cornejo, R. F., "Care of Barnyard Manure," B.S. thesis. Department of Agronomy, The Pennsylvania State University, University Park, Pennsylvania, June, 1913.

Creel, H. G., *The Birth of China.* Frederick Ungar Publishing Co., New York, 1954.

Cressey, G. B., *China's Lands and Peoples.* 2nd ed. McGraw-Hill Book Co., New York, 1951.

Derrick, B. B., *A Monograph on Lime and Its Use in Agriculture.* Department of Agronomy, The Pennsylvania State University, University Park, Pennsylvania, May, 1913.

Gordon, J. and Y. L. Djang, "The 2200-Year Hydraulic Project of West China and Its Needed Reform," *China Monthly,* New York, January, 1948.

Hopkins, D. P., *Chemicals, Humus, and the Soil.* Faber & Faber Ltd., London, 1945.

Hosie, A., "Drought in China, A.D. 620–1643," *Journal of the North China Branch of the Royal Asiatic Society,* vol. 12, n.s. (1878), pp. 51–89.

Lattimore, O., *Pivot of Asia.* Little, Brown & Co., Boston, 1950.

Lee, Mabel Ping-hua, *The Economic History of China,* Columbia Studies in History, Economics, and Public Law, XCIX.1. Columbia University, New York, 1921.

Liu, Hou-lee, "Genetic Studies of Soybeans," Ph.D. dissertation. University of Illinois, Urbana, Illinois, 1948.

Sauchell, V., *Manual on Phosphates in Agriculture.* The Davison Chemical Corporation, Baltimore, 1951.

Shaw, C. F. "Soils of China," *Soil Bulletin No. 1.* National Geological Survey of China, Nanking, China, 1930.

Shen Tsung-han, *Agricultural Resources of China.* Cornell University Press, Ithaca, 1951.

Symposium on Lime, Columbus Regional Meeting, March 8, 1939. American Society for Testing Materials, Philadelphia, 1939.

Waley, A., *Book of Songs* (*Shih ching*), translated from the Chinese. Houghton Mifflin Co., Boston, 1937.

Wang Yueh, "A Field Study on the Disposal of Garbage and Night Soil by Composting," *Science and Technology in China.* The Natural Science Society of China, National Central University, Nanking, China.

Winfield, G. F., *China: The Land and the People.* William Sloane Associates, Inc., New York, 1950.

CHAPTER 2. CLOTHING MATERIALS

Alexander, P. and R. F. Hudson, *Wool: Its Chemistry and Physics.* Reinhold Publishing Corp., New York, 1954.

Bancroft, W. D. and J. B. Calkins, "The Action of Sodium Hydroxide on Cellulose," *Textile Research,* vol. 4 (1924), p. 119.

Bodde, Derk, *China's Gifts to the West.* American Council on Education, Washington, D. C., 1942.

Brown, H. B. and J. O. Osborn, *Cotton.* 3rd ed. McGraw-Hill Book Co., New York, 1958.

Camman, Schuyler, *China's Dragon Robes.* Ronald Press Co., New York, 1952.

Casson, Lionel, "Trade in the Ancient World," *Scientific American,* vol. 191, No. 5 (November, 1954).

Charleston, R. J., "Han Damasks," *Oriental Art,* vol. 1 (1948), pp. 63–81.

Goodrich, L. C., "Cotton in China," *Isis,* vol. 34, No. 5 (1943), pp. 408–10.

Hannan, W. I., *Textile Fibers of Commerce.* Charles Griffin & Co., Ltd., London, 1902.

Harris, M., *Handbook of Textile Fibers.* Interscience Publishers, Inc., New York, 1955.

"India Believed Land of Origin of Cotton," *The Asian Student* (December 18, 1956), San Francisco, California.

Jackson, L. H., *Yarn and Cloth Calculations.* Interscience Publishers, Inc., New York, 1947.

Jaffe, B., *Chemistry Creates a New World.* Thomas Y. Crowell Co., New York, 1957.

Kaswell, E. R., *Textile Fibers, Yarns and Fabrics.* Reinhold Publishing Corp., New York, 1953.

Mathews, J. M., *The Textile Fibers: Their Physical, Microscopical, and Chemical Properties.* John Wiley & Sons, Inc., New York, 1916.

Powers, D. H., "Recent Developments of Resins for Textile Application," *American Dyestuff Reporter,* vol. 28 (1939).

Shinkle, J. H., "In Mercerization, Practice Has Kept Ahead of the Knowledge of Principles," *Textile World,* vol. 75 (1929).

Sylvan, V., "Silk from the Yin Dynasty," *Bulletin of the Museum of Far Eastern Antiquities, No. 9.* Stockholm, n.d.

Zipser, J., *Textile Raw Materials and Their Conversion into Yarns,* translated from the German by Charles Salter. Scott Greenwood & Son, London, 1901.

CHAPTER 3. DYES

Baghi, P. C., *India and China.* 2nd ed. Hind Kitabs Ltd., Bombay, 1944.

Kametaka, T. and G. Perkin, "Carthamine," Part 1, *Trans. Chemical Society (London),* vol. 97 (1910), pp. 1415–1427.

Li Chi (editor-in-chief) and others, *Ch'eng-tzu-yai· The Black Pottery Culture Site at Lung-shan-chen in Li-cheng-hsien, Shangtung Province* (in Chinese, 1934), translated by Kenneth Starr. Yale University Press, New Haven, Connecticut, 1956.

Lucas, A., *Ancient Egyptian Materials and Industries.* 3rd ed. Edward Arnold & Co., London, 1948.

Meir, G. E. and J. W. Mellor, "The Ferric Oxide Colors," *Trans. British Ceramic Society,* vol. 36 (1937), pp. 31–43.

Perkin, A. G. and A. E. Everest, *The Natural Organic Coloring Matters.* Longmans, Green and Co., New York, 1918.

Schunk, E., "On the Supposed Identity of Rutin and Quercitrin," *Journal of the Chemical Society of London,* vol. 53 (1888), pp. 262–267.

———, "The Yellow Coloring Matter of Sophora Japonica," *Journal of the Chemical Society of London,* vol. 67 (1895), pp. 30–32.

Schwarz, E. W. K., "Important Chemical Developments in the Textile Industry," *American Dyestuff Reporter,* vol. 29 (1940).

Thompson, D. V. Jr., "Medieval Color Making," *Isis,* vol. 22, No. 2 (1934–35), pp. 456–468.

Thorpe, E., *History of Chemistry.* 2 vols. Watts Co., London, 1909.

Thorpe, J. F. and C. K. Ingold, *Vat Colors.* Longmans, Green and Co., New York, 1923.

Warwicke, J., "The Coloration and Finishing of Textile with Pigments," *Textile Colorist,* vol. 61 (1939), pp. 441–443.

Watson, E., *The Principal Articles of Chinese Commerce.* Statistical Department of the Inspectorate General of Customs, Shanghai, 1930.

Watson, W., *Textile Design and Colour.* 2nd ed. Longmans, Green and Co., London, 1921.

CHAPTER 4. THE PREPARATION OF GRAINS

Anderson, J. A. and A. W. Alcock (editors), *Storage of Cereal Grains and Their Products.* American Association of Cereal Chemists, St. Paul, Minnesota, 1954.

Chapin, R. W., *The Milling and Manufacturing of Grain and Grain Products.* The American Institute of Agriculture, Chicago, 1923.

Hall, C. H., *Drying Farm Crops.* Agricultural Consulting Associations, Inc., Reynoldsburg, Ohio, 1957.

Henderson, S. M. and R. L. Perry, *Agriculture Process Engineering.* John Wiley & Sons, Inc., New York, 1955.

Hommel, R. P., *China At Work.* The Bucks County Historical Society, Boylestown, Pennsylvania, John Day Co., New York, 1937.

Hutcheson, T. B., T. K. Wolfe, and M. S. Kipps, *The Production of Field Crops.* 3rd ed. McGraw-Hill Book Co., New York, 1948.

Kirk, L. E., L. Ling, and T. A. Oxley, *Storing and Drying Grain in Canada, in the United States, in the United Kingdom.* Food and Agriculture Organization of the United Nations, Washington, 1948.

Leonard, W. H. and J. H. Martin, *Cereal Crops.* Macmillan Publishing Co., New York, 1963.

Mallory, W. H., *China: Land of Famine.* American Geographical Society, New York, 1926.

Oxley, T. A., *The Scientific Principles of Grain Storage.* Northern Publishing Co., Liverpool, England, 1948.

Singer, C., E. J. Holmyard, and A. R. Hall, *A History of Technology.* 5 vols. Oxford University Press, London, 1954–58.

Smith, E., *Storage of Fruits and Vegetables.* The American Institute of Agriculture, Chicago, 1923.

Von Loesecke, H. W., *Outline of Food Technology.* Reinhold Publishing Corp., New York, 1942.

CHAPTER 5. SALT

Graham, D. C., "Ornamented Bricks and Tiles from Western Szechwan," *Journal of the West China Border Research Society,* vol. 10 (1938), 191 ff.

Hummel, A. W. and others, Discussion of Orientalia, *Report of the Librarian of Congress for the Fiscal Year Ending June 30, 1933.* Government Printing Office, Washington, D. C., 1933.

Maynard, L. A., *Animal Nutrition.* 3rd ed. McGraw-Hill Book Co., New York, 1951.

Rudolph, R. C., "A Second-Century Chinese Illustration of Salt Mining," *Isis,* vol. 34 (April, 1952), pp. 39–41.

_____ and Wen Yu, *Han Tomb Art of West China.* University of California Press, Berkeley and Los Angeles, California, 1951.

Taylor, F. S., *A History of Industrial Chemistry.* W. Heinemann Ltd., London, 1957.

CHAPTER 6. SUGARS

Alikonis, J. J., "Carbohydrates in Confectionery," *Use of Sugar and Other Carbohydrates in the Food Industry,* Advances in Chemistry Series, No. 12, American Chemical Society (March, 1953), pp. 57–63.

Goodrich, L. C., *A Short History of the Chinese People.* Harper & Brothers Publishers, New York, 1951.

Li Ch'iao-p'ing, *The Chemical Arts of Old China.* Journal of Chemical Education, Easton, Pennsylvania, 1948.

Martin, L. F., "Sugar In Confectionery," *Use of Sugar and other Carbohydrates in the Food Industry,* Advances in Chemistry Series, No. 12, American Chemical Society (March, 1953), pp. 64–69.

Richter, G. H., *Textbook of Organic Chemistry.* 2nd ed. John Wiley & Sons, Inc., New York, 1945.

Riegel, E. R., *Industrial Chemistry,* Reinhold Publishing Corp., New York, 1942.

Slosson, E. E., *Creative Chemistry.* The Century Co., New York, 1919.

Stepanek, J. E. and C. H. Prien, "Small Chemical Industries for China, II," *Chemical and Engineering News.* American Chemical Society, vol. 28, No. 47 (November 20, 1947), p. 4063.

——— and ———, "Sugar and Insecticides Manufacture," *Ibid.,* vol. 28, No. 49 (December 4, 1950), pp. 4250–4253.

CHAPTER 7. CERAMICS

Alkins, C. and M. Francis, "Effect of Bentonite on Glaze Take-up of Mechanically Dipped Tiles," *Trans. British Ceramic Society,* vol. 42 (1943), pp. 157–162.

Andrae, W., *Coloured Ceramics from Ashur and Earlier Ancient Assyrian Wall-paintings.* Kegan Paul, Trench, Trubner & Co., Ltd., London, 1925.

Barlow, Sir Alan, "Catalogue of the Exhibition of Cleadon Wares," *Trans. Oriental Ceramic Society,* vol. 23 (1947–48), pp. 31–45.

Beech, D. G. and D. A. Holdridge, "Some Aspects of the Testing of Clays for the Pottery Industry," *Trans. British Ceramic Society,* vol. 53 (1954), pp. 103–133.

Beck, H. C. and C. G. Seligman, "Barium in Ancient Glass," *Nature,* vol. 133 (June 30, 1934, London), p. 982.

Benedetti-Pichler, A. A., "Microchemical Analysis of Pigments," *Industrial and Engineering Chemistry,* vol. 9 (March, 1937), pp. 149–152.

Brindley, G. W., *X-ray Identification and Crystal Structure of Clay Minerals.* The Mineralogical Society, London, 1951.

Bullin, L., "A Critical Survey of Tunnel Kiln Economy," *Trans. British Ceramic Society,* vol. 41 (1942), pp. 217–244.

Bushell, S. W., *Chinese Art,* vol. 1. H. M. Stationery Office, London, 1909.

Charleston, R. J., *Roman Pottery.* Faber & Faber, London, n. d.

Clark, N. O., "China Clay Research in Cornwall," *Trans. British Ceramic Society,* vol. 49 (1950), pp. 409–419.

Clems, F. H., H. M. Richardson, and A. T. Green, "The Action of Alkalies on Refractory Materials," *Trans. British Ceramic Society,* vol. 40 (1941), pp. 415–441.

Cox, W. E., *The Book of Pottery and Porcelain.* 2 vols. Crown Publishers, New York, 1944.

Dale, A. J. and M. Francis, "The Durability of On-Glaze Decoration," *Trans. British Ceramic Society,* vol. 41 (1942), pp. 245–256.

——— and ———, "The Technical Control of Glazing by Dipping and Other Methods," *Ibid.,* vol. 41 (1942), pp. 167–190.

——— and M. F. Reynolds, "Trends of Development in Ceramic Decoration Technique," *Ibid.,* vol. 50 (1951), pp. 269–283.

Davis, R. J., R. A. Green, and H. F. E. Donnelly, "Some Researches on China Clay," *Trans. British Ceramic Society,* vol. 36 (1937), pp. 173–200.

Dobson, E. and A. B. Searle, *Bricks and Tiles.* Crobsy Lockwood & Son, London, 1921.

Gad, G. M. and L. R. Barrett, "The Action of Heat on Alunite and Alunitic Clays," *Trans. British Ceramic Society,* vol. 48 (1949), pp. 352–374.

Gardner, A. W., "The Drying of Clay and Clay Ware in a Modern Pottery," *Trans. British Ceramic Society,* vol. 43 (1944), pp. 37–48.

Garner, H., "Blue and White of the Middle Ming Period," *Trans. Oriental Ceramic Society,* vol. 27 (1951–53), pp. 61–72.

German, W. L. and S. W. Ratcliffe, "Relative Translucencies of the System Bone-China-Clay-Stone," *Trans. British Ceramic Society,* vol. 53 (1954), pp. 165–179.

Goskar, T. A., "Dryers for Roofing Tiles," *Trans. British Ceramic Society,* vol. 37 (1938), pp. 62–73.

Gray, B., *Early Chinese Pottery and Porcelain.* Faber & Faber, London, 1952.

Greaves-Walker, A. F., *Drying Ceramic Products.* Industrial Publications, Inc., Chicago, 1948.

Guy, J. P., "Saggers: Past, Present, Future, Their Scientific, Humanitarian and Commercial Aspects," *Trans. British Ceramic Society,* vol. 14 (1914–15), pp. 27–37.

Harris, R. G., "Observations on Some Causes and Effects of Vitrification in Ceramic Bodies," *Trans. British Ceramic Society,* vol. 38 (1939), pp. 396–410.

———, "Saggers and Kiln Furniture in the Pottery Industry," *Ibid.,* vol. 46 (1947), pp. 247–259.

Hetherington, A. L., *Chinese Ceramic Glazes.* 2nd revised ed. P. D. and Ione Perkins, S. Pasadena, California, 1948.

———, "An Interesting and Accidental Reduced Copper On-Glaze Effect," *Trans. British Ceramic Society,* vol. 42 (1943), pp. 183–184.

———, "The Pre-Ming Wares of China," *Ibid.,* vol. 41 (1942), pp. 257–267.

Holdcroft, A. D. and J. W. Mellor, "The Action of Cobalt Salts on Clays," *Trans. British Ceramic Society,* vol. 6 (1906–07), pp. 153–158.

Hollinshead, A. D., J. Turner, and J. W. Mellor, "Cobalt and Nickel Colours," *Trans. British Ceramic Society,* vol. 14 (1914–15), pp. 167–172.

Hyslop, J. F., "The Decomposition of Clay by Heat," *Trans. British Ceramic Society,* vol. 43 (1944), pp. 49–50.

Jenyns, S., "The Problem of Chinese Cloisonne Enamels," *Trans. Oriental Ceramic Society,* vol. 25 (1949–50), pp. 49–64.

Johnson, J. A. and H. L. Steel, "Slip House Technique in General Earthenware Production," *Trans. British Ceramic Society,* vol. 39 (1940), pp. 182–198.

Kaechlin, R. and G. Migeon, *Oriental Art: Ceramics, Fabrics, Carpets,* transl. from the French by Florence Heywood. The Macmillan Co., New York, n.d.

Karlbeck, H. O., "Proto-Porcelain and Yueh Ware," *Trans. Oriental Ceramic Society,* vol. 25 (1949–50), pp. 33–48.

Laufer, B., *The Beginning of Porcelain in China.* Field Museum of Natural History Publication 192, Anthropological Series, vol. 15, No. 2 (1917, Chicago).

Lee, J. G., "Ming Blue-and-White," *Philadelphia Museum Bulletin* (Autumn, 1949).

Los Angeles County Museum, *Chinese Ceramics* (from the prehistoric period through Ch'ien Lung). A Loan Exhibition from Collections in America and Japan, March14–April 27, 1952, Los Angeles, California.

Lovejoy, E., *Burning Clay Wares.* 3rd ed. T. A. Randall & Co., Publishers, Indianapolis, Indiana, 1922.

Macey, H. H., "Clay-water Relationships and the Internal Mechanism of Drying," *Trans. British Ceramic Society,* vol. 41 (1942), pp. 73–122.

———, "The Effect of Temperature and Humidity on the Rates of Drying of Clay Shapes," *Ibid.,* vol. 37 (1938), pp. 131–150.

———, "Some Observations on the Safe Drying of Large Fireclay Blocks," *Ibid.,* vol. 38 (1939), pp. 469–475.

Martin, R. C., *Lacquer and Synthetic Enamel Finishes.* D. Van Nostrand Co., New York, 1940.

Mellor, J. W., "The Chemistry of the Chinese Copper-red Glazes," Parts I & II, *Trans. British Ceramic Society,* vol. 35 (1936), Part I: pp. 364–378, Part II: pp. 487–491.

———, "The Chemistry of the Chrome-Tine Colors," *Ibid.,* vol. 36 (1937), pp. 16–27.

———, "Cobalt Blue Colors," *Ibid.,* vol. 6 (1906–07), pp. 88–96.

———, "The Cobalt Blue Colors," *Ibid.,* vol. 36 (1937), pp. 264–265.

———, "Crazing and Feeling of Glazes," *Ibid.,* vol. 34 (1934–35), pp. 1–112.

———, "The Crazing of Glazed Wall-Tiles in Service," *Ibid.,* vol. 36 (1937), pp. 443–465.

———, "The Cultivation of Crystals on Glazes," *Ibid.,* vol. 36 (1937), pp. 13–15.

———, "The Minute Structure of Porcelain, Parian, Semi-Porcelain, and Earthware," *Ibid.,* vol. 5 (1905–06), pp. 75–92.

———, "The Scumming of Mazarine Blue," *Ibid.,* vol. 6 (1906–07).

———, "The Sulphuring and Feathering of Glazes, Part II, *Ibid.,* vol. 6 (1906–07), pp. 71–75.

Meredith, W. D., "Slip House Practice in Glazed Tile Production," *Trans. British Ceramic Society,* vol. 39 (1940), pp. 163–177.

Newton, I., "Some Colored and White Wares from Hunan," *Trans. Oriental Ceramic Society,* vol. 27 (1951–53), pp. 25–36.

Noble, W., E. Rowden, and A. T. Green, "Fuel Economy in the Firing of Intermittent Kilns," *Trans. British Ceramic Society,* vol. 40 (1941), pp. 388–414.

Norton, F. H., *Elements of Ceramics.* Addison-Wesley Press, Inc., Cambridge, Mass., 1952.

Parmelee, C. W., *Ceramic Glazes.* 2nd ed. Industrial Publications, Inc., Chicago, Ill., 1951.

———, *Clays and Some Other Ceramic Materials.* Edwards Brothers, Inc., Ann Arbor, Michigan, 1946.

Raphael, M., *Prehistoric Pottery and Civilization in Egypt,* translated by N. Guterman. Bolligen Foundation, Inc., Washington, D. C. (published by Pantheon Books, Inc.), 1947.

Ratcliffe, S. W. and H. W. Webb, "The Use of Infra-Red Drying in Pottery Manufacture," *Trans. British Ceramic Society,* vol. 44 (1945), pp. 119–134.

Rigby, A. "Potting Clays," *Trans. British Ceramic Society,* vol. 34 (1934–35), pp. 381–395.

Rowden, E., "Using Waste Heat in the Heavy-Clay Industry," *Trans. British Ceramic Society,* vol. 52 (1953), pp. 69–102.

———, W. Noble, and A. T. Green, "Fuel Economy in the Firing of Hoffman and Belgian Kilns," *Ibid.,* vol. 41 (1942), pp. 207–216.

Schofield, P. K., "Clay Mineral Structures and Their Physical Significance," *Trans. British Ceramic Society,* vol. 39 (1940), pp. 147–162.

Searle, A. B., *The Chemistry and Physics of Clays.* 2nd ed. Ernest Benn Ltd., London, 1933.

———, *The Clayworker's Handbook.* 5th ed. Charles Griffin & Co., Ltd., London, 1949.

———, *Modern Brickmaking.* D. Van Nostrand Co., New York, 1911.

Seligman, C. G., P. D. Ritchie, and H. C. Beck, "Early Chinese Glass from Pre-Han to T'ang Times," *Nature,* vol. 138 (October, 1936, London), p. 721.

Sharratt, E. and M. Francis, "The Durability of On-Glaze Decoration II," *Trans. British Ceramic Society,* vol. 42 (1943), pp. 171–182.

Shorter, A., "The Measurement of Heat Required in Fire Clays," *Trans. British Ceramic Society,* vol. 47 (1948), pp. 1–21.

Silverman, A., *Date on Chemicals for Ceramic Use.* Bulletin 118, National Research Council, National Academy of Science, Washington, D. C., 1949.

Singer, F., *Ceramic Glazes.* Borax Consolidated Limited, London, n. d.

———, "Low Temperature Glazes," *Trans. British Ceramic Society,* vol. 53 (1954), pp. 398–421.

Stiles, H. E., *Pottery of the Ancients.* E. P. Dutton & Co., Inc., New York, 1938.

Sullivan, M., "Extraction of the Royal Tomb of Wang Chien," *Trans. Oriental Ceramic Society,* vol. 23 (1947–48), pp. 17–26.

Yetts, W. P., "Notes on Chinese Roof-Tiles," *Trans. Oriental Ceramic Society*, vol. 7 (1927–28), pp. 13–42.

CHAPTER 8. CASTING

Babelon, E., *Manual of Oriental Antiquities*. H. Grevel & Co., London, 1906.

Bailey, G. L. and W. A. Baker, "Melting and Casting of Non-ferrous Metals," *Symposium on Metallurgical Aspects of Non-ferrous Metal Melting and Casting of Ingots for Working*. The Institute of Metals, London, 1949.

Campbell, H. L., *Metal Castings*. John Wiley & Sons, Inc., New York, 1936.

Davis, W., *Foundry Sand Control*. The United Steel Companies, Ltd., Sheffield, England, 1950.

Dietert, H. W., *Modern Core Practice and Theories*. American Foundrymen's Association, Chicago, 1942.

Gough, H. J. and D. G. Sopwith, "The Resistance of Some Special Bronze to Fatigue and Corrosion-Fatigue," *Journal of the Institute of Metals (London)*, vol. 60 (1937), p. 143.

Hatfield, W. H., *Cast Iron in the Light of Recent Research*. 3rd ed. Charles Griffin & Co., Ltd., London, 1928.

Herb, C. O., *Die Casting*. The Industrial Press, New York, 1936.

Johnson, F., *Heat Treatment of Carbon Steels*. Chemical Publishing Co., Inc., Brooklyn, New York, 1946.

———, *Alloy Steels, Cast Iron, and Non-ferrous Metals*. Chemical Publishing Co., Inc., Brooklyn, New York, 1949.

Katzen, R., "Copper and Copper Alloys," *Industrial and Engineering Chemistry*, vol. 46, No. 10 (October, 1954).

Pell-Walpole, W. T., "Gases in Bronze," *Mémoires du Congrès International de Fonderies*. Brussels, Belgium, 1951.

Pinel, M., T. T. Read, and T. Wright, "Composition and Microstructure of Ancient Iron Castings," *Technical Publications 882. Metals Technology*. American Institute of Mining and Metallurgical Engineers, January, 1938.

Read, T. T., "Ancient Chinese Castings," *Trans. American Foundrymen's Association*, vol. 8, No. 3 (June, 1937), pp. 30–34.

Schwartz, H. A., *Foundry Science*. Pitman Publishing Corp., New York, 1950.

Shaw, R. M., "Cast Iron," *Iron Age*, vol. 156 (January 30, 1936).

Simpson, B. L., *Development of the Metal Casting Industry*. American Foundrymen's Association, Chicago, 1948.

Symposium on Centrifugal Casting, Publication 44–37. American Foundrymen's Association, Chicago, 1944.

Symposium on Malleable Iron Melting, St. Louis, 1943. American Foundrymen's Association, Chicago, 1943.

Yang Lien-sheng, *Money and Credit in China*. Harvard University Press, Cambridge, Mass., 1952.

CHAPTER 9. BOATS AND CARTS

Cardwell, R., "Pirate-Fighters of the South China Sea," *National Geographic Magazine*, vol. 89 (June, 1946), pp. 787–796.

Chang T'ien-tse, *Sino-Portuguese Trade from 1514–1644*. E. J. Brill Ltd., Leyden, Holland, 1934.

Cohen, Morris R. and I. E. Drabkin, *A Source Book in Greek Science*. McGraw-Hill Book Co., New York, 1948.

Cressey, P. F., "Chinese Traits in European Civilization: A Study in Diffusion," *American Sociological Review*, vol. 10 (1945).

Granet, Marcel, *La Civilisation Chinoise*. Albin Michel, Paris, 1948.

Hart, H. H., *Venetian Adventurer: The Life and Times of Marco Polo*. Stanford University Press, Stanford, 1947.

Ho, K., C. S. Wan, and S. H. Wein, "Iodine Value of Tung Oil," *Industrial and Engineering Chemistry, Analytical Ed.*, vol. 7 (1936).

Hudson, G. F., *Europe and China: A Survey of Their Relations from the Earliest Time to 1800*. Edward Arnold, Ltd., London, 1931.

Lowkowitsch, J. and G. H. Warbutron, *Chemical Technology and Analysis of Oils, Fats, and Waxes*. 3 vols. Macmillan & Co., New York, 1922.

Needham, Joseph, "The Unity of Science," *Asian Horizon* (Autumn–Winter, 1949–50), pp. 55–66.

——— and Wang Ling, *Science and Civilization in China*. Cambridge University Press, vol. I, 1954; vol. II, 1956.

Pickett, J. A. and W. L. Brown, "Variations in China Wood Trees," *Circular 115, Georgia Agriculture Experiment Station* (1938).

Reischauer, E. O., *Ennin's Diary (838–847* A.D.). The Ronald Press Co., New York, 1955.

Ricci, Matthew, *China in the 16th Century: The Journals of Matthew Ricci, 1583–1610*, translated from the Latin by Louis J. Gallager, S.J. Random House, New York, 1953.

Sarton, G., *A History of Science*. Harvard University Press, Cambridge, Mass., 1952.

Worcester, C. R. G., *The Junks and Sampans of the Yangtze*. 2 vols. Statistical Department of the Inspectorate General of the Customs, Shanghai, 1947–48. Reviewed in *Isis*, vol. 42, no. 355 (1951).

CHAPTER 10. HAMMER FORGING

Boylston, H. M., *An Introduction to the Metallurgy of Iron and Steel*. 2nd ed. John Wiley & Sons, Inc., New York, 1936.

Bullens, D. K. *Steel and Its Heat Treatments*. 3 vols. 5th ed. John Wiley & Sons, Inc., New York, 1948.

Camp, J. M. and C. B. Francis, *The Making, Shaping, and Treating of Steel*. 6th ed. The United States Steel Corp., Pittsburgh, 1951.

Cook, M., W. P. Fentiman, and E. Davis, "Observations on the Structure and Properties of Wrought Copper-Aluminum-Nickel-Iron Alloys," *Journal of Institute of Metals (London)*, vol. 80 (1952), pp. 419–430.

Garland, H. and C. O. Bannister, *Ancient Egyptian Metallurgy*. Charles Griffin & Co., Ltd., London, 1927.

Hanson, D. and C. B. Marryat, "Investigation of the Effects of Impurities on Copper. Part III—The Effect of Arsenic on Copper. Part IV—The Effect of Arsenic Plus Oxygen on Copper," *Journal of the Institute of Metals (London)*, vol. 37, No. 1 (1927), pp. 121–165.

Jefferson, T. B. and G. Woods, *Metals and How to Weld Them*. The James F. Lincoln Arc Welding Foundation, Cleveland, Ohio, 1954.

Laufer, B., *Chinese Clay Figures*. Field Museum of Natural History Publication 177, Anthropological Series, vol. 13, No. 2 (1914, Chicago).

Leach, R. H., "Silver Solders," *Proceedings American Society of Testing Materials*, vol. 3 pt. 2 (1930).

Lyman, T., *Metals Handbook*, American Society for Metals, Cleveland, Ohio, 1948.

Lynes, W., "Comparative Value of Arsenic, Antimony, and Phosphorous in Preventing Dezincification," *Proceedings American Society for Testing Materials*, vol. 41 (1941).

Mantell, C. L., *Tin*. Reinhold Publishing Corp., New York, 1949.

May, R., "The Corrosion of Condenser Tubes. Impingement Attack; Its Causes, and Some Methods of Prevention," *Journal of the Institute of Metals (London)*, vol. 40, no. 2 (1928), pp. 141–185.

Needham, Joseph, *The Development of Iron and Steel Technology in China*. Second Biennial Dickinson Memorial Lecture to the Newcomen Society, 1956. Published by the Newcomen Society, London, 1958.

Nightingale, S. J., *Tin Solders*. British Non-ferrous Metals Research Association, London, 1932.

Pearson, C. E., *The Extrusion of Metals*, John Wiley & Sons, Inc., New York, 1944.

Pomp, A., *The Manufacture and Properties of Steel Wire*, translated from the German by C. P. Berthoeft. 2nd ed. The Wire Industry Ltd., London, 1954.

Pulsifer, W. H., *Notes for a History of Lead*. D. Van Nostrand Co., Inc., New York, 1888.

Read, T. T., "The Mineral Production and Resources of China," *Trans. American Institute Mining and Metallurgical Engineers*, vol. 43 (1912).

Rolling Mills, Roll, and Roll Making. Mackintosh-Hemphill Co., Pittsburgh, Pa., 1953.

Rosenhain, W. and P. A. Tucker, "Eutectic Research No. 1: The Alloys of Lead and Tin," *Philosophical Transactions of the Royal Society of London*. Series A, vol. 209 (1909), pp. 89–122.

Rose, S. T. K., *The Metallurgy of Gold*. Charles Griffin & Co., Ltd., London, 1915.

Rossi, B. E., *Welding and Its Application*. McGraw-Hill Book Co., New York, 1941.

Sauveur, A., *The Metallography of Iron and Steel*. 4th ed. McGraw-Hill Book Co., Inc., New York, 1935.

Shepard, O. C. and W. F. Dietrich, *Fire Assaying*. McGraw-Hill Book Co., Inc., New York, 1940.

Sherry, R. H., *Steel Treating Practice*. McGraw-Hill Book Co., Inc., New York, 1929.

Stoughton, B., *The Metallurgy of Iron and Steel*. 4th ed. McGraw-Hill Book Co., Inc., New York, 1934.

Teichert, E. J., *Ferrous Metallurgy*. 3 vols. 2nd ed. McGraw-Hill Book Co., Inc., New York, 1944.

Thompson, F. C., "The History of Wire-Drawing," *Metal Industry* (*London*), vol. 60 (1942), pp. 70–72.

Wilkins, R. A. and E. S. Bunn, *Copper and Copper Base Alloys*. McGraw-Hill Book Co., Inc., New York, 1943.

CHAPTER 11. CALCINATION OF STONES

Agricola, G., *De Re Metallica* (1556), translated by H. C. Hoover and L. H. Hoover. Dover Publication, Inc., New York, 1950.

Briscoe, H. T., *General Chemistry for College*. Houghton Mifflin Co., Boston, 1938, pp. 808–809.

Franke, G., *A Handbook of Briquetting*. Charles Griffin & Co., Ltd., London, 1916.

Gregg, J. L., *Arsenical and Argentiferous Copper*, American Chemical Society Monograph Series. The Chemical Catalog Company, Inc. New York, 1934.

Johnson, O. S., *A Study of Chinese Alchemy*. Commercial Press, Shanghai, 1928. Reviewed in *Isis*, vol. 12 (1929), p. 330.

Knibbs, N. V. S., *Lime and Magnesia*. D. Van Nostrand Co., New York, 1924.

Lellep, O. G., "Latest Practice in Burning Cement and Lime in Europe," *Trans. American Institute of Mining and Metallurgical Engineers*, vol. 199 (1954).

McMahon, J. E., "Millions Invested in Exhuming Ancient Oysters," *New York Times*, September 3, 1955.

Moore, E. S., *Coal*. 2nd ed. John Wiley & Sons, Inc., New York, 1947.

Strong, R. A. E. Swartzman, and E. J. Burrough, *Fuel Briquetting*. Canadian Department of Mines and Resources, Ottawa, 1937.

Wang Kung-ping, *Controlling Factors in the Future Development of the Chinese Coal Industry*. King's Crown Press, New York, 1947.

CHAPTER 12. VEGETABLE OILS AND FATS

Bailey, A. E., *Cottonseed Processing and Products*. Interscience Publishers, Inc., New York, 1948.

______, *Industrial Oil and Fat Products*. Interscience Publishers, Inc., New York, 1945.

Bauchman, W. F. and G. S. Jamieson, "The Chemical Composition of Soya Bean Oil," *Journal of American Chemical Society*, vol. 44 (1922), pp. 2947–2952.

Bogue, R. I., *The Chemistry and Technology of Gelatin and Glue*. McGraw-Hill Book Co., Inc., New York, 1922.

Burton, D., *Sulphated Oils and Allied Products.* A. Harvey, London, 1939.

Cannon, M. R. and M. R. Fenske, "Composition of Lubricating Oil," *Industrial and Engineering Chemistry,* vol. 31 (1939), pp. 643–648.

Daquin, E. L., and Associates, "Filtration-Extraction of Cottonseed," *Industrial and Engineering Chemistry,* vol. 45 (January, 1953), pp. 247–254.

Fiero, G. W., "Hydrogenated Castor Oil in Ointments, VI: Sulfated Product in Official Ointments," *Journal of American Pharmaceutical Association,* vol. 30 (1941).

Ghosh, M. N., "Oil Content of Castor Seeds as Affected by Climate and Other Conditions," *The Agricultural Journal of India,* vol. 19 (1924), pp. 81–84. Chemical Abstracts, vol. 20, (1926), 2348.

Govindarajan, V. S. and B. V. Ramachandran, "L-lysine Content of Some Oilseed Cakes," *Journal of Scientific Industrial Research* (India), vol. 11B (1952), pp. 477–479.

Hildich, T. P., *The Industrial Chemistry of the Fats and Waxes.* 2nd ed. D. Van Nostrand Co., Inc., New York, 1941.

Jamieson, G. S., *Vegetable Fats and Oils.* 2nd ed. Reinhold Publishing Corp., New York, 1943.

——— and W. F. Baughman, "The Chemical Composition of Cottonseed Oil," *Journal of the American Chemical Society,* vol. 42 (1920), pp. 1197–1204.

Klemgard, E. N., *Lubricating Greases: Their Manufacture and Use.* Reinhold Publishing Corp., New York, 1937.

Langworthy, C. F., "The Digestibility of Fats," *Industrial and Engineering Chemistry,* vol. 15, No. 3 (March, 1923).

——— and A. D. Holmes, "Digestibility of Some Vegetable Fats," *USDA Bulletin 505.* United States Department of Agriculture, Washington, D. C., February, 1917.

——— and ———, "Studies on the Digestibility of Some Animal Fats," *USDA Bulletin 507.* United States Department of Agriculture, Washington, D. C., March, 1917.

"Larger Castor Seed Production in Manchuria," *Chemical and Metallurgical Engineering,* vol. 29 (December, 1923), p. 1200.

Markley, K. S., *Fatty Acids.* Interscience Publishers, Inc., New York, 1947.

——— and W. H. Goss, *Soybean Chemistry and Technology.* Chemical Publishing Co., Brooklyn, New York, 1944.

McKinney, R. S. and A. F. Freeman, "The Analysis of Tung Fruit," *Oil and Soap,* vol. 16 (1939).

Priest, G. W. and J. D. Von Mikusch, "Composition and Analysis of Dehydrated Castor Oil," *Industrial and Engineering Chemistry,* vol. 32 (1940).

Quackenbush, F. W., H. L. Gottlieb, and H. Steenbock, "Distillation of Tocopherols from Soybean Oil," *Industrial and Engineering Chemistry,* vol. 33 (October, 1941), pp. 1276–1278.

Rose, W. G., A. F. Freeman, and R. S. McKinney, "Solvent Extraction of Tung Oil," *Industrial and Engineering Chemistry,* vol. 34 (1942).

Shough, A. H., "Castor Oil Base Hydraulic Fluids," *Industrial and Engineering Chemistry,* vol. 34 (1942).

Smith, P. I., *Glue and Gelatin.* Chemical Publishing Co., Brooklyn, New York, 1943.

Swisher, M. C. and G. W. Fiero, "The Solvent Extraction of Castor Oil Seeds," *Journal of American Pharmaceutical Association,* vol. 21 (1932), pp. 579–582.

Venkitasubramanian, T. A., "Component Fatty Acids and Glycerides of Soybean Oil," *Journal of Scientific Industrial Research* (India), vol. 11B (1952), pp. 132–134.

Wennstrom, H. E., "Sulphonated Oils: Their Manufacture, Properties, and Uses," *Oil and Fat Industries,* vol. 4 (June, 1927), pp. 225–231.

CHAPTER 13. PAPER

Anderson, J. G., "Preliminary Report on Archaeological Research in Kansu," *Memoirs of the Chinese Geological Survey,* Series A, vol. 5 (1925).

_____, *Children of the Yellow Earth.* Macmillan Co., New York, 1934.

_____, "An Early Chinese Culture," *Bulletin of the Geological Survey of China,* No. 5 (1923).

Calkin, J. B. and G. S. Witham, Sr., *Modern Pulp and Paper Making.* Reinhold Publishing Corp., New York, 1957.

Carter, T. F., *The Invention of Printing in China.* Columbia University Press, New York, 1925.

_____ and L. C. Goodrich, *The Invention of Printing in China.* 2nd ed. Columbia University Press, New York, 1955.

Grant, J., *Wood Pulp and Allied Products.* Leonard Hill Ltd., London, 1947.

Jelks, J. W., "De-inking Paper Pulp by Flotation," *Bulletin No. F10–B66.* Denver Equipment Co., Denver, Colorado.

McLean, D. A., H. A. Birdsall, and C. J. Calbick, "Microstructure of Capacitor Paper," *Industrial and Engineering Chemistry,* vol. 45, No. 7 (July, 1953), pp. 1509–1515.

Peake, C. H., "The Origin and Development of Printing in China in the Light of Recent Research," Gutenberg-Fahrb., vol. 10 (1935), pp. 9–17.

Sutermeister, E., *Chemistry of Pulp and Paper Making.* John Wiley & Sons, Inc., New York, 1929.

Witham, G. S., Sr., *Modern Pulp and Paper Making.* Reinhold Publishing Corp., New York, 1942.

CHAPTER 14. THE METALS

Addicks, L., *Silver in Industry.* Reinhold Publishing Corp., New York, 1940.

Aluminum Bronze, Bulletin 31. Copper Development Co., London, 1938.

Arbiter, N., "Oxidized Copper," *Engineering & Mining Journal,* vol. 158, No. 3 (March, 1957), pp. 80–85.

Bonnin, A. *Tutenag and Paktong.* Oxford University Press, 1924.

Butts, A., *Copper*. Reinhold Publishing Corp., New York, 1954.

Caley, E. R., "On the Existence of Chronological Variations in the Composition of Roman Brass," *The Ohio Journal of Science,* vol. 55, No. 3 (May, 1955), pp. 137–140.

Carpenter, H. C. H., "An Egyptian Axe Head of Great Antiquity," *Nature,* vol. 130 (1932, London), pp. 625–626.

Dickerman, N., *Foreign Mineral Survey: Mineral Resources of China.* U. S. Bureau of Mines. The Government Printing Office, Washington, D. C., January, 1948.

Ellis, D. W., *Copper and Copper Alloys.* American Society for Metals, Cleveland, 1948.

Ercker, L., *Treatise on Ores and Assaying,* translated by A. G. Sisco and C. S. Smith. University of Chicago Press, Chicago, 1951.

Everhart, J. L., "Aluminum Bronze," *Materials and Methods,* vol. 34 (December, 1952), pp. 119–134.

Friend, J. N., *Iron in Antiquity*. Charles Griffin & Co., Ltd., London, 1926.

Haymard, C. R., *An outline of Metallurgical Practice.* 2nd ed. D. Van Nostrand Co., Inc., New York, 1942.

Hoffman, H. D., *Metallurgy of Lead.* McGraw-Hill Book Co., Inc., New York, 1918.

Ingolls, W. R., *The Metallurgy of Zinc and Cadmium.* 2nd ed. McGraw-Hill Book Co., Inc., New York, 1906.

Laque, F. L. and C. L. Cox, "Some Observations of the Potentials of Metals and Alloys in Sea Water," *Proceedings American Society of Testing Materials,* vol. 40, no. 670 (Philadelphia, 1940).

Speller, F. N., *Corrosion, Causes & Prevention.* 3rd ed. McGraw-Hill Book Co., Inc., New York, 1951.

Stanley, R. C., *Nickel, Past and Present.* The International Nickel Co. of Canada, Toronto, 1934.

Stewart, W. C. and F. L. Laque, "Corrosion Resisting Characteristics of Iron Modified 90:10 Cupro Nickel Alloy," *Corrosion,* vol. 8, no. 8 (August, 1952), pp. 259–277. National Association of Corrosion Engineers, Inc., Houston, Texas.

Sullivan, J. W. W., *The Story of Metals.* American Society for Metals, Cleveland, Ohio, 1951.

Thornton, C. P., "Meteorites," *Mineral Industries,* vol. 27, no. 6 (March, 1958), The Pennsylvania State University, University Park, Pennsylvania.

Uhlig H. H., *Corrosion Handbook.* John Wiley & Sons, Inc., New York, 1948.

Von Zeerleber, A., *The Technology of Aluminum and Its Light Alloys,* translated by A. J. Field. Nordemann Publishing Co., Inc., New York, 1936.

Zimmer, G. F., "The Use of Meteoric Iron by Primitive Man," *Journal Iron and Steel Institute* (London), vol. 94, no. 2 (1916).

CHAPTER 15. WEAPONS

Bernadau, J. R., *Smokeless Powder, Nitro-cellulose, and Theory of the Cellulose Molecule.* John Wiley & Sons, Inc., New York, 1908.

Cook, M. A., *The Science of High Explosives,* American Chemical Society Monograph No. 139. Reinhold Publishing Corp., New York, 1958.

Cowher, Lou, W. H. Hunley, and LaDow Johnston, *How to Build a Muzzle Loading Rifle, Target Pistol, and Powder Horn.* The National Muzzle Loading Rifle Association, Inc., Portsmouth, Ohio, 1958.

Davis, T. L. and Chao Yung-tsung, "An Alchemical Poem by Kao Hsiang-hsien," *Isis,* vol. 30, no. 2 (May, 1939).

——— and ———, "Chao Hsueh-min's Outline of Pryotechnics: A Contribution to the History of Fire Works," *Proceedings of the American Academy of Arts and Sciences,* vol. 75 (1943), pp. 95–107.

——— and J. R. Ware, "Early Chinese Military Pyrotechnics," *Journal of Chemical Education,* vol. 24 (1947), pp. 522–537.

Goodrich, L. C., "Notes on a Few Early Chinese Bombards," *Ibid.,* vol. 35, pt. 3, no. 101 (1944), p. 211.

——— and Feng Chia-sheng, "The Early Development of Firearms in China," *Isis,* vol. 36 (1946), pp. 114–123, 250.

Guttmann, G., *The Manufacture of Explosives.* 2 vols. Whittaker and Co., London, 1895.

Mayers, W. F., "On the Introduction and Use of Gunpowder and Firearms Among the Chinese," *Journal of the North China Branch Royal Asiatic Society,* n. s., vol. 6 (1871), pp. 73–103.

Partington, J. R., *A History of Greek Fire and Gunpowder.* W. Heffer & Sons, Ltd., Cambridge, England, 1960.

Smith, H. DeWolf, *Atomic Energy for Military Purposes.* Princeton University Press, Princeton, N. J., 1948.

Tunis, E., *Weapons: A Pictorial History.* The World Publishing Co., Cleveland & New York, 1954.

Wang Ling, "On the Invention and Use of Gunpowder and Firearms in China," *Isis,* vol. 37, parts 3 & 4, no. 109–111 (July, 1947), pp. 160–178.

CHAPTER 16. VERMILLION AND INK

Apps, E. A., "Inks for Metal Decoration," *Metal Finishing Journal,* vol. 10, nos. 110 and 112 (February and April, 1964), pp. 71–76, 143–149.

Calcutt, J. A., "Metallic Lead as Pigment for Anti-corrosive Paints," *Corrosion Prevention and Control,* vol. 6, no. 11 (November, 1959), pp. 34–38.

Chopey, N. P., "How to Make Azo Pigments," *Chemical Engineering,* vol. 68, no. 1, (January 9, 1961), pp. 74–77.

Drogin, I., "Carbon Black as Pigments," *Color Engineering,* vol. 2, no. 3 (March, 1964), pp. 12–18, 20–32.

Ellis, C., *Printing Inks.* Reinhold Publishing Corp., New York, 1940.

Fried, E., "New Pigment-Mixture Diagram and Color System," *Optical Society of America Journal,* vol. 49 (December, 1959) pp. 1159–1168.

Hurst, G. H., *Painters' Colors, Oils, and Varnishes.* 3rd ed. D. Van Nostrand Co., New York, 1901.

Porth, V. J. Jr. and L. F. Engelhart. "Lithographic Inks and their Relation to Paper," *Tappi,* vol. 44 (July, 1961), pp. 12A–36A.

Watts, G. E., "Titanium Oxide in a Free World," *Chemical Age,* vol. 87, no. 2224 (February 24, 1962), pp. 317–318, 322.

CHAPTER 17. YEASTS

Allen, P. W., *Industrial Fermentation.* The Chemical Catalog Co., New York, 1926.

Booker, L. E., E. R. Hartzler, and E. M. Hewston, "A Compilation of the Vitamin Values of Foods in Relation to Processing and Other Variants," *USDA Circular No. 638* (May, 1942). U. S. Department of Agriculture, Washington, D. C.

Castiglioni, A., *A History of Medicine.* Knopf, New York, 1947.

Christensen, C. M., *The Molds and Man.* University of Minnesota Press, Minneapolis, 1951.

Conant, J. B., *Pasteur's Study of Fermentation.* Harvard University Press, Cambridge, Mass., 1952.

Cruess, W. V., *The Principle and Practice of Wine Making.* 2nd ed. The AVI Publishing Co., Inc., New York, 1947.

Florey, H. W. et al., *Antibiotics.* 2 vols. Oxford University Press, London, 1949.

Grumm-Grzimailo, A., "Contribution à l'histoire de l'introduction de la vigne en Chine," *Archives of the History of Science and Technology,* vol. 5 (Leningrad, 1935), pp. 499–506. (In Russian with French Summary: *Isis,* vol. 25, no. 1 (May, 1936), pp. 261–262.

Hull, W. O., W. E. Kite, and R. C. Auerback, "Modern Wine Making," Industrial and Engineering Chemistry, vol. 43 (October, 1951), pp. 2180–2192.

Large E. D., *The Advance of Fungi.* Jonathan Cape, London, 1940.

Ludovici, L. J., *Fleming, Discoverer of Penicillin.* Indiana University Press, Bloomington, Indiana, 1952.

Marriott, H. J. L., *Medical Milestones.* Williams & Wilkins, Baltimore, 1952.

Our Smallest Servants. Chas. Pfizer & Co., Inc., Brooklyn, N. Y., 1956.

Peckham, C. T. Jr., "Starch Hydrolyzates in the Food Industries," *Use of Sugar and Other Carbohydrates in the Food Industry,* Advances in Chemistry Series, No. 12, American Chemical Society (1953), pp. 43–48.

Shearon, W. H. Jr. and H. E. Weissler, "Brewing," *Industrial and Engineering Chemistry,* vol. 43, no. 6 (June, 1951), pp. 1262–1271.

"Sources for New Drugs," *Chemical and Engineering News,* American Chemical Society, vol. 35, no. 3 (January 21, 1957), pp. 16–17.

Steinhaus, E. A., "Living Insecticides," *Scientific American,* vol. 195, no. 2 (August, 1956), pp. 96–104.

Tauber, H., *Enzyme Technology.* John Wiley & Sons, Inc., New York, 1943.

Undefkofler, L. A. and R. J. Hickey, *Industrial Fermentations,* vol. 1. Chemical Publishing Co., New York, 1954.

Welch, H., *Principles and Practice of Antibiotic Therapy.* Medical Encyclopedia, New York, 1954.

CHAPTER 18. PEARLS AND GEMS

Baxter, W. T. and H. C. Dake, *Jewelry Gem Cutting and Metalcraft.* McGraw-Hill Book Co,. Inc., New York, 1942.

Bruce, R. C., "Catalogue of the Exhibition of Chinese Jades," *Trans. Oriental Ceramic Society,* vol. 23 (1947–48), pp. 46–61.

Cheng Te-k'un, "Tang and Ming Jades," *Trans. Oriental Ceramic Society, London,* vol. 28 (1953–54), pp. 23–36.

Coker, R. E., "Fresh-water Mussels and Mussel Industries in the United States," *Bulletin of the United States Bureau of Fisheries,* vol. 36 (1917–19), pp. 11–88. Government Printing Office, Washington, D. C.

Dana, E. S. and W. E. Ford, *A Textbook of Mineralogy.* John Wiley & Sons, Inc., New York, 1926.

Hansford, S. H., "Two Inscribed Jades," *Trans. Oriental Ceramic Society,* vol. 23 (1947–48).

"Kokichi Mikimoto, Pearl King, Dead," *New York Times,* September 22, 1954.

Korringa, Pieter, "Oysters," *Scientific American,* vol. 189, no. 5 (November, 1953), pp. 86–91.

Kraus, E. H. and C. B. Slawson, *Gems and Gem Materials.* 3rd ed. McGraw-Hill Book Co., Inc., New York, 1939.

Laufer, B., *The Diamond: A Study in Chinese and Hellenistic Folklore.* Field Museum of Natural History Publication 184, Anthropological Series, vol. 15, no. 1 (1915, Chicago).

______, *Jade,* P. D. & Ione Perkins in Cooperation with the Westwood Press & W. M. Hawley, South Pasadena, California, 1946.

______, *Notes on Turquoise in the East.* Field Museum of Natural History Publication 169, Anthropological Series, vol. 13, no. 1 (July, 1913, Chicago).

Palache, C., H. Berman, and C. Frondel, *The System of Mineralogy.* 2 vols. John Wiley & Sons, Inc., New York, 1944.

Pack, G., *Jewelry and Enameling.* 2nd ed. D. Van Nostrand Co., Inc., New York, 1953.

Smith, O. C., *Identification and Qualitative Chemical Analysis of Minerals.* D. Van Nostrand Co., Inc., New York, 1953.

Whitlock, H. P. and M. L. Ehrmann, *The Story of Jade.* Sheridan House, New York, 1949.

GLOSSARY

(For purposes of alphabetizing, Chinese terms are treated as single words.)

A

Abutilon fiber 茼麻

Acorn shells (*Quercus chinensis*) 栗壳

Agate 瑪瑙

A-la-ku 阿拉古

Alder wood 杞木

Alluvial soil 洲土

Alum 明礬

Alum water 白礬水

Amaxantus mangostanus seed 莧菜子

Amber 琥珀

Ang-chou 卬州

An-i 安邑

Anise 茴香

Apricot seed paste 杏仁泥

Aquilaria agrallohra 沉香

Artemisia [*Artemisia apiacea*] 青蒿

Audience bell 朝鐘

"Automatic wind" 自来風

Autumn gauze 秋羅

Aventurine, see Hsing-han Sha

Awl 錐

Awl-hoe 錐鋤

B

Bamboo eyelets 竹針眼

Bamboo fibers 竹麻

Bamboo paper 竹紙

Bamboo pipes 竹梘

Banded agate 錦纏瑪瑙

Barbarian safflower 番紅花

Bark of *Myricaeceae* 楊梅皮

Bark paper 皮紙

Basin niter 盆消（硝）

Bean jam 豆豉

Bean milk 豆漿水

Beef fat 牛油

Big-eyed *t'ung* (*Aleurites cordata*) 大眼桐

Bird gun 鳥槍

Bird pistol 鳥銃

Bitter bamboo (*Phyllostachys bambusoides*) 苦竹

Bitula japonica, hua 樺

Black 玄色

Black lentil 穭豆

Black vitriol (green vitriol) 皂礬

Blister (or raw) copper 生銅

"Blue grass" 藍草

Blue pigment 澱 (澱信)

Blue vitriol, see Gall vitriol

Board 箭

Boiling sea elixir (a reddish gem stone) 煮海金丹

Boswellia lagbora 乳香

[To] bow cotton 彈棉

Bran 麩

Bright green 豆綠色

Bright salt 光明鹽

Brine 鹵水, 鹵汁

Broad bean 蠶豆

[Broken] bituminous coal 碎煤

Brown sugar 黑砂

Buckler 盾

Buckwheat 蕎麥

Buddha's-head blue 佛頭青

Burnt-offering paper 火紙

Burnt silk 焚帛

Burweed (*Xanthium strumdrium*) 蒼耳

C

Cabbage seeds (cabbage) 菘菜子 (白菜)

Camphor seeds 樟樹子

Cane crusher 糖車

Cannon 礮

Carnelian 靺鞨

Carrying-pole lead 扁擔鉛

"Cart brain" 車腦

Cashmere goat (yü-t'iao yang) 䍧羊

Castor seed 蓖麻子

Cat's eye or tiger's eye 猫精 (睛)

Ceramics 陶埏

Ceruse 胡粉

Ch'a-ching (silver) 茶經

Chaff 糠

Chaff gold 糠金

Chan-k'ou 盞口

Chan-lu (deep vessel) 湛盧

Chang Ch'ien 張騫

Chang-chou 漳州

Chang Hua 張華

Ch'ao powder 朝粉

Che-chiang 柘漿

Che leaves 柘葉

Ch'e-li monkey 扯里猻

Ch'en 辰

Chen-chiang (Chinkiang) prefecture 鎮江府

Cheng 鉦

Cheng Ho 鄭和

Ch'en Hsüan 陳瑄

Chen-ting prefecture 真定府
Che-t'ang 蔗餳
Chi 璣
Ch'i 芑
Ch'i 臍
Ch'iang 羌
Ch'iang 鏹
Chiang-yu 醬油
Ch'iao-ch'ui 巧倕
Chiao-shih 焦石
Chicken-heart awl 雞心錐
Chieh Lake 解池
Ch'ien-li ch'uan 千里船
Ch'ien-shan 鉛山
Chien-yao 建窰
Ch'ih 尺
Chih-ku-k'ou 直沽口
Ch'i-mu Huai-wen 綦母懷文
Ch'in 秦
Ch'in-chiao (*Xanthaxylum piperitum*) 秦椒
Chin-ch'ih-wei 金齒衛
Ch'in-chou 秦州
Ch'ing-chiang-p'u 清江浦
Ching-chou 荊州
Ch'ing-fen (calomel) 輕粉
Ching-sai nu 靜塞弩
Ching-te Chen 景德鎮
Ching-yang 涇陽
Ch'ing ying 青鸎
Chin-hua 金華
Chin-kang 金剛
Chin-tan 金丹
Chin-tan ta-yao 金丹大要
Chin wood 槮木
Chiu-ching (*The Wine Classic*) 酒經
Chiu-k'ung Mountain [lit. "Nine Hollows" Mountain] 九空山
Chiu-niang (mother of wine) 酒釀
Chrysalis 蛹
Ch'u 楚
Chü 秬
Ch'ü (yeast) 麴
Ch'üan-chou 泉州
Ch'uang-tzu nu 牀子弩
Ch'ui 倕
Chu-ko Liang 諸葛亮
Chu-ko nu (Chu-ko crossbow) 諸葛弩
Chün-yao 均窰
Chu-yai 珠崖
(Ch'ü-yü) carpets 氍俞
Cinnabar (mercury sulfide) 朱砂
Cinnabar bed 朱砂牀
Clam-shell lime 蜃灰
Clean rice broth 清泔汁

Clear pale pink 水紅色

Cloud tile 雲瓦

Clove 丁香

Coal 煤炭

Coarse paper 糙紙

Cold paste 冷漿

Combed wool 搊絨

Copper coal 銅煤

Copper green [copper acetate] 銅綠

Copper-made crossbow lock 銅弩機

Core mold 模骨

Corners of eaves 榱桷

Cotton (*Gossypium indicum*) 草棉

Cotton gin or ginning machine 棉花趕車

Cotton paper 綿紙

Counter-weight lever 桔槔

Covered (a) culturing yellow mould, (b) fermentation 罨黃

Crackled ware 碎器

Crepe 縐紗

Crimson 大紅色

Crockery mortars 沙盆齒鉢

Cross beam 桄

Crouching lion 伏獅

Curers of eczema 淫瘡家

Currency coin 貨泉

Curved brick 鞠磚

Curved cover 抱同

Curved distillation tube 曲弓溜管, 曲溜

Curved pole 犂擔

Cushion (for bow-string) 墊絃

Cut money 板錢

Cut water beam 斷水樑

Cylinder wheel 筒車

D

Dalbergia hupearra tree 檀木

Damask 綾

Damask weave 綾地

Dark barley 青稞

Dark green 大紅官綠色

Deep blue 大青

Deep sky-blue 天青色

Divine-fire ball 吐焰神球

Divine-frightening cannon 神威大砲

Diviner's board 栻, 羅盤

Document paper 揭貼呈文紙

Dog-head gold 狗頭金

Double-bolted flour 重羅之麪

Double cocoon 雙繭

Dragon robe 龍袍

Dragon's mouth beam 龍口樑

Drain tile 溝瓦

Draw-boy 提花小廝

Drawer board (on a loom) 衢盤

Draw loom 花機

Driving shafts 疊助木

Dry alum 枯礬

Dryobalanops camphora 龍腦香

E

Early Chi-an 吉安早

Early common rice 秈稻米

Early Liu-yang 瀏陽早

Earth salt 土鹽

Earthen funnel 瓦溜

Earthen jars, large and small 罌缶

Egg-shell blue 蛋青

Eight-directions-rotating string-of-100-bullets cannon 八面轉百子連珠砲

Elevated cylinder wheel 高轉筒車

Elm wood 榆木

Emerald 祖母綠

Empty husk, husk 秕

Erh-ya 爾雅

Evergreen 椆

Eyelets 溜眼

F

Fang I-chih 方以智

Fang-shan 房山

Fan-yu (possibly bitumen or asphalt) 礬油

Feast candy 享糖

Felt 氈

Fertilizer wheat 肥田麦

Figured spun-silk 花綿

Figure tower (*hua-lou*) 花樓

Files 鑢

Fine bran 細糠

Fine fabric 工布

Fine split bamboo 竹絲

Fire-arrow 火矢

Fire well (natural gas) 火井

Fire yeast 火麴

First-grade vermillion 頭朱

Flail 枷

Flat awl 扁錐

Flax seeds (*Linum usitatissimum*) 亞麻子

Flood-dragon 蛟龍

Floss 浮絲

Flour bolter 麵羅

Flowering mulberry 花桑

Fo-lang-chi ("Feringi", Portuguese calivers) 佛郎機

Four-fire brass 四火銅

Fourfold paper 連四紙

Fou-yu 浮游

Fox 狐

Frames of the healds 簆耙

Fruit-cane 果蔗

Full kernels 穰

"Full sail" boats of the Yellow River 黃河滿蓬梢

G

Gallnut 棓子

Gall vitriol (blue vitriol) 膽礬

Garnet sand 紅沙

Garu-wood and musk 沈, 麝

Gauze 羅

Gem stones 寶石

General's helmet 將軍盔

Geographic Classic, The (*Shan hai ching*) 山海經

Ginger-shaped iron 薑鐵

Glaze 銹, 釉

Glaze nut 釉果

Glossy jar 甕鑑

Gluten of wheat 麪觔

Goat 山羊

Gold-back coins 金背錢

"Gold juice" 金汁

Golden-haired ape 金絲猿

Grain-flower honey 禾花蜜

Grain salt 顆盐

Grain tree 穀樹

Granular tin 錫沙

Grape blue 葡萄青色

Graphite 黑石脂

Gravel salt 砂石盐

Great commander 大將軍

Great gateway 巨闕

Green lentil flour 綠豆粉

Green lentils 綠豆

Green tally 葱符

Green vitriol 皂礬

Ground ash 地灰

Ground juice 地溲

Guide rolls 星丁頭

Gypsum 寒水石

H

Hand cranked wheels 拔車

Hand pestle 手杵

Hand spool 籰

Hao-fang 蠔房

Hardwood charcoal 火墨

Heald 篾

Hei tan 黑丹

Hemp 麻, 火麻

Hemp-quilted garment 緼袍

Hemp seeds 大麻仁

Hemp yarn 苧紗

Heng-yeng 衡陽

Hibiscus skin (*Hibiscus mutobilis*) 芙蓉膜

Ho Ch'ou 何稠

Ho-hsi 河西

Holly seeds (*Ilex pedunculosa*) 冬青子

Ho-lo-tan 呵羅單國

Honeycomb 蜜脾

Ho-p'u 合浦

Horse indigo 馬藍

Horse-teeth niter 馬牙消(硝)

Ho Sui 何遂

Ho-t'ien 和闐

Hou-kang 後崗

Hsia 夏

Hsi-an 西安 (Sian)

Hsiang Hsiu 向秀

Hsiang-yang 襄陽

Hsiao-mi 小米

Hsiao-shih 硝(消)石

Hsi-chung 奚仲

Hsien-men Tzu-kao 羨門子高

Hsien-yang 咸陽

Hsin 釁

Hsin District [i.e., Kuang-hsin] 信郡(廣信)

Hsing-han Sha [probably aventurine, sometimes called goldstone] 星漢沙

Hsing I 邢夷

Hsin-kuo-chou 興國州

Hsin stones 信石

Hsin-yang-chou 信陽州

Hsi-yang-p'ao, the occidental cannon 西洋砲

Hsüan-chi 璇璣

Hsüan-fu 宣府

Hsüan furnace 宣爐

Hsüeh T'ao note-paper 薛濤牋

Hsü Kuang-ch'i 徐光啓

Huai-an 淮安

Huai (*Sophora japonica*) flower 槐花

Huai flower bud 槐蕊

Huai-nan-tzu 淮南子

Huai River 淮河

Huang-lien (*Coptis japonica*) 黃連

Huang-lien-sha (at Hai-men) 黃連沙

Hua-yang-kuo Chih 華陽國志

Hu-chou silk (woven) 湖紬

Hulling mill 礱

Hun-chiang-lung (submarine mine) 混江龍

I

I 鎰

I-hsing 宜興

I-i 夷羿

I-ling 夷陵

Imperial wheat 御麥

I-shih 猗氏

I-ti 儀狄

Incense file 香鎈

Indigo 藍澱

Indigofera tinctoria 木藍, 槐藍, 馬棘

Indigo powder 靛花

Interior rim (of a wheel) 輔

Iron and/or steel hardening 健鐵

Iron coal (natural coke or anthracite) 鐵炭

Iron drill 鐵椎

Iron droppings 鐵落

Iron ore 鐵砂

Iron roller and trench 鐵碾槽

Iron sparks 鐵華

Irregular sugar 頑糖

Isatis tinctoria 菘藍

Ivory color 象牙色

J

Japanese lead 倭鉛

Japanese satin 倭緞

Jellied maltose 膠飴

Jen-shêng (*Panax ginseng*) 人參

Job's-tear wine 薏酒

K

Kan-chiang 干將

Kan-chu 甘藷

K'ang-mi chiu 杭米酒

Kan-ts'ao (*Glucyrrhiza glakra*) 甘草

Kao 郜

Kao-chou 高州

K'ao-kung chi 考工記

Ke hêmp 葛麻

Keng 秔

Keng 粳

Keng chih t'u (Pictorial Accounts of Agriculture and Sericulture) 耕織圖

Ke plant (creeper) 葛

Kiangsu indigo (*Indigofera kiangsu*) 吳藍

Killer-of-myriads (a toxic incendiary bomb) 萬人敵

Kiln 窯

Kiln refuse powder 窯滓灰

Knife brick 刀磚

Ko Hsüan 葛玄

Ko Hung, known also as Pao-p'u-tzu 葛洪, 抱朴子

K'uang 穬

Kuang-hsin 廣信

Ku-ku wool 孤古絨

K'un-lun vitriol 崑崙礬

Kung-sun Ch'iao 公孫僑(子産)

Kung-sun Shu 公孫述

L

La 鑞

Lake salt 池塩

Lampwick 燈心

Lanchou woolens 蘭絨

Landscape-designed agate 截子瑪瑙

Lapis lazuli 琉璃

Large French beans 刀豆

Large jar 缸

Large panicled millet 黍

Large quarto 大四連

Large rollers 巨軸

Large salt 大塩

Leather braiding

Leavening, fermentation 信

Leaven, yeast 麴

Lei Hsiao 雷斅

Lentil 小豆

Lentil wine 綠豆酒

Level trench 平槽

Liang 梁

Li Chiang 麗江

Lien 鏈, 連

Lien-chou 廉州

Li-fang 蠣房

Li Ping 李冰

Li tribesmen of Kwangtung 廣南黎人

Light blue 草白

Light green 油綠色

Light of Wu-yuan 婺源光

Lime 石灰

Lin-an prefecture 臨安府

Lin-chin 臨晋

Lin Ch'iung 臨邛

Ling-piao lu-i 嶺表錄異

Lin-t'ao Commandery [of Kansu Province] 臨洮郡

Litharge 黃丹

Litmus 紫鉚

Liu-chia Harbor at Soochow 劉家港

Liu P'i 劉濞

Long string bean 豇豆

Lo-p'an 羅盤

Lophanthus rugous 藿香

Lotus pink 蓮紅色

Lotus-seed shells 蓮子殼

Louse sesamum 壁蝨脂麻

Lowly silkworm 賤蠶

Low temperature brick 嫩火磚

Lo-yang 洛陽

Lü 莒

Lung-hsi 隴西

Lung-shan civilization 龍山文化

Lun heng 論衡

Lü-shih Ch'un-ch'iu (3rd Century B.C.) 呂氏春秋

[Lustrous] anthracite coal 明煤

M

Ma-chia-chai (Shensi) 馬家寨

Ma Chün 馬鈞

Ma-huang (*Ephedra vulgaris*) 麻黃

Malachite 空青, 曾青

Malted 糵

Maltose 飴, 餳

Malt sugar (lit. "minor sugar") 小糖

Marshes 草蕩

Mauve 藕褐色

Ma-wei 馬尾

Medicinal yeast 神麯

Mei-kuei ["round gem" probably garnet and/or mica] 玫瑰

Melon-seed gold 瓜子金

Meng hemp (*Abutilon avicenuae*) 苘麻

Meng-shan copper 蒙山銅

Mercury 水銀

Meteorite or shooting-stars cannon, The 流星砲

Method of draining 淋法

Miao 苗

Middle paper 中夾紙

Mighty-arm crossbow, The (shen-pi-nu) 神臂弩

Milk-mulberry fiber 桑穰

Millet and sorghum, different varieties of 粟粱黍稷

Min-ch'ih District 澠池縣

Mine powder 礦灰

Mined lead 出山鉛

Mixed tallow-oil 桕混油

Mo-ching 墨經

Mo-fa chi-yao (Essentials of Ink Making) 墨法集要

Mohammedan blue 回青

Mo-hsieh 莫邪

Mo-p'u (Categories of Ink) 墨譜

Mortar 臼

Mo-tzu 墨子

Mother gold 母金

Mou-i 牟夷

Mountain pomegranate flowers 山榴花

Moxa 艾

Mu-nan [probably beryl] 木難

Musical boards 雲板

N

Nardostachys jatamansi 甘松

Native sulphur 土硫黃

Natural musk 麝香

Navy-blue cloth 毛青布

"Nest of silky thread" (spun malt-sugar) 一窩絲

Nettle-hemp 枲麻

Newly hatched silkworm 乳蠶,蠶蚍

Nine-arrow heart-piercing cannon, The 九矢鑽心砲

Ningsia 寧夏

Ning-kuo Commandery [in Northern Anhui Province] 寧國郡

Nu (crossbow) 弩

Nuo 糯

O

Ochre 代(黛)赭石

Ordinary pearl 常珠

(Ordinary) wheat 小麥

Organzine rack 經耙

Orpiment 石黃

Otter fur 獺皮

Ou 甌

Oyster [shell] lime 蠣灰

P

Pa 巴

P'ai-ch'uan 柏船

Pa-tou (*Croton tiglium*) 巴豆

Pai-t'ou weng (*Anemone cernua*) 白頭翁

Pale blue 月白

Palm-fiber cloth 蕉紗

Palm leaves 貝樹葉

P'ang-lu rugs 氆氌

P'an Ku 潘谷

Pao-p'u-tzu 抱朴子

Paper-mulberry leaves 楮葉

Paper-mulberry tree (*Broussonetia papyrifera*) 楮樹

Paper sacrificial money 紙錢

Paste-impregnated paper 蠲糨紙

Pea 豌豆

Peach-bamboo 桃竹

Peach-blossom pink 桃紅色

Peacock blue 翠藍

Pei-chi-ko 北極閣

Pendant gold 帶胯金

Perilla nankinensis (*Sabra minuti flora, Sabra plebeia*) 紫蘇

Perilla ocymoides 蘇麻

Petuntse (properly pai-tun-tzu) 白不子

Picked wool 拔絨

Pierced cocoon 出種繭殼

Pig iron 生鐵

Pigweed indigo (*Amarantaceae tinctorium*) 莧藍

Pi Mou-k'ang 畢懋康

Pine-needle juice 松毛水

Ping 邴

P'ing-yang (Shansi) 平陽

Pin-t'ieh 鑌鐵

Plane 鉋

Poison-mist divine-smoke cannon, The 神烟砲

Polygonum tinctorium 蓼藍

Pongee 絹

Poppyseed 金罌子

"Pot-bottom" silk 鍋底綿

Potter's wheel 陶車

Pounding mill (water-powered) 水碓

Powdered coal 末煤

Powdered salt 末塩

Powdered smartweed 辣蓼末

Powdered white brass 白銅末

Po-wu chih 博物志

Pressed ware 印器

Pulley well 轆轤

Pulp tank 抄紙槽

Purple powder 紫粉

Pyrolusite 無名異

Q

Quartz crystal 水晶

Quartz crystal prism 火齊珠

Quilted clothing, quilting 挾纊

R

Rain or snow-water 天露

Ramie, rhea, or China grass 苧麻

Rape-flower honey 菜花蜜

Rape seed 芸薹子（菜子）

Receiving hole 鴨嘴

"Red-barbarian cannon" 紅夷砲

Recumbent brick 眠磚

"Red-flower" agate 錦紅瑪瑙

Red kaolin 赤石脂

Red vitriol 紅礬

Red wine-mash 紅酒糟

Red yeast 丹麴

Reed (*Miscanthus sacchari* flowers) 荻

Reeling machine （繅）絲車

Reeling silk 治絲

Refined copper 熟銅

Revolving awl 旋錐
Resounding bronze 響銅
Resurrected paper 還魂紙
Rice stalks 稻稭
Rigid rods (on a loom) 衢脚
Roasted ore 青砂頭
Rock salt 崖盐
Rock sugar, rock candy 凝冰, 冰糖
Roller (to husk) 碾
Rolling stone 碾石
Rosa banksia 木香
Rosin 松香
Rough woolens 褐
Round awl 圓錐
Rounded or wheeled ware 圓器
Ruby or rubellite 喇子
Running pearls 走珠
Rushes 蒲草

S

Sable 貂
Safflower cakes 紅花餅
Safflower juice 紅花汁
Saggers 匣鉢
Salted mud 盐泥
Salt frost 盐霜
Salt pan 盐盆
Saltpeter 硝
Saltpeter earth 土消(硝)
San 饊
Sandalwood 檀木
Sandless clayey earth 無沙粘土
Sandy bottom 沙脚
"Sandy" pots and pans 沙鍋沙礶
Sandy soil 夾沙土
Sappanwood 蘇木
Sapphire 瑟瑟珠
Satin 緞
Scarlet 猩紅
Scoop chisels 剜鑿
Screen 簾
Sea brine 海鹵
Sealing wax coins 火漆錢
Sea salt 海盐
Second Commander 二將軍
Second-grade vermillion 二朱
Seeds of Job's tears (*Coix lachryma*) 薏苡仁
"Seeth" water of natural indigo 靛水
Seiaena sehlegeli 石首魚
Separation rods 交竹
Serge 褐子
Sesamum 胡麻, 芝(脂)麻
Shao-chiu 燒酒
Shao-chou 韶州
Shao-hsing wine 紹興酒
Shao powder 韶粉

Sheep's fat jade 羊脂玉

Sheep's head 羊頭

Sheepskin 羊皮

Sheet of silkworm eggs 蠶紙

Shelf (for silkworms) 箔

Shell fortress 螺城

Shells of wood-oil seeds [*Aleurites cordata*] 桐殼

Shield 楯, 傍牌

Shih (*tan* after Sung Dynasty) 石

Shih-mo 石墨

Shining pearls 滑珠

Short plank 短枋

Shu 蜀

Shui ching chu 水經注

Shu-mi chiu 黍米酒

Shu-mi lu-chiu 粟米爐酒

Shuo wen 說文

Shu River 蜀江

Side brick 側磚

Sieve 篩

Silk (woven cloth) 紬

Silken gauze for the bolter 絲織羅地絹

Silk-mulberry fiber paper 桑穰紙

Silk wadding (絲)綿

Silkworm bath 蠶浴

Silkworm moth 蠶蛾

Silver pink 銀紅

Silver rust 銀銹

Silver sparkles 銀花

Single cocoon 獨蠶繭

Sinker lead 釣腳鉛

Sizing 過糊

Skein frame 絡篤

Sky blue 天藍

Sky-lights 天窗

Slot 齒

Small bark paper 小皮紙

Small-grained millet 粟

Small-leafed *Polygonum tinctorium* 小葉莧藍

Small panicled millet 稷

Small roller 小碾

Smartweed [*Polygonum posumbu*] 馬蓼

Smelly coal 臭煤

Smithsonite 爐干石

Smoked Chinese plums 烏梅

Snail-shell pearls 螺珂珠

Snake-head awl 蛇頭錐

Soft jade 軟玉

So-li 琑里

Solid weave 實地

Solution of ox glue 牛膠水

Solution of red earth in water 紅土水泥

Sorghum 梁

Soy bean 大豆，黄豆

Spin 紡

Spinning wheel 紡車

Split bamboo 篾

Spooling silk 調絲

Square frame brick 方墁磚

Ssu-chiao Mountain [Four-cornered Mountain] 四角山

Standard vat 標缸

Starch 小粉

Starching and smoothing process 漿碾

Stone green [verdigris made from malachite] 石綠

Stone marrow lead 石髓鉛

Stone mill 磑

Stop-foot beam 挽脚樑

Strawcocks (for silkworms) 山

Straw salt 蓬盐

Strip 升

Stripping 絲匡

Strong pan 牢盆

Sublimation of mercury 升湏

Submarine mine 混江龍

Suckling lamb 乳羔

Sugar-cane 糖蔗，荻蔗

Sulphur stones 石亭脂

Sun-dried salt 大晒塩

Sung Ying-hsing 宋應星

Superior paper 紅上紙

Supernatant liquid of manure 清糞水

Su-shen tribesmen 肅慎

Swan's down 天鵝絨

Sweating the green 汗青

Sweet-wine 醴

Szechuan stone 巴石

T

Tacking 搶風

T'ai-chou (Chekiang) 台州

T'ai-o (large beam) 泰阿

Tallow tree seeds [*Stillingia sebifera*] (烏)桕子

Tan (*shih* until Sung Dynasty) 石

T'ang 餳

T'ang Ching-ch'uan 唐荊川

T'ang pen-ts'ao 唐本草

T'ang-shuang p'u 糖霜譜

Tan-sha 丹砂

Ta-t'ung 大同

Ta-t'ung 大通

Tax boats 課船

Tea indigo 茶藍

Tea (or *camellia*) seed [木+茶]子

Teng-t'an pi-chiu (Necessary Knowledge for a Military Commander) 登壇必究

Thin gauze 紗

Thin-glue gauze 清膠紗

Thousand-year jade-containing rocks 千年璞

Thread 縷

Thread-passing rod 送絲干(竿)

Three-barrel pistol 三眼銃

Throwing apparatus 經具

Tidal mound 潮墩

T'ieh-li-mu 鐵力木

T'ien-kung k'ai-wu 天工開物

Tiger-spot bean 虎皮豆

Tiller 關門棒

Time vitriol or chicken droppings vitriol 時礬, 鷄屎礬

Ting-ning 丁寧

Ting-tripods 鼎

Topaz 酒黃

Tou 斗

Tough cocoon 棘繭

Treadles 踏輪

Treadle wheel 踏車

Tree cotton (*Ceiba pentandra*) 木棉

Tree leaf salt 樹葉塩

Tribute boats 漕舫

Tributes of Yü 禹貢

Trough 槽梘

Ts'an-tou 蠶豆

Tsao-chiao, pods of the *Gleditschia sinensis* 皂角

Ts'ao-chün 草菌

Ts'ao-wu (*Aconitum*) 草烏

Tseng 甑

Tsin 晉

Ts'un 寸

Ts'ung-ling 葱嶺

Tsun-hua 遵化

T'u 稌

Tu-chiang-yen 都江堰

Tu-chung (*Eucommia ulmoides*) 杜仲

Tu-lan sands 獨攬沙

T'ung-jen 銅仁

T'ung oil (*Aleurites cordata*) 桐油

T'u Po-chü 涂伯聚

T'u-shu chi-ch'eng 圖書集成

Turnip seeds 萊服子

Twilled fabric 文成斜現

Two-colored agate 夾胎瑪瑙

Two-fire brass 二火銅

Tz'u-chou 磁州

Tzu-ma 紫麻

Tz'u Mountain 茨山

U

Ultramarine ("Buddha's-head blue") 佛頭青

Unborn lamb 胞羔

Uncanny jade 玉妖

Unfashioned iron 毛鐵

Used iron 勞鐵

Using-the-wind beam 使風樑

V

Valve(?) 消息

Vanquisher crossbow, The (k'o-ti nu) 克敵弩

Varnish tree (*Rhus verniciflua*) 漆樹

Vegetable tallow (*Stillingia sebifera*) 桕皮油

Venetian sumach (*Rhus cotinus*) 蘆木

Vermillion 銀朱

Vinegar dregs 醋滓

W

Waist loom 腰機

Wang Chen 王楨

Wang Chin 王璡

Wang Mang 王莽

Warp 經

Warp beam 的杠

Warping 穿經

Water-drip tile 滴水瓦

Water sulphur 水硫黃

Water-turned wheel 水轉翻車

Watery or milky agate 漿水瑪瑙

Wax [insect] eggs 臘子

Wax [tree] seed 臘種

Weaving borders 牽邊

Weft; tram 緯絡

Weights, or rigid rods (on a loom) 衢脚

Well salt 井盐

Wen-chou (Chekiang) 溫州

Western barbarian wheat 西番麦

Western Regions 西域

Western sugar 洋糖

Wheat-beard niter 芒消(硝)

Wheat-husk gold 麩麦金

Wheat yeast 麦麹

Whirring dart 嚆矢

White bean 白藊豆

White clay 堊土

White [wheat] flour 白麪

White hammered paper 白硾紙

White-Jade River 白玉河

White lead 胡粉
White sugar 白霜, 糖霜
White wax 白臘
Wild silk 野蠶絲
Wild silkworm 野蠶
Willow-catkins alum 柳絮礬
Winch 大關車
Window gauze paper 窗紗紙
Wine-mash 酒糟
Wine yeast 酒母
Winnowing fan 颺(扇)
Winnowing machine 風車
Winnow riddle 簸
Wo 瓮
Wood ash 柴灰
Wooden cask 楻桶
Wood red 木紅色
Woolen 絨
Woolly sheep 蓑衣羊
Woven bamboo tray 篾盤
Wrapping paper 包裹紙
Wrought iron 熟鐵
Wu 吳
Wu 巫
Wu-ching tsung-yao 武經總要
Wu-li hsiao-shih 物理小識
Wu-pei chih 武備志
Wu-pien 武編

Y

Yang [pure positive] (純)陽
Ya-tsao (*Gleditschia japonica*) 牙皂
Yang-shao civilization 仰韶文化
Yang-t'ao vine 羊桃藤
Yao-chiu 藥酒
Yellow berberine 黃蘗
Yellow soy bean 黃豆
Yellow vitriol 黃礬
Yellow yeast 黃麴
Yen 燕
Yen-ching (Peking) 燕京
Yen-hsiao 盐消
Yew nut 榧子
Yin [pure negative] (純)陰
Ying-ch'ing 陰青
Yin-p'ing lead 陰平鉛
Yü-chou 禹州
Yü-nieh 羽涅
Yü-t'iao yang 喬芀羊
Yü-t'ien 玉田
Yü-tien 圩田
Yü-t'ien, known as Ho-t'ien in Sinkiang Province 于闐(和闐)

APPENDIX A

Summary of Chinese Dynasties[1]

DYNASTY		DATES
The legendary Five Rulers		
1. T'ai-hao 太昊	Fu-hsi 伏羲 P'ao-hsi 庖羲	ca. 2852–2738 B.C.
2. Yen-ti 炎帝	Shen-nung 神農 Lieh-shan 烈山	ca. 2737–2696 B.C.
3. Huang-ti 黃帝	Yu-hsiung 有熊 Hsuan-yuan 軒轅	ca. 2697–2596 B.C.
4. Shao-hao 少昊	Chin-t'ien 金天	ca. 2597–2512 B.C.
5. Yao, Shun 堯舜等		ca. 2513–2196 B.C.
Hsia kingdom 夏		ca. 2197–1766 B.C.
Shang-Yin kingdom 商殷		ca. 1765–1122 B.C.
Shang	商	ca. 1765–1400 B.C.
Yin	殷	ca. 1401–1122 B.C.
Chou dynasty 周	(Feudal Age)	1121– 220 B.C.
Early Chou period		1121– 796 B.C.
Spring and Autumn (Ch'un-ch'iu) period	春秋	770– 403 B.C.
Warring States (Chan-kuo) period	戰國	403– 220 B.C.
Ch'in dynasty 秦	(First Unification)	221– 205 B.C.

[1] A detailed description of Chinese dynasties can be found in *A Chinese-English Dictionary* by R. H. Mathews (Shanghai, 1931).

DYNASTY		DATES
Han dynasty 漢		205 B.C.–A.D. 220
Former Han (Ch'ien Han)	前漢	205 B.C.–A.D. 8
Hsin interregnum	新莽	A.D. 9– 24
Later Han (Hou Han)	後漢	A.D. 25– 220
Three Kingdoms (San Kuo) 三國	(First Partition)	A.D. 220– 265
Shu (Han)	蜀	A.D. 220– 265
Wei	魏	A.D. 220– 265
Wu	吳	A.D. 222– 280
Tsin dynasty 晉	(Second Unification)	A.D. 265– 430
Western	西晉	A.D. 265– 316
Eastern	東晉	A.D. 317– 430
Northern and Southern dynasties (Nan Pei Ch'ao) 南北朝	(Second Partition)	A.D. 420– 588
Liu-Sung dynasty	劉宋	A.D. 420– 478
Ch'i dynasty	齊	A.D. 479– 501
Liang dynasty	梁	A.D. 502– 556
Ch'en dynasty	陳	A.D. 557– 588
Northern (T'opa) Wei dynasty	北魏	A.D. 386– 534
Western (T'opa) Wei dynasty	西魏	A.D. 535– 557
Eastern (T'opa) Wei dynasty	東魏	A.D. 534– 550
Northern Ch'i dynasty	北齊	A.D. 550– 577
Northern Chou (Hsienpi) dynasty	北周	A.D. 557– 581
Sui dynasty 隋	(Third Unification)	A.D. 589– 618
T'ang dynasty 唐		A.D. 618– 906
Five dynasties 五代	(Third Partition)	A.D. 907– 960
Later Liang	梁	A.D. 907– 922
Later T'ang (Turkic)	唐	A.D. 923– 935
Later Tsin (Turkic)	晉	A.D. 936– 946

DYNASTY			DATES
Later Han (Turkic)	漢		A.D. 947– 950
Later Chou	周		A.D. 951– 959
Liao (Ch'itan Tarter) dynasty	遼(契丹)		A.D. 916–1125
West Liao (Qāra-Khitāi) dynasty			A.D. 1126–1211
Hsi Hsia (Tangut Tibetan) state	西夏		A.D. 1038–1227
Northern Sung dynasty	北宋	(Fourth Unification)	A.D. 960–1126
Southern Sung dynasty	南宋		A.D. 1127–1279
Chin (Jurchen Tartar) dynasty	金(女真)		A.D. 1115–1234
Yuan (Mongol) dynasty	元		A.D. 1279–1368
Ming dynasty	明		A.D. 1368–1644
Ch'ing (Manchu) dynasty	清		A.D. 1644–1911
Republic	民國		A.D. 1912–

APPENDIX B

The Twenty-Four Solar Terms

	SOLAR TERM	APPROXIMATE SOLAR DATE	APPROXIMATE LUNAR DATE	ZODIACAL POSITION OF THE SUN
Hsiao-han	小寒 (Slight cold)	Jan. 6	Early Dec.	Capricorn
Ta-han	大寒 (Severe cold)	Jan. 21	Middle Dec.	Aquarius
Li-ch'un	立春 (Spring begins)	Feb. 5	Early Jan.	Aquarius
Yü-shui	雨水 (Rain water)	Feb. 19	Middle Jan.	Pisces
Ch'i-chih	啓蟄 (Excited insects)	Mar. 5	Early Feb.	Pisces
Ch'un-fen	春分 (Vernal equinox)	Mar. 20	Middle Feb.	Aries
Ch'ing-ming	清明 (Clear & bright)	Apr. 5	Early Mar.	Aries
Ku-yü	穀雨 (Grain rains)	Apr. 20	Middle Mar.	Taurus
Li-hsia	立夏 (Summer begins)	May 5	Early Apr.	Taurus
Hsiao-man	小滿 (Grain fills)	May 21	Middle Apr.	Gemini
Mang-chung	芒種 (Grain in ear)	Jun. 6	Early May	Gemini
Hsia-chih	夏至 (Summer solstice)	Jun. 21	Middle May	Cancer

	SOLAR TERM	APPROXIMATE SOLAR DATE	APPROXIMATE LUNAR DATE	ZODIACAL POSITION OF THE SUN
Hsiao-shu	小暑 (Slight heat)	Jul. 7	Early Jun.	Cancer
Ta-shu	大暑 (Great heat)	Jul. 23	Middle Jun.	Leo
Li-ch'iu	立秋 (Autumn begins)	Aug. 7	Early Jul.	Leo
Ch'u-shu	處暑 (Limit of heat)	Aug. 23	Middle Jul.	Virgo
Pai-lu	白露 (White dew)	Sep. 8	Early Aug.	Virgo
Ch'iu-fen	秋分 (Autumnal equinox)	Sep. 23	Middle Aug.	Libra
Han-lu	寒露 (Cold dew)	Oct. 8	Early Sep.	Libra
Shuang-chiang	霜降 (Hoar frost descends)	Oct. 23	Middle Sep.	Scorpio
Li-tung	立冬 (Winter begins)	Nov. 7	Early Oct.	Scorpio
Hsiao-hsueh	小雪 (Little snow)	Nov. 22	Middle Oct.	Sagittarius
Ta-hsueh	大雪 (Heavy snow)	Dec. 7	Early Nov.	Sagittarius
Tung-chih	冬至 (Winter solstice)	Dec. 21	Middle Nov.	Capricorn

APPENDIX C

The Equivalence of Chinese Weights and Measures in Metric Units

(Data taken from Wu Ch'eng-lo's *Chung-kuo tu liang heng shih*, or History of Chinese Weights and Measures, Shanghai, 1937)

DYNASTY	DATE	ONE CH'IH OR (CHINESE) FOOT [a] IN CENTIMETERS	ONE CHIN OR (CHINESE) CATTY IN GRAMS	ONE LIANG OR (CHINESE) OUNCE IN GRAMS	ONE SHENG [g] OR (CHINESE) PECK IN MILLILITERS
Huang-ti	After 2697 B.C.	24.88	—	—	—
Yü	2254–2204 B.C.	24.88	—	—	—
Hsia	2204–1765 B.C.	24.88	—	—	—
Shang	1765–1121 B.C.	31.10	—	—	—
Chou	1121– 220 B.C.	19.91	228.86	14.93	193.7
Ch'in	349– 205 B.C.	27.65	258.24	16.14	342.5
Former Han	205 B.C.–A.D. 8	27.65	258.24	16.14	342.5
Hsin Mang	A.D. 9– 24	23.04	222.73	13.92	198.1
Later Han	A.D. 25– 220	23.04	222.73	13.92	198.1
Wei	A.D. 220– 265	24.12	222.73	13.92	202.3
Western Tsin	A.D. 265– 273	24.12	222.73	13.92	202.3
Western Tsin	A.D. 274– 316	23.04	222.73	13.92	202.3
Eastern Tsin	A.D. 317– 430	24.45	222.73	13.92	202.3
Former Chao	A.D. 318– 319	24.19	—	—	—
Liu-Sung	A.D. 420– 478	24.51	—	—	—

South Ch'i	A.D. 479– 501	24.51	334.10	20.88	297.2
Liang & Ch'en	A.D. 502– 588	24.51	222.73	13.92	198.1
Liang	A.D. 502– 557	23.20 [b]	—	—	—
Liang	A.D. 502– 557	23.55 [c]	—	—	—
Ch'en	A.D. 557– 588	23.55	—	—	—
Later Wei & West Wei	A.D. 386– 557	29.51	222.73	13.92	—
Later Wei & East Wei	A.D. 495– 550	29.97	—	—	396.3
North Ch'i	A.D. 550– 577	29.97	445.46	27.84	396.3
North Chou	A.D. 557– 566	29.51 [d]	—	—	157.2
North Chou	A.D. 566– 581	26.68 [e]	250.56	15.66	210.5
North Chou	A.D. 577– 581	24.51 [f]	250.56	15.66	210.5
Sui	A.D. 581– 606	29.51	668.19	41.76	594.4
Sui	A.D. 607– 618	23.55	222.73	13.92	198.1
T'ang	A.D. 618– 906	31.10	596.82	37.30	594.4
Five dynasties	A.D. 907– 960	31.10	596.82	37.30	594.4
Sung	A.D. 960–1279	30.72	596.82	37.30	664.1
Yuan	A.D. 1279–1368	30.72	596.82	37.30	948.8
Ming	A.D. 1368–1644	31.10	596.82	37.30	1073.7
Ch'ing	A.D. 1644–1911	32.00	596.82	37.30	1035.5

a: The "Book of Measures and Calendars" in the *History of Sui Dynasty* classifies the various types of Chinese rulers, dating from the Wei and Tsin dynasties to the Sui dynasty, into 15 different grades in accordance with their length.

b & c: The new legal ruler and the sun-dial measuring ruler, respectively.

d, e, f: The market ruler, the jade ruler, and the iron ruler, respectively.

g: Ten *Sheng* make 1 *tou* and ten *tou* make 1 *tan*. Up to the Sung dynasty the character for *tan* was pronounced *shih;* thereafter it has been commonly pronounced *tan*.

APPENDIX D

Transmission of Certain Techniques From China to the West[1]

	INVENTION OR DISCOVERY	CHINA		EUROPE		APPROXIMATE MINIMUM TIME-LAG, IN CENTURIES
		PERIOD OF EXPERIMENT	FIRST PRECISE DATE	PERIOD OF EXPERIMENT	FIRST PRECISE DATE	
1	Square-pallet chain-pump	1st century B.C.; probably mentioned in 83	189	17th century?	1672	15
2	Edge-runner mill	1st century B.C.	170	15th century? [2]	1607	13
3	Edge-runner mill with water-power drive		400	15th century?	1607	11
4	Trip-hammer mill		4th century B.C.			
5	Trip-hammer mill, with water-power	2nd to 1st century B.C.	20	15th century? [3]	1607	14
6	Rotary Winnowing-machine, with crank-handle		40 B.C.	850, crank	late 18th century	14
7	Rotary fan for ventilation		180	15th century?	1556	12
8	Blowing-engines for furnaces and forges, with water-power	2nd to 1st century B.C.	31		13th century	11
	(idem. crank-drive type)		1310		1757	4

NOTES: 1. The table is reproduced from *A History of Technology,* by Charles Singer, E. J. Holmyard, A. R. Hall, and T. I. Williams, vol. 2, (1956), pp. 770–771, which in turn was amplified from table 8 (p. 242) in *Science and Civilization in China,* vol. I (1954), by Joseph Needham with the research assistance of Wang Ling. All dates are A.D. except where otherwise indicated. The periods in Column IV attempt to allow for considerable doubt and obscurity, and therefore are frequently less than the number obtained by subtracting the date in Column II from that in Column III.

2. The Chinese edge-runner mill had a roller revolving on a plane surface, and differed considerably from the split-ball-and-cup arrangement of the classical trapetum.

3. Water-driven fulling-mills were known in Europe in the thirteenth century and perhaps earlier, but it is not clear whether these were of the Chinese tilt-hammer type, or vertical-lift stamp-mills. The tilt-hammer was however associated with the development of the blast-furnace.

INVENTION OR DISCOVERY	CHINA		EUROPE		APPROXIMATE MINIMUM TIME-LAG, IN CENTURIES
	PERIOD OF EXPERIMENT	FIRST PRECISE DATE	PERIOD OF EXPERIMENT	FIRST PRECISE DATE	
9 Piston-bellows, for continuous blast		4th century B.C.	15th century?	16th century	14
10 Draw-loom for figured weaves	2nd century B.C.	Ca 100 B.C.		4th to 5th century	4
11 Silk-working machinery: reeling machine	Before 100 B.C.	1st century B.C.		All introduced about the end of the 13th century	3–13
flyer; twisting and doubling		1090			
water-power applied	13th century	1310		14th century	
12 Wheel-barrow	1st century	231		Ca 1200	9–10
13 Sailing-carriage (first high land speeds)		552		1600	11
14 Wagon-mill, grinding during travel		340		1580	12
15 Efficient draught-harness for horses:					
breast-strap	4th century B.C.	2nd century B.C.	6th century?	Ca 1130	8
collar		3rd to 7th century	9th century?	Ca 920	6
16 Cross-bow (individual weapon)		3rd century B.C.	Known in Roman period, but not widely used before the Middle Ages	11th century	13
17 Kite		Ca 400 B.C.		1589	12
18 Helicopter top (spun by cord)		320		18th century	14
19 Zoetrope (lamp-cover revolved by ascending hot air)		180		17th century	Ca 10
20 Deep drilling (for water, brine, and natural gas)	2nd century B.C.	1st century		1126	11
21 Iron casting	4th century B.C.	2nd century B.C.	Cast iron yielded only by accident in ancient and medieval times	13th century	10–12

INVENTION OR DISCOVERY	CHINA		EUROPE		APPROXIMATE MINIMUM TIME-LAG, IN CENTURIES
	PERIOD OF EXPERIMENT	FIRST PRECISE DATE	PERIOD OF EXPERIMENT	FIRST PRECISE DATE	
22 Concave curved iron mouldboard of plough		9th century [B.C.?]		Ca 1700	25 [?]
23 Seed-drill plough, with hopper		85 B.C.	Known in antiquity	Ca 1700	14
24 Gimbals	First century B.C.	180		Ca 1200	8–9
25 Segmental arch bridges		610		1345	7
26 Cable suspension-bridges		1st century B.C.			
27 Iron chain suspension-bridges		580	Proposed 1595	1741	10–13
28 Canals and rivers controlled by series of gates		3rd century B.C.		1220	17
29 True lock-gates and chambers		825		1452	7
30 Ship-building:					
stern-post rudder		8th century		1180	3
water-tight compartments		5th century		1790	12
31 Rig:					
efficient sails (mat-and-batten principle)		1st century B.C.		19th century	18
fore-and-aft rig		3rd century		9th century	6
32 Gunpowder:	8th century	Ca 850		13th century	4
as an igniter for an incendiary weapon		919	(Greek Fire used in the 7th century)		
rockets and fire-lances		Ca 1100	Described Ca 1300	15th century	3–4
projectile artillery,		Ca 1200		Ca 1320	1
explosive grenades and bombs		Ca 1000		16th century	4–5
33 Magnetism:					
lodestone spoon rotating on bronze plate	1st century B.C.	83			
floating magnet		1020		1190	4
suspended magnetic needle	10th century	1086			

INVENTION OR DISCOVERY	CHINA		EUROPE		APPROXIMATE MINIMUM TIME-LAG, IN CENTURIES
	PERIOD OF EXPERIMENT	FIRST PRECISE DATE	PERIOD OF EXPERIMENT	FIRST PRECISE DATE	
compass used for navigation	11th century	1117			2
knowledge of magnetic declination		1030		Ca 1450	4
theory of declination discussed		1174		Ca 1600	4
34 Paper:		105		1150	10
printing with wood or metal blocks	6th century	740		Ca 1400	6
printing with movable type		1045 (earthenware) 1314 wood			
printing with movable metal type	Ca 1340	1392 (Korea)		Ca 1440	1
35 Porcelain	1st century	3rd to 7th century		18th century	11–13

INDEX